Cave and Karst Systems of the World

Series Editor

James W. LaMoreaux, P.E.LaMoreaux and Associates, Tuscaloosa, AL, USA

More information about this series at http://www.springer.com/series/11987

Mladen Garasic

The Dinaric Karst System of Croatia

Speleology and Cave Exploration

Springer

Mladen Garasic
Committee for Karst, Croatian Academy
of Sciences and Arts (HAZU)
University of Zagreb
Zagreb, Croatia

ISSN 2364-4591 ISSN 2364-4605 (electronic)
Cave and Karst Systems of the World
ISBN 978-3-030-80586-9 ISBN 978-3-030-80587-6 (eBook)
https://doi.org/10.1007/978-3-030-80587-6

This Springer imprint is published by the registered company Springer Nature Switzerland AG
The registered company address is: Gewerbestrasse 11, 6330 Cham, Switzerland

Preface

One of the most important things I have realized throughout the years of my scientific life was the fact that just because I do not see or know enough about the processes around me it does not mean they do not exist or occur. This is the main driver for developing new ideas and understanding. The underground world of karst is full of mysteries that keep pushing us towards the new discoveries. Exploring the less known or unknown phenomena attracts the scientific interest.

The idea about this book was evolving inside me for years. For over fifty years, I have been entering the underground realm of karst. Whether as an amateur, a professional or a scientist, I have visited thousands of caves throughout the world and have written hundreds of scientific articles about them. Therefore, I felt the need to write a book which would succinctly describe the cave phenomena and their research in a Classic Karst environment—The Dinaric Karst. I focused on the Croatian part of Dinaric Karst where I started practising speleology and where I observed some of the phenomena which are rare or still undiscovered in other parts of the world.

Over a thousand caves in the Croatian part of the Dinaric Karst did not have a natural entrance from the surface until recently. They were subsequently discovered during construction works such as tunnelling, building roads, bridges, quarries and hydropower plants. Some of the rare and previously unknown processes that created them were discovered in those caves—caverns. I have visited all continents and got to know the underground world of around eighty countries, thus realizing the Dinaric Karst is special in a lot of ways.

During my stay in New Mexico, Ph.D. Jim Lamoreaux, the editor of the respectable scientific journal *Environmental Earth Science* by Springer, introduced the idea of publishing a series of books about karst and caves of the world. I feel the right moment has come to give the readers a summary about the scientific understanding of a unique part of karst that they will be able to compare to other types of karst in the world. *The Dinaric Karst System in Croatia—Speleology and cave exploring* will be part of the series of books *Cave and karst System in the World* by Springer.

This book is meant for speleologists, geologists, experts in geoscience and students of geoscience as a new source of data and as a useful handbook. I did not want to repeat the information already published in similar books. I simply wanted to point out some things that were not mentioned in the scientific literature before or were not emphasized enough. This is not a book about all the caves in Croatian Karst nor the list of caves in Dinaric Karst. This is a summary of the current speleological understanding about a small, very important and interesting karst area in the world.

I would like to thank all the people that helped me develop my knowledge throughout the years, collecting data and publishing this book. Special thanks goes to Public institution (PI) "Žumberak—Samoborsko gorje Natural Park", Dugopolje Tourist board, Fužine Tourist board, PI "Učka Natural Park", PI "Natura Jadera", PI "Paklenica National Park", PI " Mljet National Park", PI "Priroda Dubrovačko—Nevetvanska" Dubrovnik, PI "Priroda" Rijeka, PI "Priroda" of Šibenik—Knin County, Silvio Legović, Jama Baredine, Institute for Quaternary Paleontology and Geology of Croatian Academy of Sciences and Arts (HAZU), academician Ivan Gušić, Prof. Ph.D. Meho Saša Kovačević, mr.sc. Slobodan Kolbah, Prof. Ph.D. Maša

Surić, Prof. Ph.D. Tomislav Malvić, Prof. Ph.D. Josipa Velić, Ph.D. Kristina Pikelj, Prof. Ph. D. Čedomir Benac, Neven Korač, mag.ing.geol., Ph.D. Boris Vrbanac, Ph.D. Jadranka Mauch Lenardić, Marija Brajković, dipl.ing.geol., Prof. Ph.D. Darko Mayer, Ph.D. Josip Terzić, Ph. D. Petra Konrad Kovač, Ph.D. Natalija Matić, Ph.D. Vesna Štamol, Milica Bjelić, dipl.iur., Croatian Environmental Protection Agency, PI "Sjeverni Velebit National Park", Damir Lacković, dipl.ing.geol., Ph.D. Neven Cukrov, Ph.D. Gordan Lukač, Zlatko Marasović, Ph.D. Dubravka Kljajo, Ivana Maras, Donat Petricioli, Vladislav Brnčić.

Aside from the majority of photographs I have taken personally, some of the best speleophotographers like Boris Krstinić, Dinko Stopić, Alan Kovačević, Tihomir Kovačević, Gordan Polić, Željko Marunčić, etc. allowed me to use some of their photographs in this book.

The works of three Croatian experts of speleology and geology Dr. Josip Poljak, academician Dr. Mirko Malez and Dr. Srećko Božičević, who were my predecessors in many researches of Croatian Karst motivated me to continue with their research and to inspire the future generations to do the same.

I would also like to thank the hydrogeologists and geologists who expertized in solving the problems in Croatian Karst and whose works are used in this book: Željko Babić, Željko Blagus, Dinka Borčić, Božidar Biondić, Ranko Biondić, Luka Bojanić, Srećko Božičević, Franjo Fritz, Ivan Gušić, Milan Herak, Mladen Juračić, Vladimir Jurak, Mladen Kuhta, Antun Magdalenić, Mirko Malez, Darko Mayer, Ante Pavičić, Josip Poljak, Andrej Stroj, Josip Terzić, Kosta Urumović, Željko Vulić ...

I would like to thank posthumously my friends speleologists who helped me with the research: Tomislav Marinčić, Mladen Šebian, Ljubiša Kalinić, Boris Resimić, and Drago Opašić.

I would like to thank all the members of DISKF, especially Tihomir Kovačević, Tihomir Jukica, Alan Kovačević, Boris Watz, Ivan Krpina, Tomislav Gospodinović, Davor Garašić, Toni Terzić, Krešimir Ajduković, Ivan Drnić, Goran Kresnik, Mislav Krvavica, Nebojša Anić, Tomislav Jech, Hrvoje Jelković, Tomislav Jech, Edo Aganović, Miljenko Garašić, Miroslav Hranjec, Zlatko Piškorić, Nevio Putinja, Eugen Ujčić, Vedran Putinja, Mladen Raškaj, Tibor Perok, Slobodan Meničanin, etc. who helped me with the research of many dangerous caverns and caves.

I am grateful and thankful for the international help from the members of the UIS bureau Ph.D. Fadi Nader, Ph.D. George Veni and academician Ivan Gušić for advises and review of this book.

A special thanks goes to "Jama Baredine" and my friend Silvio Legović that helped me finance the English translation and the proof reading of this book. I would like to thank Davor Garašić and Eleni Ktisti for the English translation, editing and proof reading.

Years of working on this book must have affected my family so I am sincerely grateful for the understanding and support during the time of writing. Without the support of my family, this book would have never been written.

Zagreb, Croatia Mladen Garasic

Acknowledgements

Photographs: The author of all photographs printed in this book is Mladen Garasic except if state otherwise. Other authors are: Srećko Božičević (Fig. 6.7), J. P. Bresser (Fig. 3.35), Mark Burkey (Fig. 4.26), Luigi Casati (Fig. 3.33), Alan Kovačević (Fig. 3.30), Tihomir Kovačević (Figs. 3.16 and 7.12), Boris Krstinić (Figs. 1.1, 1.4, 1.12, 2.14, 3.19, 3.46, 3.50, 4.12 and 9.8), Damjan Krstinić (Fig. 2.9), Željko Marunčić (Figs. 3.9 and 10.35), Gordan Polić (Figs. 3.42, 4.27, 4.44 and 9.13), Tonći Rađa (Fig. 10.32), Krzysztof Starnawski (Fig. 3.37), Dinko Stopić (Figs. 3.5, 3.6, 3.52, 3.53, 4.13, 9.12, 10.31 and 10.41), Goran Šafarek (Fig. 3.29), Boris Vrbek (Fig. 4.6).

Contents

About the Author

Prof. Mladen Garasic, Ph.D. is one of the world's well-known caver and speleologist with more than 56 years of continuous activity in international caving and speleology.

He has Ph.D. in geology, hydrogeology, and geological engineering and ScD. in earth sciences.

Born in Zagreb, Croatia, in 1951, Mladen Garasic graduated in geology and karst hydrogeology in 1977. He earned his master's degree in 1981, and doctorate in geosciences and geological engineering in 1986. He is a scientist and a professor of geology, karst hydrogeology, applied geology, engineering geology and speleology at the University of Zagreb and Technical University of Graz (Austria). He has authored more than 360 scientific and professional papers and about 800 speleological and geological reports. He serves as a Karst committee secretary for the Croatian Academy of Sciences and Arts (HAZU), he is the member of UNESCO World Heritage Team for the Dinaric Karst, International Association of Hydrogeologists, International Association for Engineering Geology and the Environment, etc.

He started with caving in 1963 and was one of the founder and president of several caving clubs in Croatia. He served as the first president of the Croatian Speleological Federation from 1990 to 2010 (elected five times on general Assemblies of CSF).

Since 1993 to 2018, he has served as Croatia's delegate to the International Union of Speleology—UIS (24 years) and to the European Speleological Federation—FSE beginning in 2009 (8 years).

Prof. Garasic has conducted research in, and explored and visited more than 5150 caves in 80 countries on all continents. He has led many speleological expeditions in the longest and deepest caves in Croatia, Europe and the world. He also studied about 1050 caves without natural entrances, discovered by tunnels and quarries, and evaluated their hydrogeology and engineering geology.

He was the youngest member ever in Croatia who passed exam for cavers in caving organizations. He is cave diver and cave rescuer. He was founder of Zagreb Speleological School in 1971, school that has one of the longest tradition and every years teach 20–30 new cavers. He was editor and one of the authors of the first Manual for cavers on Croatian language (1976). Cave photographer (he has taken more than 300,000 photos from the caves) and bibliophile of speleological literature (collected more than 30,000 speleo editions).

He was one of rare Croatian scientists who wrote and passed doctorate (1986) on speleology field. He is member of many organizing and scientific international, regional and national committees on congresses and conferences (speleology, karstology, geology, hydrogeology, geotechnics, etc...). He and his team projected and constructed the longest bridge in the world (58 m long) ever made in big cave chamber in the highway's tunnel (2006–2008).

He was invited speaker for more than twenty times the entire world and attended 12 International Congresses of Speleology. He was the first and the only member of NSS from Croatia (Yugoslavia) tens of years which he began in 1977. He is member of Geological Society of America for years too.

He is experienced instructor in caving, made several exploring and teaching expeditions (Africa, Asia, Middle East, South America, Caribes...) and got some awards for these jobs. He is editor and coordinator of UIS Caver's International Dictionary written on 29 languages (!), speleological Bulletin Spelaeologia Croatica (1990–2007), cave diver, cave rescuer (1971–1979).

Croatia is small country with well-known karst, with four caves deeper than 1000 m (the deepest is Lukina jama -1432 m), and the longest inside vertical cave channel in the world (-518 m), and the second entering vertical (-553 m) cave channel in the world, deep karst river spring (Una river dive to -248 m), etc. All these are in world well-known Dinaric Karst, with several UNESCO World heritage localities (Plitvice Lakes, Dubrovnik, Split, Trogir, Zadar, etc.). The first speleological organization in Croatia was founded on 20th of February 1892—the fourth in the world.

Mladen Garasic is Adjunct Secretary of International Speleological Union (UIS) for periods 2013–2017 and 2017–2021, and Vice President of European Speleological Federation (FSE) for period 2015–2019.

For his professional and scientific speleological works, he has been proclamated as honorary member of several speleological federations and associations in Croatia and the World.

List of Figures

List of Tables

1 Introduction to Karst and Geology of Croatia

Abstract

The Dinaric Karst of Croatia was formed in a tectonically active area, where several geotectonic plates of variable size are in direct contact. The African tectonic plate, with its smaller Adriatic plate, extends northwards towards the Alps. The Adriatic plate subducts under the Dinaric plate, which is the base part of the Dinaric Mountains. Therefore, this is a seismically palaeo and neotectonically active area with many earthquakes. Since these area is composed mainly of soluble carbonate rocks (mostly Mesozoic and Cenozoic ages) that are up to 8 km thick in this area, with favourable hydrogeological conditions, the karstification processes are very prominent. A brief overview of the geology of Croatia also indicates the complexity of hydrogeological processes in this area.

Dinaric Karst is the area where the first scientific exploration and research on the surface and underground began. In this area, the first theories about the movement of the groundwater in karst were born. Therefore, it is not surprising that one part of the Dinaric Karst is called Classic Karst. Dinaric Karst is the birthplace of speleology and cave explorations as well as the modern hydrogeology of the karst (Ford and Williams 2007; White 1988).

It is important to mention that the word karst originated from the area of Dinaric Karst. This area was the birthplace of some other internationally accepted terms such as polje, ponor, dolina, etc., which describe certain karst phenomena and relief.

"Dinaric Karst" was named after the Dinara Mountain which is the highest mountain in Croatia (Fig. 1.1). Dinara Mountain spreads from north-west to south-east which suggests the significant era of karst genesis in this part of Europe (Fig. 1.2). The largest area of Dinaric Karst is located in Croatia, but there are areas of it in Italy, Slovenia, Bosnia and Herzegovina, Montenegro, Serbia and Albania. Almost all of the Dinaric Karst area under the Adriatic Sea is located in Croatia.

Unlike the term "Dinaric Karst", the terms "Dinarides" or "Dinaric Alps" represent the area consisting both of karst and non-karst.

1.1 Karst Reliefs

Josip Roglić a world-known Croatian geographer, geomorphologist, the second president of Croatian Speleological Society (later Federation), defines karst as a set of phenomena related to flow of water and forms of relief in carbonate rocks. In one of his subsequent works, Roglić (1974) states: "Karst is a specific form of relief and circulation of water on soluble rocks (limestones), but karst also appears in saddle (gypsum) terrains. Water drainage and the creation of karstic relief are mutually causal and functionally linked and need to be observed together."

Milan Herak wrote in the well-known university textbook *Petrography and Geology (1966)*: "For karst it is characteristic that underground water communication is stronger than on the surface (Bögli 1980; Bakalowicz 2005) and is related to hollows of different dimensions which gradually expand due to the solubility (corrosion) of limestone and dolomite." Karst areas in Croatia are very widespread and the karst morphological and hydrological particularities are mostly well developed. Their historical significance is also great because the foundations of the karstic doctrine are set up precisely in these terrains. In the classification of karstic fields, Herak and Stringfield (1972) uses a tectogenetic approach and on the basis of specific morphological and hydrogeological features. He distinguishes between two types of karst. "Epirogenetic" (deep) karst with tabular and bored basin and "orogenetic" (accumulated) karst with oblique, bored, dissected basin.

Hydrologist Bonacci (1987) in the book *Karst Hydrology* states: "Karst represents an area composed of a special surface and underground relief and a surface-underground hydrographic network resulting from the circulation of water

M. Garasic, *The Dinaric Karst System of Croatia*, Cave and Karst Systems of the World,
https://doi.org/10.1007/978-3-030-80587-6_1

Fig. 1.1 Dinara Mountain is the highest mountain in Croatia

Fig. 1.2 Dry riverbed of the karst river and Dinara Mountain

and its aggressive chemical and physical action on fractures, fissures and cavities in layers of soluble rocks such as limestone, chalk, dolomite, gypsum and salt. Karst inherent soluble rocks are located on or near the ground surface. The karstification process is the result of the physical and chemical action of water dissolving and transferring dissolved solids from the rock mass."

Srećko Božičević in his work *Fenomen krša* (1992), writes: "From the geological point of view, karst is a characteristic form of relief on the limestone-dolomite substrate, predominantly on the rocky carbonate surface, formed by the long run of precipitation and flowing water. The karst as a morphologically-hydrogeological peculiarity of the relief is characteristic or specific to a series of irregular spiked or toothed peaks on a rough stone surface where we do not regularly find surface water flows."

Garašić (1986, 1991), in several papers, points out the essential conditions for forming the karst: "water-soluble rocks, intensive tectonic activity (cracking breakdown) and hydrogeological conditions (a significant amount of water acting on the primary rocks)". The karst is a specific form of relief that develops on soluble rocks (limestone, dolomite, salt, gypsum). We primarily associate it with limestone and dolomite (carbonates, $CaCO_3$ i $MgCO_3$) because of their extreme susceptibility to chemical (corrosion) and mechanical wear (erosion, abrasion) and high dispersal. The karst is characterized by specific hydrological and hydrogeological relations. These relationships are defined by absence of surface water (Fig. 1.3). Namely, surface water flows are rare and largely allogeneic, with the expressed abundance of water in the underground (Fig. 1.4). In the karst, underground runoff is always higher than surface runoff, and there are significant differences between the topographic (surface) and hydrogeological (underground) boundaries between the basins. The fractural porosity is characteristic of karst, while the phreatic water does not exist in the classical sense. It is located in several levels, depending on capillary elevation, hydrogeological dynamics of groundwater, rocks water flow, flow rate (slope, quantity, etc.). It is a fractural secondary porosity, as opposed to the primary or intrinsic porosity that occurs in cluster sediments (pebbles, sands, etc.).

International Glossary of Hydrology (1992) states that the karst areas are covered with limestones and dolomites characterized by special topography generated by the underground solution as well as the sinking of surface water.

Fig. 1.3 Absence of surface water (Bijela voda)

1.2 Geology of Croatia in General

Croatia geographically belongs to the Mediterranean countries of Europe. From a geological point of view, a part of Croatia (along the Adriatic Sea) is under the influence of geological changes caused by the shifting of the African Geotectonic Plate, and the second, inner part is under the influence of Eurasian Geotectonic Plates (Figs. 1.5 and 1.6). Recently, the assumption about the geosyncline development of the Adriatic plate has been rejected, but the thickness of the carbonate complex is estimated to be about 7–10 km (Herak 1987; Vlahović et al. 2002; Velić et al. 2003).

The structure and composition of the relief of Croatia are determined by its long geological history. From the earliest Precambrian to Cenozoic (i.e. Holocene) era, the relief of Croatia was influenced by the endogenous forces and exogenous processes in the triangle area of Asia–Europe–Africa. All three basic groups of rocks are represented in the relief of Croatia; sedimentary rocks (95% of the relief of

Fig. 1.4 Water from the riverbed infiltrates the underground

Croatia), metamorphic rocks (2–4% of the relief of Croatia) and igneous rocks (1% of the relief of Croatia). Sedimentary rocks are mostly clastic (sandstones, conglomerates, marl, breccias) and organogenetic rocks (limestones and dolomites). Metamorphic rocks are mostly marble, schist, gneiss, while igneous rocks are mostly andesite and granite. The oldest rocks ones are found in the cores of mountains Papuk, Psunj and Moslavačka gora. From the Mesozoic era, the carbonate rocks (dolomites and limestones) prevail in the karst, mostly in Banovina, Kordun, Gorski Kotar, Velebit, Lika and Dalmatia. The youngest rocks belong to the Cenozoic era (clastic sediments) (Mihevc and Zupan 1996). They can be found in the area of Pannonian and Peripannonian basin (Fig. 1.7).

1.2.1 Tectonics and Seismic Activity of Croatia

In Croatia, earthquakes occur in areas of contact between the smaller tectonic units. The cause of the earthquakes in the coastal part of Croatia is the subduction of the Adriatic platform under the Dinarides. Subduction is the result of the movement of the African tectonic plate towards the Euro-Asian tectonic plate (Fig. 1.8).

In the north-western continental part of Croatia, the causes of earthquakes are the compression processes that occur due to the movements of the Dinarides and the Alps. In the area of the Central Slavonian Mountains, the earthquakes are caused by the different movements of certain mountain masses. The margins of this area are predominantly active. The Republic of Croatia is located in the seismically active area. This is particularly true for the coastal area, especially for southern Dalmatia and the north-western part of Croatia. In the coastal area, from north-west to south-east, the earthquake zone extends from the border with Slovenia to the area south of Senj. Seismic activity is less present in the area between Velebit and Bukovica. Further to the south-east the area of strong seismic activity continues all the way to the south of Dubrovnik with a few breaks between Šibenik and Split.

In the Adriatic Sea, the seismic activity of the central and southern parts is stronger. The most prominent activity can be traced south of the island of Lastovo. In the western part of continental Croatia, a zone extending from the border with Slovenia to the west of Karlovac is also prominent. The zone then continues through Žumberak and Medvednica Mountains to Kalnik and the western part of Bilogora. This zone joins the active Zagreb area belt which extends all the

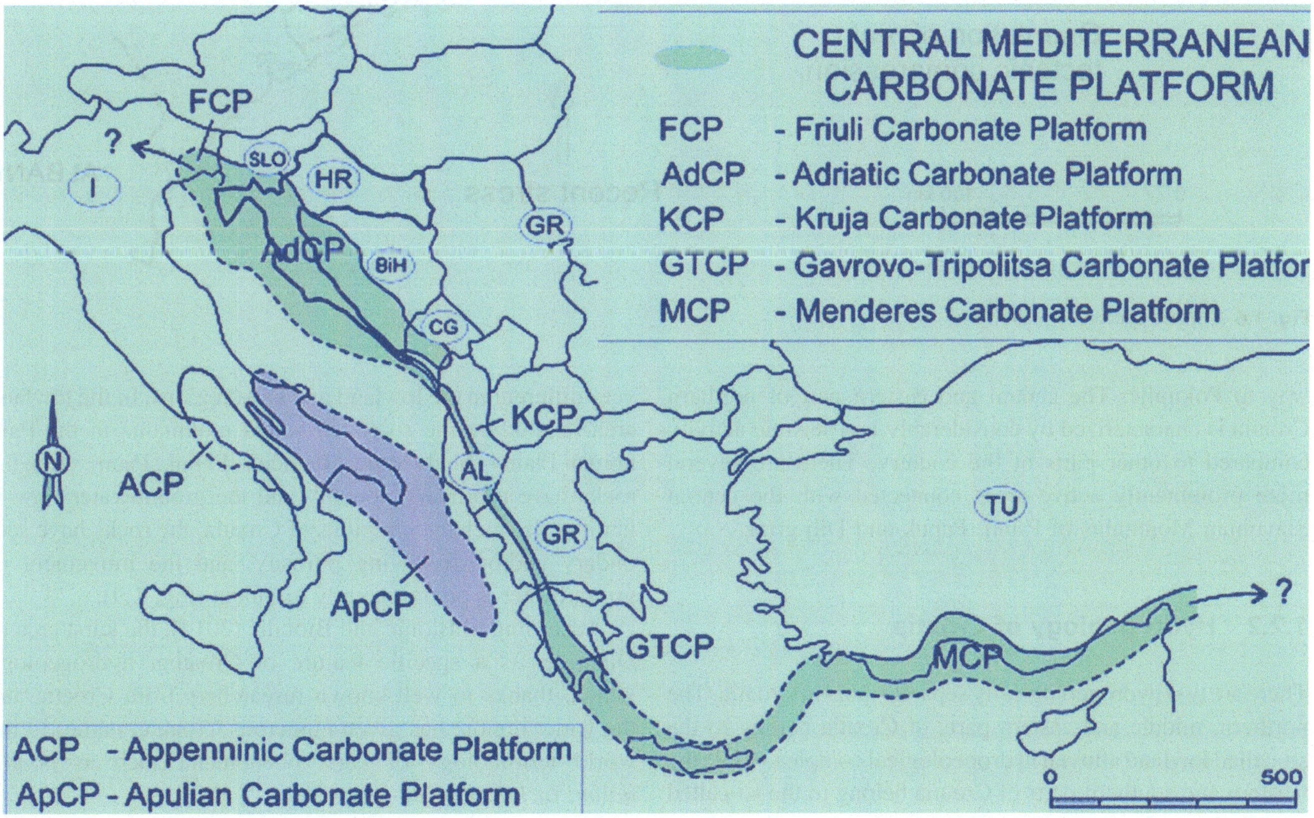

Fig. 1.5 Central Mediterranean carbonate platform

Fig. 1.6 Dinaric carbonate platform

way to Pokuplje. The central and eastern part of northern Croatia is characterized by considerably less seismic activity compared to other parts of the country. There are several more prominently active areas connected with the central Slavonian Mountains of Psunj, Papuk and Dilj-gora.

1.2.2 Hydrogeology of Croatia

There are two hydrogeologically separate units in Croatia. The northern, middle and eastern parts of Croatia belong to the so-called lowland alluvial hydrogeological complex, while the western and southern parts of Croatia belong to the so-called mountain or karst hydrogeological complex. Quantities, movements and chemical composition of groundwater are very different in the lowland and karst regions. In the lowland area (excluding the so-called island mountains in the Pannonian Plain—Medvednica, Ivančica, Papuk, Psunj, etc.), the rocks have a primary porosity and the groundwater flow is laminar. In the karstic regions of Croatia, the rocks have secondary and/or dissolving porosity, and the movement of groundwater is predominantly turbulent (Fig. 1.9).

According to Bionić and Biondić (2014), the karst area of Dinarides is a specific feature of Croatian hydrogeology which, thanks to well-known researchers from Croatia and the wider region, has given numerous Croatian names to the world terminology of karst phenomena such as *ponor*, *dolina* or *krško polje*.

The terminology derives from continuous extensive research of karst fields of the Dinarides since the middle of the nineteenth century. The first organized researches were carried out in the border regions of Croatia and Slovenia (between Trieste and Kvarner Bay), where the term "karst" was created after the name of the area. Therefore, it is not surprising that many world researchers of the Dinarides use the name "Classic Karst".

The Dinaric Karst in Croatia covers half of the territory of the country. If the area of Adriatic Sea is included, then about two thirds of the total territory of Croatia is karst. Dinarides stretch from Slovenia through Croatia and Bosnia and Herzegovina to Montenegro. In Croatia, the south-western part of the Karlovac depression, Istria, Gorski Kotar, the coastal area, Lika, Ravni Kotari, Dalmatia, the mainland of Dubrovnik and all the islands of our Adriatic region belong to the Dinarides.

The main features of the Dinaric Karst are specific landscapes developed in rocks masses of predominantly carbonate rocks, a general lack of constant surface water flow, karst fields with numerous swallow-holes, underground karst phenomena in the form of caverns and underground channels, grand karst springs and high speeds of underground streams through karstic reservoirs of the vast surface and underground distribution. Given that the system of karstification and underground flow is mainly related to fracture and fault systems, karstic aquifers are extremely heterogeneous, making it extremely difficult to extrapolate data and model underground streams. This also points to the particularity of each individual water system and suggests they should be observed separately. Precipitation can infiltrate the underground by diffusion or through swallow-holes. The rapid and violent reaction of karstic aquifers to large amounts of precipitation indicates rapid infiltration of precipitation in the underground, rapid propagation of long-range hydraulic pressures and relatively low retention capacity of karst aquifers. The result is a great discharge on karst springs that can reach $Q_{\min}:Q_{\max} = 1:100$ and above.

In the past, it was believed that the level of the sea was also the level of karstification or depth to which karstic channels were developed. Submarine springs (vrulja) were understood to be formed due to hydraulic erosion and corrosion in the final phase of underground water runoff under high pressure. Nowadays, it is well known that during Quaternary, the sea levels of Mediterranean and Adriatic seas were up to 120 m lower than today (Fig. 1.10). This allowed the karstification to reach the lowest recorded sea level during that period. The rising of the sea level at the end of Quaternary resulted in the rising of specifically lighter freshwater systems as the seawater penetrated the deep karst channels and fissure systems of today's coastal areas. This created the problem of the active relationship between the fresh and salt water in the karst underground and the deltas of the rivers that flow from the mountainous area of Dinarides into the sea. Solving different types of problems in karst aquifers of the Dinarides, with regard to the abovementioned basic characteristics, requires appropriate selection of hydrogeological exploration methods, starting from the determination of geological–structural relationships, through the assessment of hydrogeological characteristics of the rocks by: tracing underground streams, determining catchments, conducting hydrological studies, water balancing, hydrogeochemical research, geophysical research, exploratory drilling, etc.

The basic climate elements for the formation and dynamics of water resources are precipitation, air temperature and wind. Precipitation regimes, along with the geological structure of the terrain, directly affect the formation of groundwater and surface water, while air temperatures and wind affect evapotranspiration processes and the quantitative relationships of water on the terrain surface and in the karst underground (Fig. 1.11). The high degree of fluctuation of these parameters is a consequence of the greatly indented relief of the karst area of the Dinarides. The Dinaric carbonate platform, with its tectonic elevation in relation to the Adriatic Sea, has the function of a climate barrier to cyclonic activities from the north-west and vice versa.

The magnitude of rainfall that amplitudes in the Dinaric region is best seen from the map of climatic rainfall zones. The Dinarides area is divided into 12 out of a total of 13 climate rainfall zones covering the entire country (Gajić-Čapka et al. 2003). Average annual rainfall in Croatia varies from about 300 mm on the outer islands to 3500 mm on the outer mountain ranges of Gorski Kotar and Lika). On the islands and the coastal areas of central and northern Dalmatia and in western Istria, rainfall of around 800 to 900 mm falls annually. Towards the inland, rainfall increases up to a maximum of 3000–3500 mm per year in the mountainous areas of Gorski Kotar and Velebit. In the interior of Croatia towards the eastern Slavonia, precipitation is gradually reduced to 600–700 mm. The karst area of the Dinarides can be considered very rich in rainfall, and this reflects on the development of very abundant aquifers in both the Adriatic and Black Sea basins. However, it is an uneven annual precipitation pattern occurring mainly in the autumn, winter and early spring periods.

The problem are the summer droughts, which leave some areas of the Dinaric Karst with no precipitation for several months. On the one hand, high rainfall intensities in short periods of time cause large flood waves and on the other hand, long dry periods result in considerable lack of water in most areas of the Dinaric Karst.

Air temperature is one of the basic elements for determining the climatological characteristics of an area. A map of isotherms of mean annual air temperatures in Croatia (Zaninović et al. 2008) was made on the basis of measured data from 161 climatological and main stations.

The lowest average annual temperatures (2–3 °C) were measured in the highest mountains of the Dinaric region

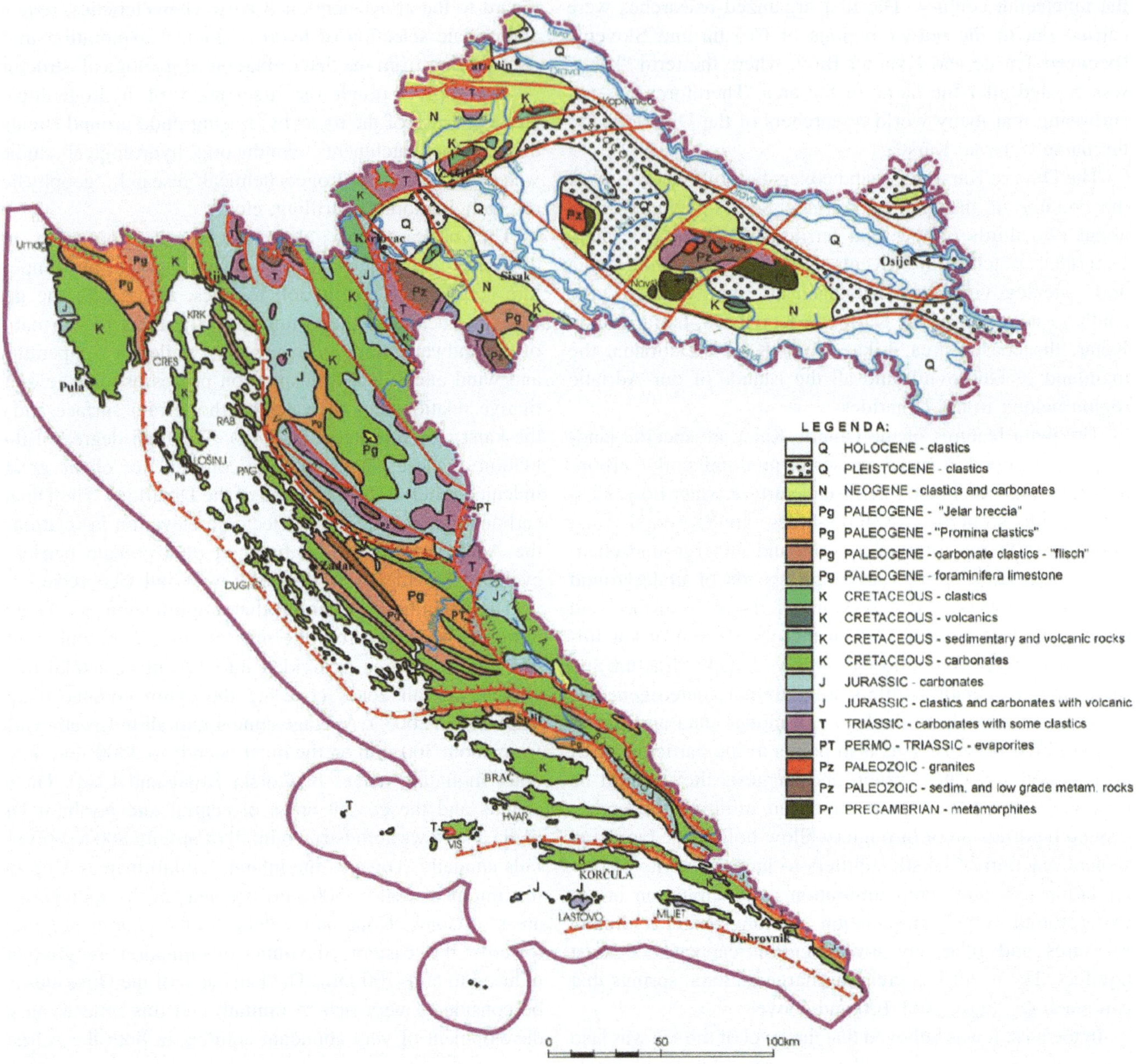

Fig. 1.7 Geological map of Croatia

(Risnjak, Snježnik, Velebit, Dinara, etc.). The highest average annual temperatures were measured in the coastal area of the Adriatic and the islands (13–17° C), and in the interior of the Dinaric Karst area (Gorski Kotar, Lika), mean annual temperatures of 5–11 °C were measured. Air temperature is one of the most important parameters for the calculation of the water balance using empirical formulas and directly affects the groundwater temperature.

Wind is also an important climatological element in water resources studies especially for the area of the morphologically very indented mountain and coastal area of the Dinarides. There, apart from regional climatic influences, are locally conditioned movements of the air masses. Wind speed values were modelled on the basis of measurements at five stations representative of different climatic zones in Croatia. The highest average annual wind speeds are associated with the mountainous areas of Velebit (Fig. 1.12), Licka Pljesevica and Dinara, especially in the mountain pass zones due to the strong Bora wind. In coastal parts of the mountains and islands, the impact from Bora is still strong, but significantly reduced compared to the mountainous areas. The Dinaric region is dominated by two winds: Jugo, from the south-east, affecting all of the Mediterranean area, and Bora, a local wind blowing from the mountains towards the sea.

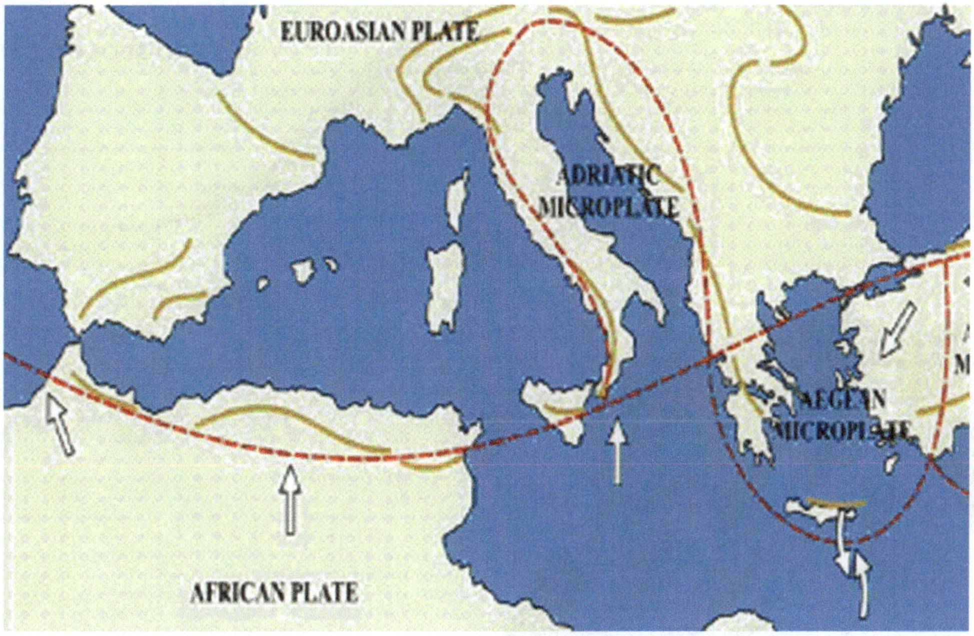

Fig. 1.8 Movements of the African tectonic plate

According to the Krippen classification, Dinarides belong partly to the moderately warm humid climate type and partly to the Mediterranean type (Šegota and Filipčić 2003).

Croatia is located in the northern temperate zone with the impacts of the Atlantic Ocean (western winds) and the Sahara depreciated by the Mediterranean Sea and the Adriatic. In the Dinaric highlands, indented relief and the formation of local climatic conditions play an important role.

The knowledge of the geological structure of the karst area of the Dinaric Mountains developed from the mid-nineteenth century to the present day. F. Koch's geological maps from the beginning of the twentieth century are well known and were used in various geological projects until the creation of the Basic Geological Map M 1:100,000 (OGK). The basic geological map of the Republic of Croatia, drawn up by numerous experts from the Croatian Geological Institute in Zagreb in the second half of the twentieth century, has become the basis of various studies for the assessment of the environment, groundwater research, mineral resources and many other disciplines related to geology. A good overview of the geological structure of the Dinarides is given in the Interpreter of the Geological Map of the Republic of Croatia M 1:300,000 by the Croatian Geological Institute. In the last few years, a dozen new OGK M 1:50,000 sheets have been produced, mostly in karst areas (islands Vis, Biševo, Brač, Hvar, Mljet, Šolta, Čiovo, Drvenik, Dugi otok and Konavle area). When it comes to the structural-geological composition of the Dinarides, it is necessary to start with a mobilist approach (Herak 1986, 1991), based on plate tectonics and subduction effects in plate margin regions. The end result of subduction is directional stress, which results in folding of rock masses, switching of folds and their movement. Accordingly, Dinarides are a macrostructural form of very complex geological relationships typical of the margins of tectonic plates. The karst area of the Dinarides is part of the Alpine mountain range resulting from the convergent movement of the African Plate towards the Eurasian Tectonic Plate (Matičec 2009). The approximation of the above tectonic plates resulted in considerable contraction of the wide area of the former Neotethys, in which there were several carbonate platforms separated by deep furrows. One of these platforms was the Dinaric carbonate platform of the Triassic to Cretaceous sedimentary cycle. The carbonate deposits formed during this period make up the largest part of the karst area of the Dinarides (Brlek et al. 2012).

The base of the carbonate mass of the Dinarides was built mainly from the clastic deposits of Carboniferous and Permian which are stratigraphically marked as the younger Palaeozoic (Velić et al. 2009). These rocks are of a limited surface spread and mainly construct cores of anticlines. The oldest rocks in the karst areas of the Dinarides are of the Carboniferous age, and they were discovered in Lika between Štikada and Sv. Rok. They consist predominantly of shale with limestone, sandstone and conglomerate inserts. They are deposited in the areas of relatively shallow sea with a pronounced impact from the land. Similar deposits from the younger Carboniferous age were also detected in the area of Gorski Kotar near Lokvarsko jezero and Lake Bajer near Fužine. Permian age deposits have a larger surface spread than those of Carboniferous age. In the

Fig. 1.9 Hydrogeological map of Croatia

area of Lika, they can be found along the north-eastern slopes of Velebit and in the area of Velika Paklenica on the south-west side of the mountain. In the older part of this lithostratigraphic unit clastic deposits with carbonate rocks are dominating, while the younger part is dominated by dolomites and limestone with clastic insertions. In Gorski Kotar, most of the clastic Palaeozoic rocks are of the Permian age (Mrzle Vodice, Čabar, Brod na Kupi). Shales and sandstones with characteristic appearances of barite in the youngest part of the deposits prevail in the lithological composition. Permian deposits are in tectonic contact with older Carboniferous deposits. The Permian age strata have quite a large spread in other parts of the Dinarides too, where they also construct cores of the anticlines. In the Kordun area, these are mostly shales and sandstones, and in Dalmatia they are evaporites. Evaporites can be found in the Sinjsko Polje, Vrličko Polje, Petrovo Polje, Kosovo Polje and valleys of Zrmanja and Una rivers. The gypsum, anhydrite, dolomite and limestone with smaller occurrences of clastic and magmatite prevail in the lithological composition.

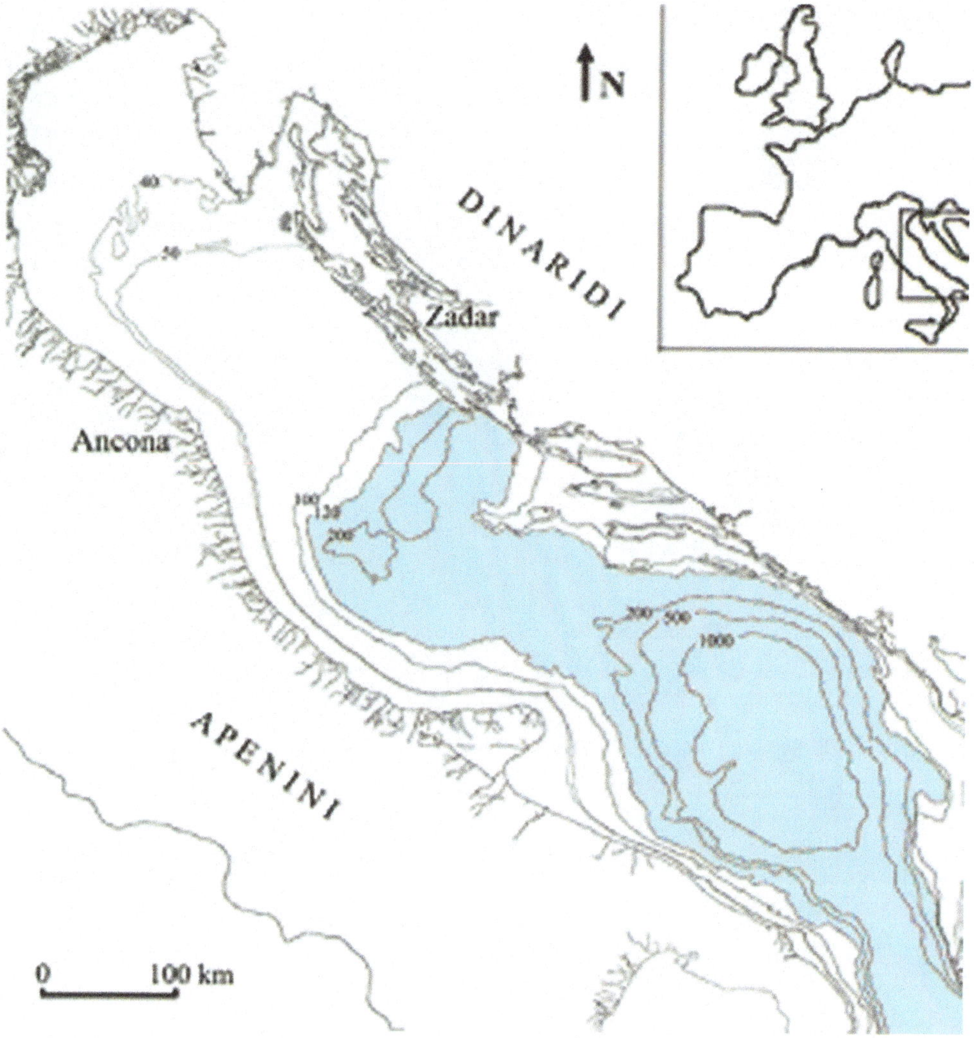

Fig. 1.10 25,000 years ago the levels of Mediterranean Sea and Adriatic Sea were up to 120 m lower than today

The beginning of the Mesozoic era was marked by the sedimentation changes resulting from the beginning of the formation of shallow sea carbonate shelf of the Dinarides. The lowering of water level of the deposition basin continues from the era of the young Palaeozoic and stabilizes during the Triassic era with the dominating carbonate sedimentation during most of the Mesozoic. The lowering of the water level processes in basins were characterized by the appearance of red clastites at the transition from Permian to the lower Triassic. These are the so-called Sajski klastiti. They are characterized by their intensive red colour and can be found the Gorski Kotar region in the north-west of the Dinarides through the area of Lika to Petrovo Polje and Kninsko Polje in the Dalmatian Zagora. Lower Triassic deposits are mainly covered by the central Triassic deposits consisting mostly of carbonate rocks. These are mostly thicker layered limestones and dolomites that are missing in Gorski Kotar where they are mostly eroded. Only minor occurrences of limestone conglomerates point to the edge of the sedimentation basin during mid-Triassic period. Within the carbonate rocks of Ravni Kotari and Dalmatinska Zagora, there are clastic and igneous rocks (Fužinski Benkovac, Senjska Draga, Donje Pazarište, Knin, etc.). The sedimentation of the upper Triassic deposits generally cover the deposits of mid Triassic era in continuity. The first part consists of clastites and then a thick series of dolomite whose occurrences are registered throughout the entire Dinaric Karst area of Gorski Kotar in the north-west to the Konavoski region in the south-east. After the period of shallow sea deposition during the upper Triassic, the beginning of Jurassic initiated a deep differentiation of basin deposition (Velić and Vlahović 2009). The shallow sea sedimentation continued through the entire Jurassic era on the terrain south of Žumberak and Karlovac—the Dinarides area, and at the same time in the Pannonian area, deep-sea sedimentation began. That was the result of tectonic events

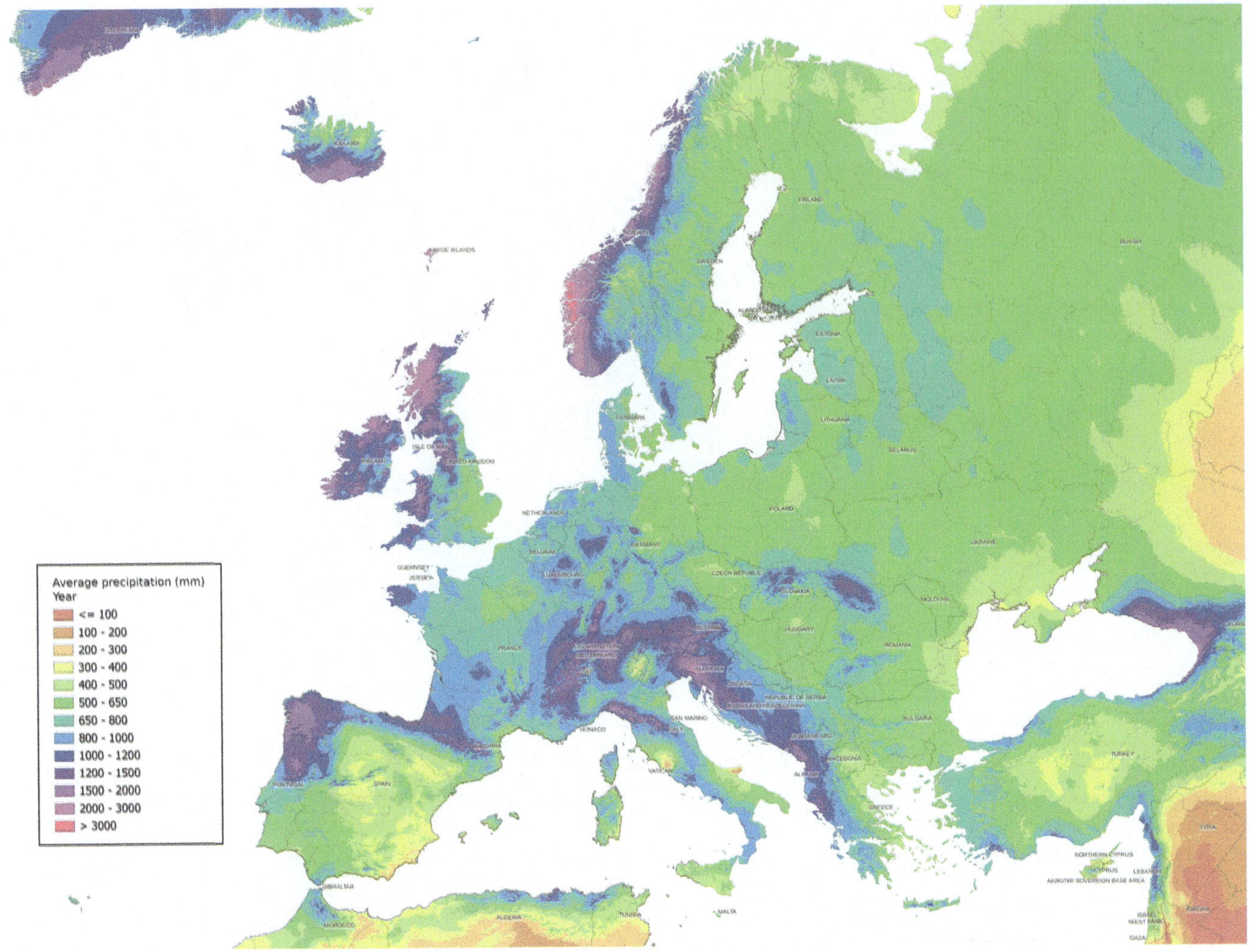

Fig. 1.11 Map of average year precipitation in Europe

during the Jurassic period. For the Dinarides area changes of sedimentation conditions in the area of shallow sea carbonate platform south of Žumberak and Karlovac and the marginal area of the deep-sea basin where mostly clastic sediments were deposited are particularly interesting. In the shallow sea ambience of the carbonate platform of the Dinarides, the transition from Triassic to the lower Jurassic —Lias—is continuous and marked by the gradual appearance of limestones within the dolomite in the lower Lias, the emergence of layered limestones with the presence of shellfish "Lithiotis problematica" in the Middle Lias and laminated limestones in the upper Lias.

In the middle Jurassic era—Dogger—a thick layered limestones with a very small amount of dolomite prevails in the lithological composition. Because of the variable sedimentation conditions on the shallow sea carbonate platform, certain lithological differences along the Dinarides were also detected. Towards the end of Dogger era, the pack of limestone deposits is in many areas interrupted by the appearance of limestone breccia as a result of a locally intensive lowering of the sea level. The carbonate deposits of the upper Jurassic—Malm age have the largest surface spread in the Dinarides area. Their main feature is the great facies diversity due to the greater depth differentiation of the sedimentation basin, from the shallow sea over the deep-sea pelagic and lagoon to the reef type. Great facies diversity means rotation of different lithological members such as layered and laminated limestones with cherts (Lemesh deposits), reef limestones, dolomites and layered massive dolomites.

In the north-eastern basin edge of the Dinarides, lithological differences during the Jurassic period were far greater than on the carbonate platform. This is an ophiolite–sedimentary rock complex dominated by metamorphic and igneous rocks created during the tectonic breakdown of the labile belt of the carbonate platform during lower Jurassic era. The shallow sea sedimentation on the carbonate platform continues in the Cretaceous period, with certain deviations in the north-west and central part of the platform due to the emerging of the land on a part of the platform. Most of

Fig. 1.12 Velebit Mountain is the longest mountain in Croatian Dinaric Karst

the carbonate rocks of the lower Cretaceous age were constructed from various types of limestone and diagenetic dolomite, mostly at the transition from Jurassic period to Lower Cretaceous period and from lower to upper Cretaceous period. In the areas of emerging of the land, the complex of carbonate deposits begins with the limestone breccias made of large components that gradually turn into laminated limestones with dolomite particles. The carbonate series of lower Cretaceous age ends again with poorly stratified limestone and dolomitic breccias indicating a transition to the upper Cretaceous. In general, the depositions of the carbonate rocks of the lower Cretaceous age took place mainly in shallow sea conditions which the footprints and remains of the entire skeletons of the dinosaurs in the area of the Istrian peninsula confirm.

The transition between the lower to the upper Cretaceous is marked by dolomite breccias resulting from the emergence of the sea on the most of the carbonate platform. Gradual palaeogeographic changes are characteristic of Upper Cretaceous. These changes indicate disintegration of what used to be relatively stable platform under the influence of sedimentation tectonics. A fairly indented platform surface was created with a relatively high degree of environmental differentiation. The shallow parts of the platform were occupied by the rudist communities, resulting in the deposition of limestone with the fossil remains of the rudists. The older part of the upper Cretaceous is marked by the occurrences of dolomite alternating with limestone, followed by a thick series of so-called rudist limestone. The deeply differentiated platform at the end of Upper Cretaceous experienced considerable emergence, so the greatest part of the platform was dry land and the deeper parts secured the continuity of sedimentation towards Cenozoic. The land parts of the platform are characterized by the development of karst forms with numerous sinkholes and caverns filled with terra rossa and bauxite. It is a palaeo karst in Croatia.

The deeper parts of the platform with the remains of the sea gradually turned to swamps with a continuous deposition from the upper Cretaceous to Palaeogene (Cenozoic). These are so-called *Liburnian* deposits with traces of coal. At the beginning of the Eocene, by the transgression of the sea, the true marine sedimentation was re-established throughout most of the platform. All of the so-called foraminiferal limestones (*miliolid, alveolina, nummulite*) typical of the shallow turbulent sea environment were settled. In the middle of the Eocene on the carbonate platform, the formation of elongated basins of the Dinaric orientation (north-west–south-east) occurred. Basins were longitudinally gradually

Fig. 1.13 Zrmanja River canyon

deepened, and the increasing amounts of incoming terrigenous material altered the lithological characteristics of the sediment. The foraminiferal limestones gradually turned to marl and then to clastic deposits with predominantly clay and sandy components (flysch). At the end of Middle Eocene, the whole area of the Dinarides was tectonically extremely active. The elongated deposition basins moved to the central part of the Dinarides in Ravni Kotari and the so-called *Promina* deposits were created.

With the end of precipitation of the flysch and the Promina deposits (Babić and Zupanič 2007), the marine sedimentation on the carbonate platform ended. Intense compressive tectonics in Oligocene and Miocene resulted in the formation of large amounts of coarse-grained clastic material that was lithified and turned into breccia (Jelar deposits). The rise of the Dinarides at the end of Palaeogene and the beginning of Neogene resulted in the separation of the Pannonian and the Adriatic Depression (Tethys and Paratethys) and the formation of two separate sedimentation areas. The Dinarides generally entered the land phase of development, but intense tectonics at the end of the Palaeogene and the beginning of Neogene formed the longitudinal depressions of the Dinaric line of prostration and depression transversal to the structures of the Dinarides along the strong transversal faults. The longest known caves in the Dinarides in Croatia have been found in these areas (*Crnopac Cave System, Munižaba, Cerovac Caves, Muda labudova, Duša* in the Crnopac area on the southern Velebit with a total length of over 70 km, etc.). Freshwater lakes emerged in depressions, where carbonated and clastic deposits with occasional shades of coal were deposited during Neogene. It is believed that the lakes were larger than today's large karst fields, but sediments today mostly fill the bottom of large karst fields such as Kninsko Polje, Sinjsko Polje, Zrmanja river valley (Fig. 1.13) and other karst fields in the Dinarides area. Remains of lake sediments exist even in the great depressions in the Adriatic belt, such as the depression of the Vrana Lake on the island of Cres. Lithologically, those sediments are marls, sandstones and less often limestone inserts. Sediments of the Neogene ages in the Dinarides play an important hydrogeological function as barriers to the movements of groundwater.

The Quaternary period (Pleistocene and Holocene) is the youngest geological period referring to the last 1.75–2.5 million years of the Earth's development. At the beginning of the Quaternary, the Dinarides generally had their present morphological forms and were ready for further predominantly land development. The climate was very variable at relatively short intervals, so during the Quaternary there were four ice ages (Gunz, Mindel, Riss and Wurm), three intermittent and one later ice age, resulting in strong surface and underground erosion processes in predominantly

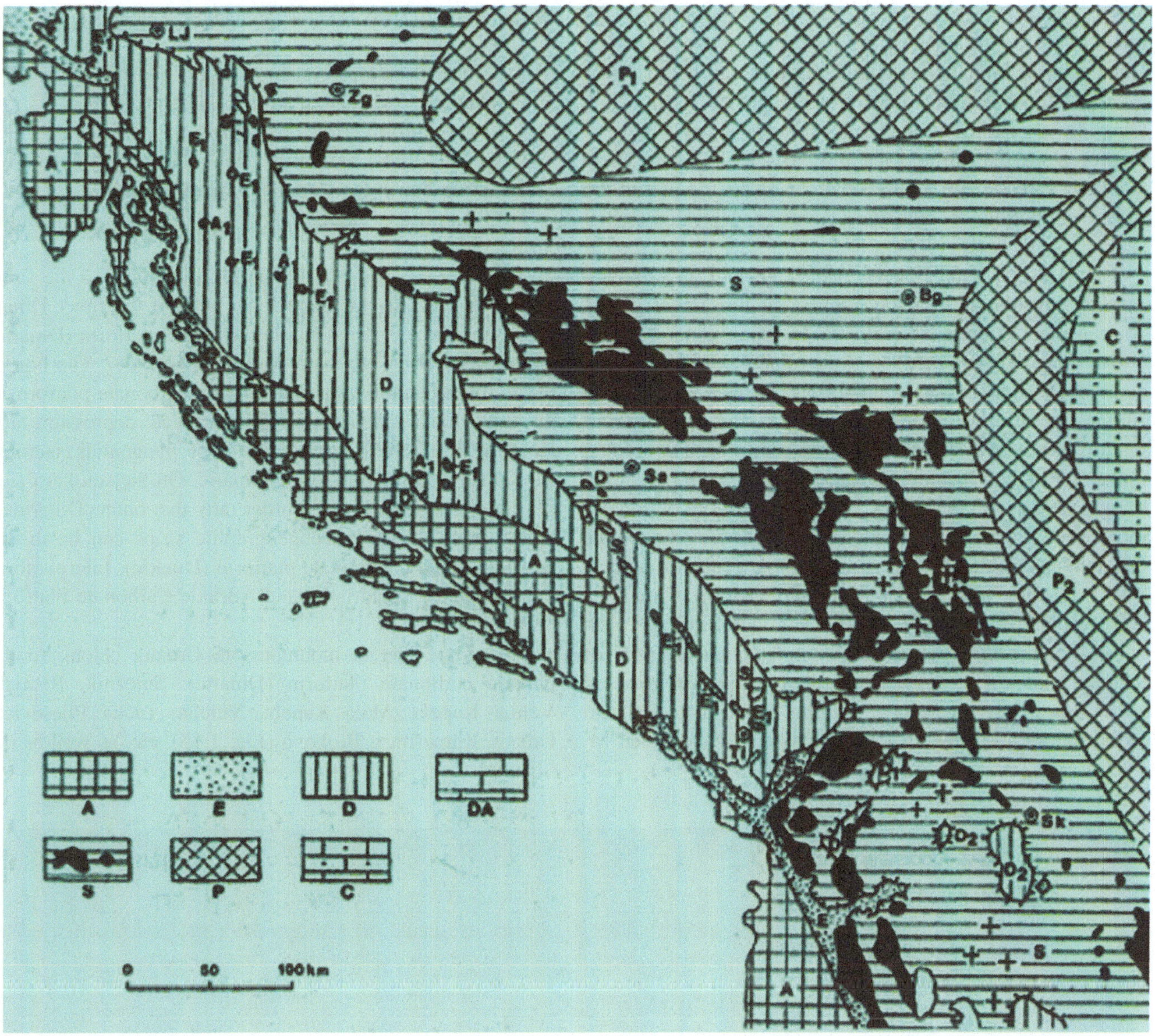

Fig. 1.14 Geotectonic units in Dinarides (A—The Adriatic carbonate platform (Adriatic); D—Dinaric carbonate platform (Dinaric), E—Interplatform space (Epiadriatic), DA—The border area of the Dinarides and Alpides, S—Inner Dinarides complex (supradinaric), P—Prealpine structural complexes (Pannonian structures)

carbonate rocks of the Dinarides. At the beginning of Quaternary strong karstification of carbonate rocks and the creation of the largest part of today's surface and underground karst forms began. By the aggressive action of the water, the tectonically formed fracture and fault systems have been expanded to the dimension of large caves and pits (e.g. the previously mentioned area of Velebit). Moreover, the formerly locally formed basins were connected to each other, enabling tens of km of groundwater flows within the dimensions of the basins (1000 km^2). This resulted in strong concentrations of underground streams and occurrences of large karst springs. According to the type and spatial distribution of sediments, it can be concluded that during the ice age, the Dinarides were covered with glaciers that descended to the sea or other depressions where glacial lakes were formed on Grobničko Polje and some other karstic fields in the Dinarides (Marjanac and Marjanac 2004). Glacial lakes have largely generated the development of karstic processes due to the concentration of water at high altitudes. Quaternary sediments were not very wide spread due to the indivisibility of relief. Lake sediments in karst fields, where gravel, sand and clay predominate, have the greatest surface and depth distribution. There are also large sediments of Aeolian origin. However, in the upper part of the Dinarides, these are mostly small occurrences of erosive remnants of these deposits, while in the Adriatic they are visible on the northern islands (Susak, Mali Lošinj, Unije and Srakane). In the Adriatic Depression and the north-east of the Dinarides, Quaternary sediments have higher surface and depth distribution. Fluvial origin sediments in the karstic areas of the Dinarides are mainly limited to karst fields and river valleys that run along the Adriatic Sea or the Black Sea basin. There is a large spread of terra rossa that covers a large part of the slopes and bottoms of sinkholes, as well as talus and eluvial sediments and rock-creeps.

Dinarides are a part of the Alpine geological–structural assembly resulting from the convergent movement of the African plate in comparison to the Eurasian plate (Matičec 2009). The convergence of these plates resulted in the contraction of the former Neotethys where there were several carbonate platforms separated by deep basins. One of these platforms was the Dinaric carbonate platform. The mutual movement of continental plates and formed platforms at the marginal parts of the plates is the foundation of today's mobilistic approach to the interpretation of tectonic events throughout Croatia, as well as in the area of the Dinarides. The originator of the mobilistic approach in Croatia is Academician Herak (1986, 1991). The mobilistic approach is based on the effects of subduction in the boundary areas of tectonic plates (Fig. 1.14). The end result is the directed stress that results in folding, shifting folds, tangential and radial mutual movements of rock masses. Academician M. Herak distinguished several geotechnical belts in the territory of Croatia:

A—The Adriatic carbonate platform (Adriatic).

D—Dinaric carbonate platform (Dinaric).

E—Interplatform space (Epiadriatic).

DA—The border area of the Dinarides and Alpides.

S—Inner Dinarides Complex (supradinaric).

P—Prealpine Structural Complexes (Pannonian Structures).

The largest part of the Dinarides, according to the same author, belongs to the geotectonic belts of the Inner Dinarides (supradinaric), the Dinaric carbonate platform (Dinaric) and the Adriatic Carbonate Platform (Adriatic). The border of the inner Dinarides and the Dinaric carbonate platform is on the south-western edge of Karlovac depression and Kordun with the characteristics of tangential tectonic supradinaric in relation to the Dinaric. On the southern side of the above-mentioned border are the outer Dinarides, within which three palaeogeographic strips can be distinguished: Dinaric carbonate platform (Dinaric), Interplatform Pelagic Band (Epiadriatic) and Adriatic Carbonate Platform (Adriatic).

All of the largest mountains in Croatia belong to the Dinaric carbonate platform (Dinaric): Snježnik, Risnjak, Velika Kapela, Mala Kapela, Velebit, Lička Plješevica, Dinara, Kamešnica, Biokovo (Fig. 1.15), etc. as well as the

Fig. 1.15 Biokovo Mountain is one of the largest mountains on Dinaric carbonate platform in Croatia

great karst fields in this mountain area. During Mesozoic over the Dinaric, a large mass of carbonate rocks was deposited in the shallow sea environment. The total thickness of those sediments reaches to the depths of several thousand metres. The structural complexity of the Dinaric is evident in the area of Gorski Kotar where the clastic rocks of the Palaeozoic base of the carbonate platform are laid over the carbonate rocks of the Dinaric. Such occurrences can be found in other areas of the Dinaric as well. The south-western edge of the Dinaric carbonate platform (Dinaric) along the Adriatic Carbon Platform (Adriatic) is also of an overthrust character. However, it is not a very pronounced zone of overthrust, but of a system of allochthon forms in relation to the peripheral parts of Adriatic with frequent occurrences of tectonic shafts.

The Adriatic coastal area from Istria in the north-west to the Dubrovnik seaboard in the south-east and the islands belong to the Adriatic carbonatic platform (Adriatic). Adriatic has similar lithostratigraphic characteristics as Dinaric, but it is an area of pronounced subduction effects manifested by strong seismic activity (Aljinović 1981, 1986). One of the features of the Adriatic Sea is the deposition of clastic sediments in the final stage of the sedimentation cycle (Herak 1986, 1991)—the deposits of Eocene Age flysch.

While the academician M. Herak emphasizes the existence of a few sedimentation environments and the considerable mutual tangential shift of these entities, a group of authors from the Croatian Geological Institute in 2009. Adhered to the view of the existence of a single carbonate platform, occasionally of deeply differentiated conditions with pronounced sedimentation tectonics. Recent Dinarides morphology is largely a result of tectonics that can be identified on sediments from the end of Cretaceous with activity during Palaeogene to the end of Neogene and the beginning of Pliocene (Matičec 2009). The orientation of the Dinarides structure (north-west–south-east) is interpreted by deactivation of the compression and extension pressures in the upper Cretaceous. The carbonate platform was so differentiated to the extent that some parts were underground conditions, some under shallow sea conditions, and in some parts of the platform deep-sea conditions were formed. At the end of Upper Cretaceous, almost the entire platform was mainland. In the upper Eocene and Oligocene, sedimentation ceases and the rise of the Dinarides begins as a consequence of spatial contractions. Along the regional direction of the strain (north–south-western) folding, reverse faulting, and overthrusting of the rock masses occurred by the north-west–south-east direction, which is defined as the Dinaric direction of provision. In what extent the contraction tectonic movements were intensive is best shown on the overthrust of the Palaeozoic rock base of the carbonate platform over the carbonate rock platforms, which is visible in Gorski Kotar. Neotectonic movements begin in the Miocene and continue until today. They are characterized by a change of stress from the direction north-east–south-west to the direction north–south. These movements activated existing structures and faults created by the earlier tectonic movements, but they also created new faults, very often of a regional spread, diagonal to earlier structures and faults.

The causes of the deformation of the recent structural complex and the occurrence of the earthquakes are tectonic movements caused by the mutual movement of some parts of the Dinarides (Prelogović et al. 1995, 2004) especially so-called Adriatic microplate (Adriatic) in the north–north-west direction. The consequences are reverse shifts in compression zones, especially in areas of previously formed reverse faults in the contact zones between Dinaric and Adriatic. Thus, for example, for the north Adriatic area shifts and rotation of the Adriatic microplane (Istria), which exerts pressure on the Adriatic, are important. Movements of the Adriatic are opposed by the parts of the Dinaric that are under pressure from the north and the consequence is plunging of rock masses into the depths.

The hydrogeological characteristics of rocks are evaluated according to the collected data on the lithological composition of the rocks, the genesis of the explored part of the Dinarides, the degree of deformation of the rocks on the surface of the terrain and the karst subterranean, the apparent velocities of the underground streams, the position of springs and swallow-holes in certain basins, the state of the rock cores in exploration and exploitation wells and other elements which, according to the researcher's assessment, are important for the formation and movement of groundwater and surface water.

The term 'carbonate rock' encompasses the complex of different lithologic members from pure limestone to dolomite of different lamination. Although the entire lithological complex of carbonate rocks falls into the range of water-permeable rocks with the possibility of forming aquifer systems, hydrogeological features are graded from good water-permeable rocks (predominantly limestone) to poorly water-permeable rocks (mostly dolomites). Accordingly, the higher the content of the limestone the water permeability is greater, and with the greater content of dolomite the water permeability decreases. The name "clastic rocks" also includes the complex of rock of different lithological composition and different degree of water permeability; however, the differences are related to the range between poorly water-permeable and water-resistant rock masses. Carbonates and clastic rocks construct the base rock mass of the Dinarides and their hydrogeological function in the formation of the basins and the network of underground streams depends on their water permeability and their position in the structural composition of the Dinarides. The third group of deposits was created by erosion processes during the youngest geological period of the Quaternary. They are variable lithological structures and are relatively thinner in relation to the rocks of the basic structures of the Dinarides. It is mainly about unbound and semibound deposits whose water permeability depends on the content of the clay component that reduces water permeability. Given the underground water flows in the deep karst subterranean and the barriers built from the clastic rocks, deposits of Quaternary age do not have a major impact on the flows in the karst subterranean, but as surface cover deposits they affect the infiltration rate of the precipitation to the underground (Garašić 1997).

Carbonate rocks can generally be divided into two groups of different degrees of water permeability. The first group is good water-permeable carbonate rocks that make up the majority of water systems with exceptionally fast underground streams, especially in heavy fault zones with deep impact of the karstification processes. A great part of the rock complex of Mesozoic age, starting from Lias to the upper Cretaceous, has all the features of a highly water-permeable medium. Carbonate rocks from the beginning of Tertiary share the equal hydrogeological features (foraminiferal limestone). The porosity of the rock is fractural or secondary. The average porosity of limestones is about 3%, which is very low compared to the interstitial porosity deposits, but the large mass of limestones and the interconnection of karstic canals in the underground provide rapid flow of large amounts of groundwater during rainy periods. Concentrated underground streams are mainly tied to tectonically damaged zones within limestones. In the second group of poorly water-permeable carbonate rocks are the rocks in which the dolomite prevails over limestone. These are primarily the dolomites of the upper Triassic that stretch along the entire Dinarides and are part of the underlying mountain created after the formation of the Dinaric carbonate platform. The sedimentation conditions on that platform were quite different so regardless of the generally shallow sea environment, there were conditions for the dolomitization processes and the formation of clastic sediments. Major occurrences of dolomite are related to the beginning of Jurassic (Lias) and Malm when a significant differentiation of the Dinaric platform occurred. Dolomites also have a fractural porosity, but due to lower water solubility than limestone, dolomites generally have no developed karstic forms. Caves are predominantly of a fractural type. So-called Jelar deposits in which rough limestone breccias prevails among the lithogenic members (Velebit), but with a considerable share of the clastic component, these rocks are commonly attributed to the hydrogeological status of the variable water permeability.

The clastic rocks are generally evaluated as watertight because the lithological composition is dominated by clay sediments. It is a lateral and vertical change of shale, silt, sandstone, marl limestone, conglomerate and breccia that give this rock mass a variable water permeability with the possibility of forming smaller aquifers with small springs. However, as a whole, depending on depth penetration, they are barriers to the movement of groundwater in the predominant karst carbonate medium of the Dinarides. In the Dinarides, there are generally two stratigraphic clastic rock types. These are mostly clastic rocks of the Palaeozoic and Triassic ages in Gorski Kotar, Lika and Dalmatia building the cores of anticlines. They mostly form the basis of the carbonate mass of the Dinaric platform and mark the beginning of sedimentation on that platform. Another stratigraphic type consists of clastic deposits of Palaeogene (flysch) whose occurrences in the coastal Adriatic, in Istria (Babić and Zupanič 2007) and on islands have an important hydrogeological function of the barrier to movement of the groundwater. There are numerous large karst springs formed on the fault contacts of the carbonate mass of Dinarides as aquifers and clastic flysch deposits as barriers. Equivalent to the "Jelar" deposits on the Velebit mountain range are the "Promina" deposits in Ravni Kotari, which due to the high content of the fine clastic component have similar hydrogeological characteristics as the older flysch. Sediments of variable water permeability and relatively small thickness include the complex of Quaternary deposits that emerged during the mainland phase of the Dinaric platform development. These are lake, fluvioglacial and alluvial deposits of karst fields, terra rossa, most common cover on carbonate rocks, rock-creep and rock-creep breccias, talus and eluvial deposits, etc. The porosity of these deposits is interstitial. The thickness varies from place to place, but not so much that they can significantly influence the formation and movement of groundwater in the karst media except in the marginal areas of karst fields where lake deposits can have a barrier function in relation to the deep streams of groundwater towards karst fields.

At the very beginning of the presentation of the method of formation and dynamics of groundwater and surface water in the Dinarides, it is necessary to emphasize the crucial significance of the geological structures. The cause of the former extreme perceptions of the groundwater movement (theories of different authors) in the karst lies in the fact that the karstic issues have long been outside the interests of geologists and that different karstic theories have arisen before geological relations have been sufficiently explored. Today, the situation is significantly different because in the twentieth and twenty-first centuries, numerous tracing of underground streams have been made for different purposes. The research methods have progressed significantly, especially regarding the preparation of geological substrates, hydrogeochemical research methods, different types of measurements and of course the application of GIS. All this has considerably improved the spatial analysis as well as the water resource analysis. Hydrogeological units are defined according to the available data on spatial layout of rocks of different hydrogeological characteristics within geological–structural forms, underground stream tracing data, hydrogeochemical parameters in groundwater, spatial distribution of springs and swallow-holes, water balance of individual units and other available parameters obtained by different research methods. From the geographic data, it is clear that some of the Dinarides water runs towards the Adriatic Sea, and part towards the Sava River and across the Danube to the Black Sea (Fig. 1.16).

Determining the watershed for large karst aquifers such as the Dinarides karst area has always been a major problem for researchers because of the stated hydrogeological characteristics of karstic aquifers (Fig. 1.17). Given that in most cases it is the underground watersheds of highly variable hydrological flow conditions, it is common for the watersheds in karst to say that they are zonal with the possibility of forming underground streams in the distribution area in different directions during different hydrological conditions. The first hydrogeological watersheds between the Adriatic

Fig. 1.16 Map of river basins in Croatia

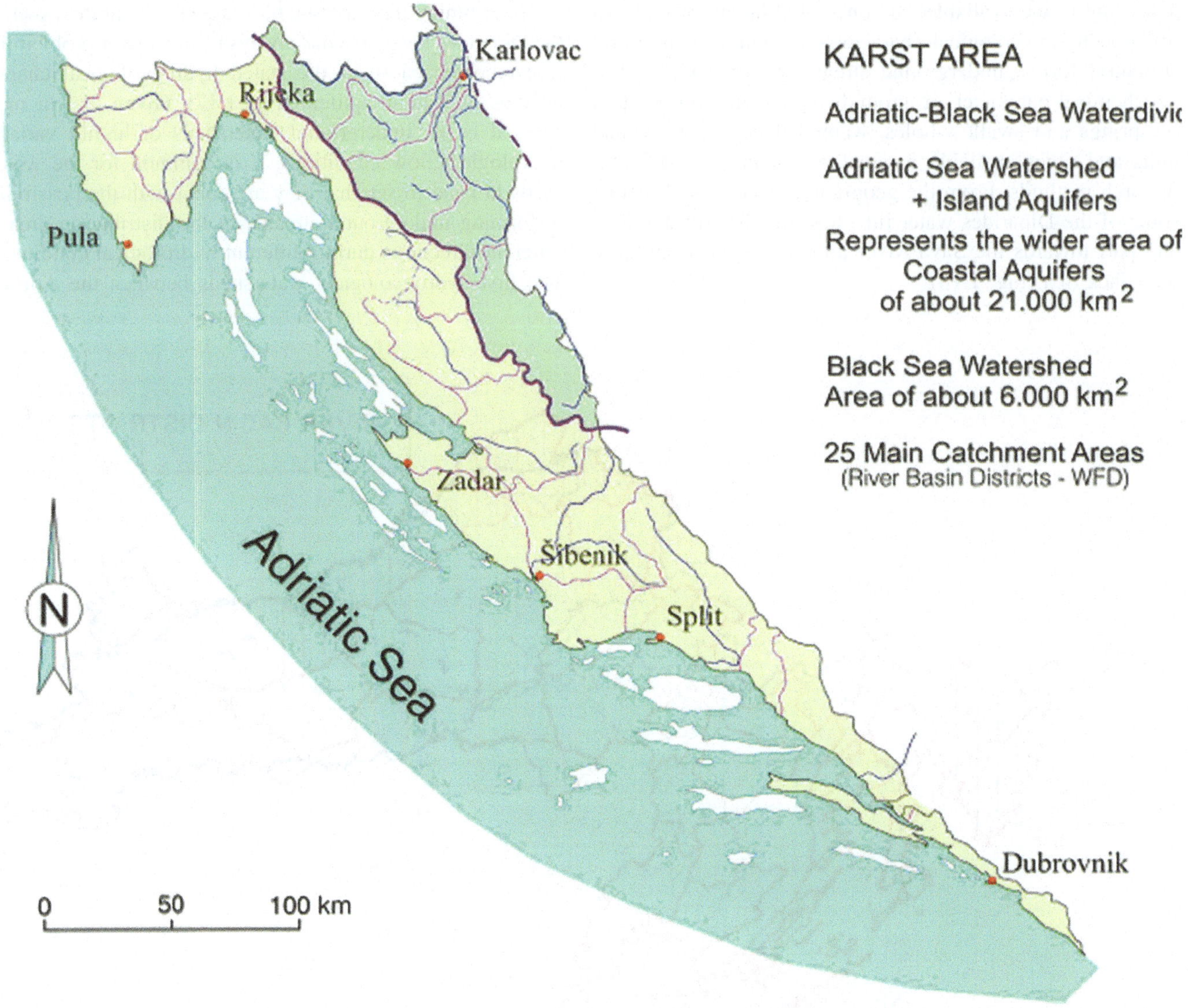

Fig. 1.17 Borders of basins

and Black Sea basins were determinate for the springs in the town of Rijeka, and the first vision of the entire watershed between the Adriatic and Black Sea basins on the territory of the Republic of Croatia according to the experiences of the researchers came shortly afterwards. Generally speaking, the watershed between the Adriatic and Black Sea basin extends from Slovenia to the north-west across the mountainous area of Gorski Kotar, Velika Kapela, the part of Mala Kapela, Ličko Sredogorje, the area of Bruvno and the southern Velebit to the Pošitak mountain, where it leaves the territory of Croatia and crosses into the territory of neighbouring Bosnia and Herzegovina. The inclusion of certain areas in one or the other basin has been confirmed by numerous tests and traces of groundwater performed on both sides of the watershed.

In the north-western part of the Dinaric carbonate area in Gorski Kotar, the watershed between the Adriatic and Black Sea basins separates water source systems in the coastal area of the Kvarner Bay in the Adriatic Sea basin from the springs of the Čabranka and Kupa Rivers in the Black Sea basin. The watershed from the Slovenian border to Gerovski Kraj is related to the occurrences of poorly water-permeable Dolomite of Triassic age and large mass of the entirely watertight clastic rocks of Palaeozoic age in the peak of the Black Sea basin. In the hinterland of the original spring of the Kupa River, the structures of anticlines of Gorski Kotar with the waterproof clastic rocks of the Palaeozoic age tectonically separate and open a spacious carbonated mountain area for the development of the zonal watershed. It is interesting to note that the Kupa river spring in the Black Sea basin and the Rječina river spring in the Adriatic basin separate on the equal altitudes (325 m above sea level), indicating a uniform development of karstification on both sides of the watershed and the possibility of moving of the watershed in the vast mountain range of Gorski Kotar. The watershed site was confirmed by tracing of the underground streams from the Gomanac area in the Adriatic basin and a swallow-hole in the border region of Slovenia to the Black Sea basin.

From the area of Crni Lug to the south-east to Lič Polje near Fužine, the watershed is again linked to the occurrences of poor water-permeable dolomites of the upper Triassic age and completely watertight clastic rocks of Palaeozoic age. Part of the area built out of watertight clastic sediments drains to the Adriatic basin. This is confirmed by tracing the swallow-holes of the Ličanka watercourse and a swallow-hole near the Lepenica watercourse towards the Adriatic coast as well as the swallow-hole along the Zagreb-Rijeka highway towards the Black Sea basin. The Lič Polje near Fužine watershed between the Adriatic and Black Sea basins is linked to the spread of the anticlinal form of the mountain areas of Burni Bitoraj and Velika Kapela, which is considerably lowered than the structural form near Fužine, where the core of the anticlinal form on the surface of the terrain is constructed by the watertight clastic rocks of Palaeozoic age. In Burni Bitoraj and Velika Kapela mountain range, watertight rocks are at a much greater depth, but with regard to the anticline form and the watertight core, they are a good foundation for forming of watershed. In the Velika Kapela zone, the watershed is the closest to the seashore, so the Adriatic basin is the narrowest in that area. In the area of watershed, tracing was only possible from the swallow-hole zone of Drežničko Polje, which established an underground link with the spring of Zagorska Mrežnica near Ogulin and belonging of the area to the Black Sea basin. From the height of Brinje, the Adriatic coast is spreading towards the mountain area of Mala Kapela, including the Brinjsko Polje, for which the connection with the source Jurjevska Žrnovnica in the coastal part of the Adriatic is established by tracing. The positions of springs and swallow-holes oriented towards the Adriatic region point to the affiliation to the Adriatic basin. On the Black Sea side of the watershed a series of karst fields (Jezerane, Crnac Polje) were developed, from which the connection with the spring of Zagorska Mrežnica in the Black Sea basin was established by tracing. From Brinjsko Polje, the watershed stretches across the area of Plitvice Lakes along the larger occurrences of Jelar deposits, which is the reason for the shallow karstification and practically surface transitions from the Adriatic to the Black Sea basin. Tracing of the swallow-holes in Dabarsko Polje from the north-eastern side of the watershed, connection with springs in Plaški area in the Black Sea basin has been established. On the south-eastern side of the watershed is Gacko Polje with springs in the area of Ličko Lešće and Sinac and swallow-hole zones of Švica, Brlog and Hrvatsko Polje. Tracing of the swallow-holes in Gacko Polje has been associated with numerous coastal springs from Novljanska Žrnovnica (Denić-Jukić et al. 2007) in the north-west to Karlobag in the south-eastern part on the nearly 100-km-long coastline. The watershed of the Adriatic and Black Sea basins from the mountain range between the Plitvice Lakes and the Gacka River Basin built to a great extent from the dolomite passes to the Mala Kapela mountain range and further to the Kozjan area, where due to the appearance of watertight clastic rocks of Palaeogene age it has all the characteristics of the surface watershed between Gacka springs and Krbavsko Polje (Biondić et al. 2010). There were several underground traces performed in this area due to the interest of defining the Plitvice Lakes National Park basin. Tracing from Brezovačko Polje and Homoljačko Polje area confirmed the connection with springs of Plitvice Lakes, and tracing from Vrhovinsko Polje and polje near Turjanski and Kozjan confirmed the connection with the Gacka spring area.

From Kozjan, the watershed passes to the area of Ličko Sredogorje and extends along the south-western edge of Krbavsko Polje towards the area of Udbina built mainly from the water-permeable clastic deposits of the lower Triassic age. The result is the occurrence of numerous small springs along the south-western edge and the swallow-holes on the north-eastern side of the field. The tracing of the swallow-holes in Krbavsko Polje confirmed the affiliation with the Black Sea (river Una) and tracing of the swallow-holes in Gornja Ploča in Ličko Sredogorje along the Jadova River towards the Adriatic (Ričica River). Ličko Sredogorje is gradually transformed from the mountain region into the plateau area of Bruvno, where a network of the surface runoff with the watercourses of Jadov, Ričica and Otuc on the northern edge of Velebit (Štikada, Gračac) has been developed. There, the waters sink and reappear at the springs along the Zrmanja River on the lowest step of the basin. The watershed between the Adriatic and Black Sea basins on this area is associated with the geologic structure called the Kremen from where it descends towards the Bruvno Plateau, part of which belongs to the Black Sea basin and the other part of the Adriatic Sea basin. It is interesting to note that Čemernica Mountain, although built exclusively from the watertight Triassic age clastic rocks, is not a hydrogeological barrier because watertight rocks have been overthrusted so the underground flows towards the Una

River spring are possible through the younger karstified rocks underneath the overthrust. From the Bruvno, the watershed connects to the structure of Velika Popina, the final part of Velebit, and further to the mountainous area of Poštak, crossing the state border to the neighbouring country of Bosnia and Herzegovina. South-east of this point, the whole of area of Croatia belongs to the Adriatic Sea basin.

The mountain area on the south-west side of the watershed towards the **Black Sea basin**, i.e. the coastal area and the karst hinterland from Istria in the north-west to Dubrovnik and Prevlaka in the south-east, as well as all the islands on the eastern side of the Adriatic belongs to the Adriatic basin. This is an enormous supply area built predominantly of karstified carbonate rocks. The presence of aquifers with the interstitial porosity is negligible. When they do exist, their balance of water is bound to the deep karst aquifers in the base (Denić-Jukić and Jukić 2003). Basic features of the Adriatic basin are the extensive water catchment zones in mountain areas with very high precipitation, long underground flows and very complex spring conditions commonly found on contact with watertight barriers or under the influence of the sea. The area of the Adriatic basin in the Dinarides consists of extremely complex geological structure with large amounts of precipitation and carbonate rocks susceptible to the dissolving processes, which is the basis for the differences in formation and the flow of surface and groundwater. During the youngest Quaternary geological period, numerous surface and underground drainage systems have been formed as hydrogeological units of equilibrium balanced water resources. By the end of the 1960s, hydrogeologists considered that due to the high degree of heterogeneity of the water bearing rocks, it is impossible to separate and balance hydrogeological units or basins of individual springs or groups of springs in the karst area of the Dinarides, regardless of the large number of tracing conducted on the underground streams. Hydrologists have mostly used surface basins for their calculations, which caused great confusion in the estimates of the amount of runoff and water losses in karst basins. Significant decisions have been made on the use of water resources on the basis of hydrological measurements on open watercourses, especially for the purpose of building accumulation lakes for the energy use of water resources. First attempts to allocate hydrogeological basins and calculate water balances for individual springs or groups of springs in the karst region of the Dinarides are made for springs in the wider area of Rijeka. It was stimulated by the need for rational use and protection of drinking water sources from pollution, which was impossible to do without knowing the spatial layout of the basins. During the early seventies, hydrogeological research was extended to the Lika, Dalmatia and Istria regions. Today, it is inconceivable to establish the protection of spring water or water resources management system without knowing the basins, or how the EU Water Framework Directive states, without knowing the spatial arrangement of groundwater bodies. Groundwater bodies are conceived as basic water units for preventing water degradation, protection and improvement of the ecosystem status as well as the establishment of a sustainable long-term water use protection system. As a part of regulatory alignment with the EU, Croatia has set up groundwater bodies for its entire territory. For the karst area of the Dinarides, a total of 17 groundwater bodies have been isolated, of which 12 are in the Adriatic basin and 5 in the Black Sea basin.

Groundwater bodies are defined by grouping natural hydrogeological basins for easier water resource management organization and the assessment of qualitative and quantitative status of water resources.

Part of the Dinarides on the north-eastern side of the watershed towards the **Adriatic basin** belongs to the Black Sea basin. In Croatia, the Black Sea basin includes mountain ranges of Risnjak and Klek in Gorski Kotar and Mala Kapela, Lička Plješevica, Kremen and Čemernica in Lika. In geological terms, this is an area of exterior and interior Dinarides and the border area of these two megastructural units. Part of the karst area of the Dinarides in Croatia belonging to the Black Sea basin drains two main rivers that join the Sava River. They are the rivers Una and Kupa with the tributaries of Dobra, Mrežnica and Korana. Those are the main aquifers in the mountain range of the outer Dinarides, and the great karst springs in the border area of the outer and inner Dinarides. From the hydrogeological point of view, two areas of different hydrogeological features are distinguished, spatially directly depending on the layout of the geological structures. It is the mountainous area of the structural unit of the outer Dinarides with the development of deep karstic underground forms on relatively flattened karst plateau, so-called shallow or fluviokarst located north-east of the mountain area. River Kupa and its tributaries drain a large part of Gorski Kotar and the western part of Lika, including Plitvice Lakes. River Una drains the eastern part of Lika and hence in the karst area of the Black Sea basin, basins of the river Una, Kupa, Dobra, Mrežnica and Korana can be distinguished as part of the drainage system of the river Kupa (Fig. 1.18).

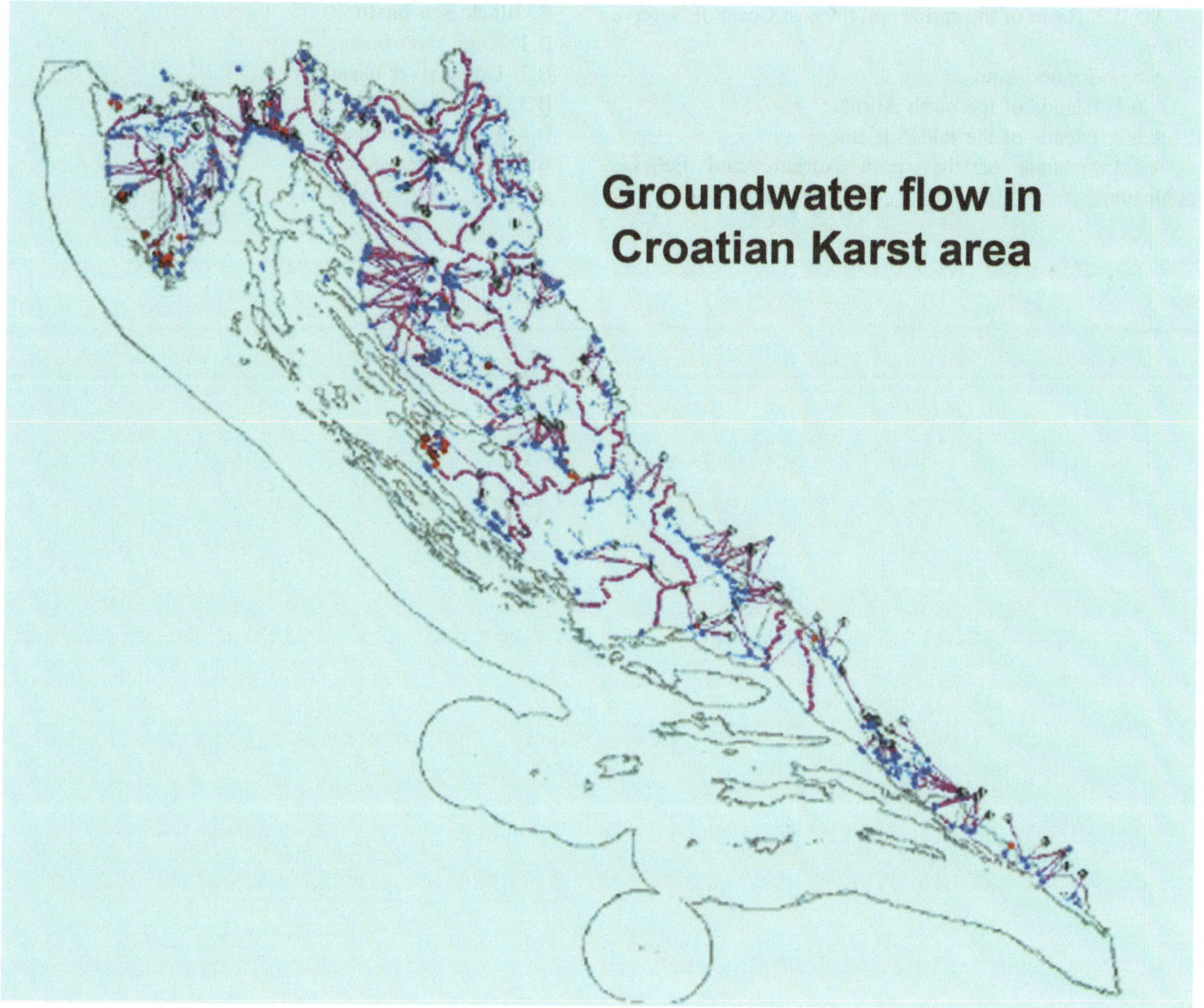

Fig. 1.18 Groundwater movements in Croatian karst

Hydrogeological basins in the Dinaric Karst in Croatia (Biondić and Biondić 2014):

A. Adriatic basin

A.1. Area of Istria.

A.1.1. Dragonja river basin.

A.1.2. Mirna river basin.

A.1.3. Raša river basin, south and west Istria.

A.2. Rijeka bay area.

A.2.1. Coastal springs of the upper part of Rijeka bay basin.

A.2.2. Rijeka city spring basin.

A.2.3. Bakar bay spring basin.

A.3. Area of Lika.

A.3.1. Coastal springs of Lika and south part of Hrvatsko primorje basin.

A.3.1.1. Gacka river basin.

A.3.1.2. Lika river basin.

A.3.2. Zrmanja river basin.

A.4. Area of Ravni Kotari and Krka river basin.

A.4.1. Bokanjačko Blato basin.

A.4.2. Vranskoga jezero lake basin.

A.4.3. Krka river basin.

A.5. Area of Dalmatia and Dubrovnik seaboard.

A.5.1. Cetina river basin.

A.5.2. Neretva river basin.

A.5.2.1. Basin of the springs on the west Coast of Neretva River.

A.5.2.2. Basin of the springs on the east Coast of Neretva River.

A.6. Adriatic islands.

A.6.1. Islands of the north Adriatic.

A.6.2. Islands of the mid-Adriatic.

A.6.3. Islands of the south Adriatic and Pelješac peninsula.

B. Black Sea basin

B.1. Kupa river basin.

B.2. Dobra river basin.

B.3. Mrežnica river basin.

B.4. Korana river basin.

B.5. Una river basin.

2 The Dinaric Karst System in Croatia

Abstract

Research in the Dinaric Karst area has been conducted for several centuries. Many scientific discoveries about karst originated here. The Dinaric karst in Croatia is special due to the thickness of carbonate masses which can reach up to several kilometres. With constantly active tectonics in these sediments and great depths of impermeable substrates, karst processes create mostly vertical or steep cave channels. In some places caves deeper than 1400 m were explored. Some of them feature over 500 m deep vertical shafts. Areas along the Adriatic coast are special due to the appearance of vruljas or underwater freshwater springs. Different types of speleogenesis in this area are presented. Morphological characteristics and hydrogeological functions for each karst area in Croatia (inner, central and outer part of the karst area) are mentioned. Caves over 55 kms long, the deepest active karst springs (over 250 m deep) and ponors (over 500 m deep) show the great potential of this area. A brief overview of the tradition and history of research of caves in Dinaric Karst of Croatia is presented.

2.1 Introduction

"Planine" (Mountains) is the first novel written in Croatian language and printed in 1569. It is the first known written record of the name Dinara.

"Dinaric Alps" are described in Kozzen's Geographical Atlas first published in 1861. In 1869 in the "Atlant" atlas figured both "Dinara Gora" (D. Mountain) and "Dinarske planine" (D. Mountains). The entry "Dinara" in popularly used Meyers Konverssations-Lexikon (1875) included»Dalmatischen« (Dalmatian) or»Dinarischen Alpen« (Dinaric Alps). From the same period 1878. is "Prirodni zemljopis Hrvatske" (Natural Geography of Croatia) by Klaić where both expressions are used "Dinarske planine" (plural) and "Dinara planina" (singular). In "Karst" from 1893 (Cvijić 1893), the main part of Dinaric Karst as we see it nowadays was a part of Adriatic (Jadranski) Karst. Lot of work and research is waiting scholars to explain the etymology and to find first apparitions of the name Dinara on maps. All this is mentioned in papers of Kranjc (2004, 2010).

The Dinaric Karst is a world-renowned area of specific geomorphology (Bognar 2006) in which the largest number of different karst phenomena can be found. From the smallest ones, measured in centimetres to the largest karst fields measured in tens of kilometres. The definition of karst is complex because it includes:

(a) Materials (rocks), predominantly carbonate and to lesser extent evaporites; and
(b) factors affecting materials, namely tectonics, i.e. rock crushing, mechanical wear, i.e. erosion by the action of water, ice, wind and chemical wear or tear or corrosion also by the action of water.

There are many references in the literature, and one of the simplest ones was given by Roglić (1974b): "*Karst is a set of phenomena related to the flow of water and forms of relief in carbonate rocks*". Karst also exists in non-tectonized sequences of carbonate rocks. Tectonic disturbance/rock damage only accelerates their karstification.

Karstification is not just a recent process, it also took place during the geological past. Today's Dinaric Karst relief was mostly shaped in the youngest geological periods, Neogene and Quaternary (Bognar et al. 2012). Therefore, this text on the general characteristics of the Dinaric Karst will give an overview of its geological history (Zupan-Hajna 2010; Mihevc 2010).

The Dinaric Karst is a world-famous natural landmark not because its inhabitants and nature lovers want it or like to talk and hear about it, but for a number of irrefutable facts, of which the most important are three:

M. Garasic, *The Dinaric Karst System of Croatia*, Cave and Karst Systems of the World,
https://doi.org/10.1007/978-3-030-80587-6_2

- Exceptionally large thickness of carbonate rocks, which in some areas is over 8,000 m and tectonics can make it reach up to 12,000 m;
- Development of all possible forms of karst phenomena;
- Karstification processes that were in repeated stages from the youngest Carboniferous till today.

The Karst Dinarides (Mihevc and Prelovšek 2010) stretch from the Italian province of Friuli in the north-west across Slovenia to Croatia, Bosnia and Herzegovina, Montenegro and Serbia to the north-west Albania (Nicod 2003, 2003a).

When it comes to the geographical boundaries of the Dinaric Karst, experts and scientists do not share a common opinion, so there are different data in the literature (Cerkvenik et al. 2018;). The criteria for determining the boundaries of the Dinaric Karst as a world geomorphologic feature should, in principle, be based on at least two determinants—(1) a thickness exceeding several hundred metres and (2) the development of typical karst phenomena (sinkholes, swallow-holes, caverns, caves, pits, etc.). According to this criteria, the Karst Dinarides are sensu stricto and include the eastern part of Friuli and Carso Triestino in Italy, southern Slovenia, a part of Croatia from Žumberak, Karlovac and Kordun to the outer islands of the Adriatic and the associated seabed, south-western Bosnia and Herzegovina, Montenegro and north-western Albania.

Karst Dinarides can be observed in the following areas:

Italy: The far north-eastern part of Italy in the vicinity of Udine and slightly north, in the province of Friuli, to the border with Slovenia near Tolmin. The second area is to the south, the Carso Triestino, which is actually the south-western part of the Slovenian Karst.

Slovenia: South of the line Tolmin–Idrija–Ljubljana–Višnja Gora–Trebnje–Boštanj–Krško–Gorjanci (Žumberak)–Plešivica (Gams 1974, 2003, Jurkovšek et al. 2013).

Croatia: West, south-west, south and south-east of the line Plešivica–Kostanjevac–Krašić–Karlovac–Veljun–Cetingrad–Tržac.

Bosnia and Herzegovina: South-west and south of the Tržac line (in Croatia along the border)–Bosanska Krupa–Bušević–Sanski Most–Vrhpolje–Bronzani Majdan (Banja Luka)–Piskavica (Banja Luka)–Dragočaj (Banja Luka)–Jagare–Gornji Borak–Bešpelj–Travnik–Jajce–Bugojno–Prozor–Konjic–Neretvom do Glavatićeva–Ulag–Gradina (On north-east foot of Bjalašnica)–Gacko—border with Montenegro (towards Piva Monastery). In addition, there is karst in eastern Bosniain the area between Sarajevo, Olovo, Zvornik, Srebrenica and Goražde (Milanović 1981, 2006).

Crna Gora: South from Piva Manastir–downstream of the Piva River to Plužine–Žabljak–north to Tara Canyon–Tara to Đurđević Tara–south and south-west with the entire Kučanjevica mountain range and north-western Sinjajevina to Šavnik–south-east to Međurječje on Morača–Tara spring under Maglić–and south-east to the Albanian border.

Albania: From the border with Montenegro (Tara River spring) to the sea and south-east to the Drima Valley and the tributary on the Skadar–Firëz–Tropojë stretch, or geologically speaking, to the Cukali Zone.

Adriatic Sea bed: The borders of the Karst Dinarides in the Adriatic extend even beyond the so-called outer islands. They extend approximately to the midpoint of the international waters belt, and they narrow down south of the island of Korčula towards the Montenegrin coast (which is a consequence of tectonic structure) (Fig. 2.1).

2.2 Geology of Dinaric Karst in Croatia

Geological maps show that the recent geological structure of the Karst Dinarides in Croatia is extremely intricate, since it covers rocks and deposits from younger Carboniferous to the present day. Geographically, this area was located at about 10° south latitude at the time of deposition of the first carbonates in the younger Carboniferous while today it is located between 42° and 46° north latitude. In other words, it traversed over 6,000 km. It would also periodically move back to the south occurred (Fig. 2.2).

The geological history of the Dinaric Karst in Croatia can be divided into six periods or, in terms of facies, into six depositional megasequences. From older to younger these are: (1) younger Carboniferous-older Permian, (2) older Permian-middle Triassic, (3) younger Triassic (younger Norian)-older Jurassic (older Toarcian), (4) younger Toarcian-younger Cenomanian/older Turonian, (5) older Turonian-end Cretaceous, (6) Eocene, (7) Oligocene–Miocene and (8) Pliocene–Holocene. The boundaries between megasequences are marked by important geological events, such as interruptions of deposition due to the uplift of the platform and deforestation or deepening of depositional environments.

(1) **Younger Carboniferous-Older Permian**. The first occurrences of the shallow sea limestones in the Croatian part of the Karst Dinarides are found as inserts within the Upper Carboniferous clasts and were discovered on the surface of 1 × 1.5 km in Lika in the Ričica formation. Minor occurrences of limestone can be found in Lower Permian clasts, but only in two places in Raduč and Rizvanuša. They were certainly spread on a larger surface, but eroded and encrusted

Fig. 2.1 Relief of Dinaric Karst

during the period of entrenchment that engulfed the Velebit-Lika area in the middle part of the older Permian.

(2) **The older Permian-Middle Triassic** was the land regime period. At that time, red clastic deposits, Brušane sandstones and siltites with lenses of Košna conglomerates precipitated. They are located on the foothills of Velebit in Lika and are best preserved in Brušane and the surrounding area.

In the Younger Middle Permian, extremely shallow, tidal plain environments were established with the deposition of early diagenetic dolomites in the Velebit-Lika region by ingression of the sea. Their thickness at Brušane and in Velika Paklenica is over 1000 m. They are known as the Schwagerina and the Mizzia dolomites. In Brušane, three zones of dark grey and black limestones, which are the original hydrocarbon rocks, have been developed in the dolomites. A sequence of those limestones at Velnička Glavica in Brušane is a protected natural monument. Inserts of red siltites, sandstones and conglomerates in the dolomites correspond to the Upper Permian–Gredenian deposits. At the same time, evaporites precipitated in eastern Lika, Una valley, Butišnica valley and Dalmatia, from Kninsko Polje through Kosovo Polje, Petrovo Polje and Vrličko Polje to Sinjsko Polje. Evaporites are mostly represented by gypsum and anhydrite. In Vrlika, karst forms such as scraps were also developed in gypsum.

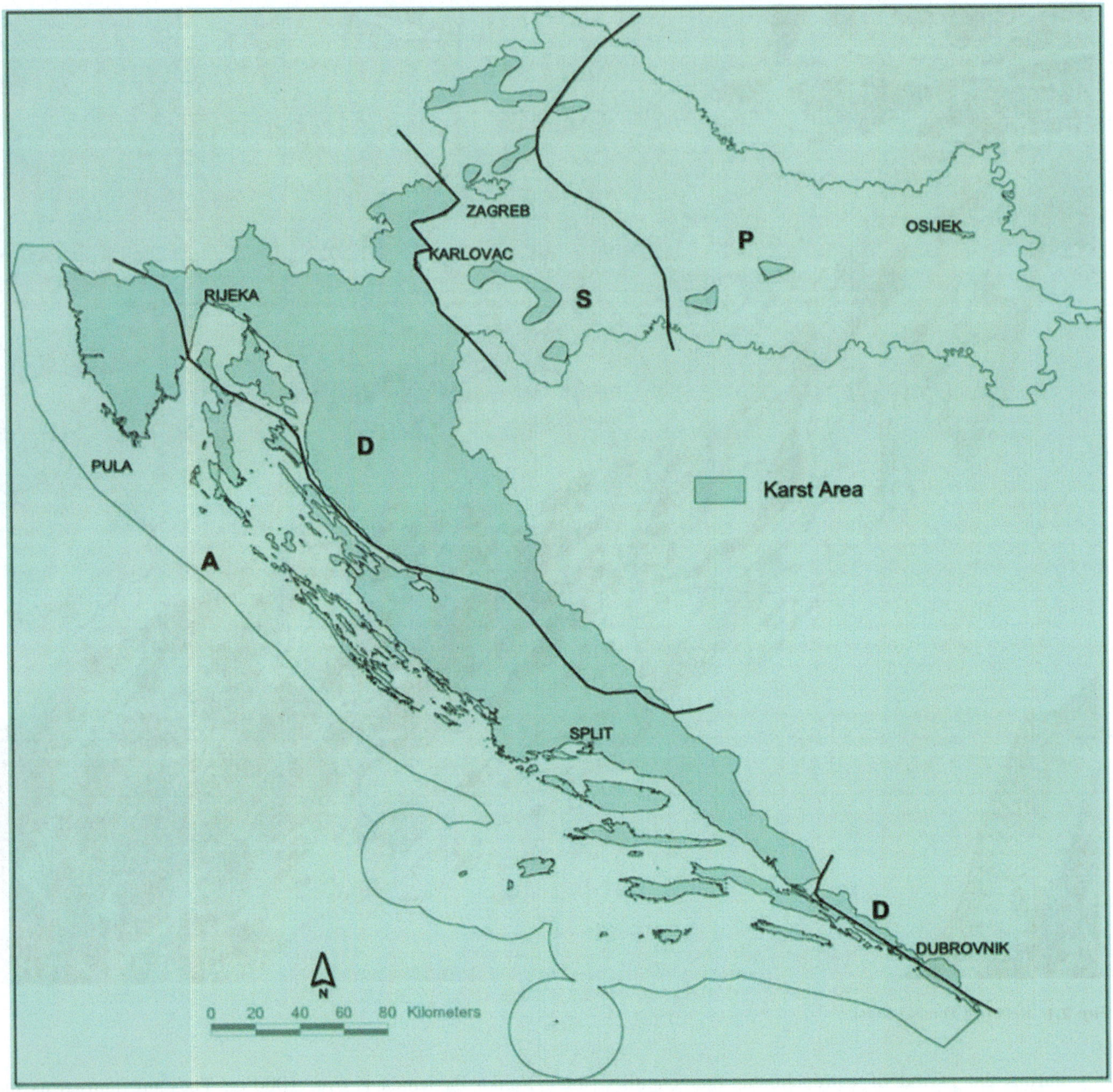

Fig. 2.2 Map of Croatian Karst (A—Adriatic, D—Dinaric, S—Supradinaric, P—Panonic)

In contrast to the areas described in Gorski Kotar, clastic deposits of dark shales and sandstones, sometimes of a flysch-like features, were continuously deposited during the Permian period.

Similar to the Upper Permian, the Lower Triassic deposits also show lithological differences. In Gorski Kotar, Ogulin area and Velebit, the Lower Triassic is represented by sandy dolomites (impacts of land) and mica sandstones. On the other hand, in eastern Lika and Dalmatia, along the aforementioned dolomites and sandstones, clay limestones with a rich cephalopod fauna are found in the Lower Triassic deposits.

Pure carbonate limestones were deposited in the shallow sea during the Middle Triassic. Lush wildlife has developed in such shallow-water environments. Limestone algae from the diplopora group predominated, so these Middle Triassic deposits are known in the literature as diploporic limestones. They were found in all areas of the Lower Triassic deposits except Gorski Kotar. In the Middle Triassic of the Karst Dinarides, occurrences of mostly basic, rarely neutral volcanic rocks are also known, along with clastic (shales and sandstones) and pyroclastic deposits (tuffits and tuffs), near Fužine, in Senjska Draga, Donje Pazarište, Baške Oštarije, Kosovo and Petrovo Polje, the Cetina Valley, Zelovske staje near Neorić, in the vicinity of Sinj, in Komiža and on the islands of Brusnik and Jabuka. Dark shales in volcanoclastic Ladinik deposits in Donje Pazarište are original but ripe hydrocarbon rocks.

In the younger part of the Middle Triassic, similar events took place as in the Lower Permian, so the area of the Karst Dinarides was elevated and turned into land. Intense erosion and karstification began. In karstified paleorelief, in valleys, sinkholes, etc., wine-red siltites, tuffite siltites and sandstones were deposited with lenses of colourful conglomerates, known as Rabelj sediments. They were discovered in Gorski Kotar, near Ogulin, in Senjska draga, Velebit from Štirovača (where they are over 200 m thick) to Gračac, Bruvno area, south-eastern Lika and northern Dalmatia). The youngest are those of Ladinian, Carnian and older Norian ages. However, there are also Carnian carbonate-clastic deposits, limestones, clayey limestones and siltites near Karlovac (Barilović) and in Komiža on the island of Vis, where Carnian evaporites are found, like on the island of Palagruža and elsewhere in the Adriatic.

(3) **The younger Triassic-older Jurassic**. Similar to the Lower Permian, shallow-water depositional environments have been established in most of the karst area due to the sea ingression during the younger Norian. The precipitates are predominantly dolomite in the literature known as the Main (Haupt) dolomite with alteration of the layers of early and late diagenetic varieties. These events resulted in a relatively stable period of shallow sea sedimentation, which lasted until the younger part of the Lower Jurassic–Toarcian.
The first limestone layers on the Main Dolomite mark the beginning of Jurassic deposits, characterized exclusively by carbonate sedimentation. For the older Lower Jurassic, the alteration of limestones and dolomites is significant, and for the middle part, Lithiothis limestones are important. In the lower Lower Jurassic, in the Toarcian, deepening of marine environments was identified, and bioturbated Stained Limestones were deposited.
Toarcian was the time when the large shallow-sea shallows in southern Tethys (Southern Tethys Megaplatform according to (Vlahović et al. 2005) fell apart into several smaller carbonate platforms, the largest of which was the Adriatic Carbonate Platform (ACP). It existed until the end of the Cretaceous, characterized by the intensive shallow sea carbonate sedimentation, mainly of limestones. Its boundaries throughout the Jurassic and Cretaceous were generally stable, and because of the high thickness of deposited carbonate sediments, they could also be considered as the boundaries of the Karst Dinarides.
(4) **Younger Toarcian-Younger Cenomanian**. Continuous and stable shallow sea sedimentation resulted in the deposition of thick carbonate sequences. They are widespread throughout the ACP.
The most beautiful example of the continuous succession of Jurassic deposits is found on the Mali Alan pass in the bare and "gentle" karst along the historic Master Road, ranging from the Triassic–Jurassic border to the end of the Kimmeridgian. This is considered a classic profile of the shallow-water Jurassic in the Karst Dinarides.
In terms of karstification, the thick sequence of Middle Jurassic limestones from which the White and Samar rocks are built (one of two strict nature reserves in Croatia) are especially stand out. Locally and occasionally there were shorter periods of land emergence and karstification in the Upper Jurassic, in Kimmeridgian (Biokovo), occasionally with bauxite deposition (Dinara, western Istria,). In contrast, in the central parts of the ACP tectonics caused deepening, and dark limestones with cherts precipitated in the wider area of Velika Kapela, and from Bihać towards Knin and Sinj in the wide Lemeš sediments, the original hydrocarbon rocks with rich ammonite communities (Vlahović et al. 2005). Periods of land emergence also occurred in the Lower Cretaceous (e.g. western Istria, Svilaja, Lastovci) and bauxite deposition occurred in the Cenomanian (Dugi otok).
Lower Cretaceous carbonates are found in all parts of the ACP. Significant to them are the Orbitolina limestones of the younger Barremian-Middle Cenomanian range. In the vicinity of Tounj and in Istria in the Lower-Aptian limestones, along with Orbitolina, the first Rudist communities are also important. They develop explosively at the beginning of Cenoman. In the literature, often all Upper Cretaceous deposits are referred to collectively as Rudist limestone.
In younger Cenomanian and older Turonian, a global rise in sea level occurred, which was reflected by deepening on ACP and deposition of thinner layered fine-grained limestones or massive limestones with pelagic fauna (Gušić and Jelaska 1990). From this facies several well-preserved fish remains (e.g. eastern Istria, Prapatnica near Trogir, island Mali Drvenik) and ammonite (Medulin) have been found in several places.
(5) **Older Turonian—the end of the Cretaceous**. As a consequence of partial uplift due to tectonic movements in the Cenomanian, some parts of the ACP became

mainland (western Istria, Kvarner islands), and elsewhere carbonate deposition continued into the younger Cretaceous, in some places (Brač, Mosor, Biokovo, Konavle) almost to the very end of Cretaceous. It was a period of gradual disintegration of the ACP during which carbonate sedimentation ceased at the end of Cretaceous with complete conversion to mainland due to strong regional, peri-Mediterranean tectonic activity (Cvetko et al. 2001). The period of land regime with strong karstification and deposition of bauxite in paleo sinkholes in Cretaceous limestones started (e.g. Istria, Cres, Obrovac,).

The Jurassic and Cretaceous carbonate deposits described are found in almost all parts of the Karst Dinarides in Croatia from the border with Slovenia to Konavle, with the exception of the youngest ones mentioned in this section. It should be pointed out that Jurassic deposits were found only on the closely related islands in the southern Adriatic, Lastovo, Kopište and Mljet, where only the Upper Jurassic limestones and dolomites were found.

(6) **Eocene**. The long terrestrial regime that began in the Upper Cretaceous continued throughout most of the Paleocene with the aforementioned paleokarstification and the occasional deposition of bauxite. In the youngest Paleocene and older Eocene by ingression of the sea, shallow water environments have been established and foraminiferal limestones are deposited with rich communities of miliolida, alveolina and nummulite foraminifera. Gradually, with the increasing tectonic movements in the mid-Eocene, the folding of the Dinaric orientation began, that is, with the folds of providing north-west–south-east orientation. The anticlinal parts became land, with their karstification and erosion, deposition material was formed in the synclinal parts which remained under the sea with prevailing basin sedimentation. Clastic sediments are deposited with predominantly flysch features. This model of deposition continued towards the south of Croatia and into the older Oligocene. However, the end of the foraminiferal limestones deposition also ended the original shallow sea deposition of carbonates.

(7) **Oligocene–Miocene**. Clastic—flysch marine sedimentation continued in the south of Croatia (e.g. Konavle) in the remaining smaller beds. In today's Ravni Kotari (Fig. 2.3), and in other parts of Dalmatia (Drniš, Sinj, Imotski), in the Oligocene, Promina sediments are deposited, characterized by a regressive facies sequence, with high carbonate content from marine environments with limestone deposition, known as Benkovac stone, to river deltas (marls, sandstones, conglomerates) and fluvial plains with the deposition of

Fig. 2.3 Karst of Bukovica and Ravni Kotari

enormous quantities of layered colourful Promina conglomerates.

All remaining area of the Karst Dinarides is exposed to karstification and occasional deposition of bauxite in the recesses of paleorelief. These were high-quality bauxites with Promina sediments layered above them.

The Oligocene is a period of strong tectonic activity and continued uplift in the Karst Dinarides. Freshwater, lake sedimentation of marls, clays, sands, etc. took place in the then formed fields. Such deposits are known in the Ogulin region near Dubrava and Perjasica, on the island of Pag, in Krbavsko Polje, in the Kninsko Polje and Peterovo Polje, and especially in Sinjsko Polje.

After the deposition of Promina sediments ceased in Miocene, they were and still are exposed to karstification, as were all other parts of the Karst Dinarides. However, parallel with the deposition of Promina sediments and partly after, in the Kvarner region, especially in Velebit and Lika, in the midst of the most intense neotectonic movements, Velebit limestone breccias were created. A tectogenetic facies unique in world proportions found only on Velebit is a valuable scientific knowledge. They are of exceptional importance for Croatian geology, geomorphology, the Karst Dinarides and the karst landscape in general, since they form the most beautiful karst forms and landscapes. They build, among other things (going from south to north): the Crnopac massif, Tulove grede, the lower parts of the canyons of Mala Paklenica and Velika Paklenica with Anića kuk, Bojinac, Stap, Stapina, Dabarki kukovi, Veliki Kozjak, Begovački kuk, the strict nature reserve of Hajdukčki and Rožanski kukovi, etc. There are also many smaller but well-known forms such as Zavratnica Bay, Strogir Stup and Kameni most near Živi bunari in Northern Velebit, a group of attractive pillars in the Tulove grede system (Fig. 2.4), and the Rapavac window in South Velebit, etc.

(8) **Pliocene–Holocene**. Through the Pliocene, Pleistocene and finally through the Holocene, karstification and relief formation of the Karst Dinarides continued. Pleistocene erosion and Holocene warming played a significant role in this.

The action of the ice, and in particular the movement of the glaciers, directly influenced the formation of the relief. Thus, after the melting of glaciers and ice, today's topography of the top parts of the entire Velebit was formed. There are also moraine deposits, for example in North Velebit in

Fig. 2.4 Tulove grede system on Velebit Mountain

the coves of Mirovo, Tudorevo and Lubenovac. In South Velebit, they are found on Struga, Oglavinovac, Javornik, Malo and Veliko Rujno, where the thickest moraine deposits of Rujanska Kosa are also found. It is similar, but to a lesser extent on Bukovac in Biokovo.

Holocene ice melt has raised sea levels globally and up to 100 m in the Adriatic area. Many of the former hills have become islands in the central and northern Adriatic, thus submerging countless karst forms.

This section will list only the stages of karstification through the geological past and the areas where it can be seen today:

- Older Permian–karstification of the Upper Carboniferous and Lower Permian limestones, Ričica and Rizvanuša;
- Younger Ladinian–older Norian–Karstification of the Middle Triassic diploporic limestones, Vratnik, Velebit from Štirovača to Gračac (typical locality Bađek), Bruvno;
- Kimmeridgian–karstification of Oxfordian and older Kimmeridgian limestones, Rovinj, Vrsar, Biokovo (Ravna Vlaška);
- Tithonian–Berriasian–karstification of Kimmeridgian-Lower Tithonian limestones, Dinars (Bravčev Dolac);
- Upper Aptian–Lower Albian–karstification of Lower-Aptian and Barremian limestones, western Istria;
- Cenomanian–karstification of older Cenomanian limestones, Dugi otok (Luka);
- Cenomanian–Santonian–karstification of Cenomanian limestones, west of Karlovac;
- Upper Cretaceous–Eocene–karstification of the Cenomanian–Coniacian rudist limestone, entire area of the Karst Dinarides in contact with foraminiferal limestones;
- Cretaceous rudist limestones and Eocene foraminiferal limestones (Fig. 2.5) in contact with Promina deposits, Ravni Kotari, Dalmatian Zagora;
- All land areas emerged from Oligocene to today.

Some statistical data about Croatia connected with karst:
Coastal area: 33,200 km^2.
Land area 56,594 km^2.
Territorial sea area 31,067 km^2.
Carbonate rock surface 29,356 km^2.
Land area: 89,810 km^2.

Fig. 2.5 Eocene foraminiferal limestones in Zagora

Coastline: 5,835 km.

Mainland coastline length: 1,777 km.

Coastline in islands: 4,058 km.

Number of islands: 1,246 (67 populated).

Highest point: Dinara (Sinjal) 1,831 m.a.s.l.

Almost the entire surface of the Adriatic Sea seabed in Croatia is carbonate facies of approximately 30,000 km^2, which means that the area of rocks subject to karstification in Croatia is approximately 60,000 m^2. At least 66% (according to some authors over 70%) of the territory of Croatia is karst. Some authors, however, believe that karst in Croatia covers only about 45%. Differences in these calculations result from the fact that most of the information did not explicitly determine, in geological sense, what types of rocks may be subjected to the process of karstification (paleo and recent karst). The dominant role in future calculations, or rather, accurate forecasts of the percentage of karst surface in the Republic of Croatia should be given exclusively by geologists who use speleologenesis as a benchmark for reaching such conclusions.

The 1997 Environmental Report in the Republic of Croatia states that the carbonate rocks on which the karst develops cover 29,356 km^2 or 52% of the total territory of the Republic of Croatia. Karst areas are presented on a separate map and the table. The data in the table is adapted to the changes of the borders of Croatian counties, and a correction of the data for karst areas of Sisak-Moslavina County was made. According to the data published in the Environmental Report, compact karst space covers 1,488 km^2 of Sisak-Moslavina County. The GISDATA documentation, however, found that carbonate rocks, on which karst forms can develop, cover only 369 km^2, or 8.3% of the total area of the county. In addition, they do not belong to a compact karst area but an isolated one. Minor changes were also made in the numerical indicators for the County of Zagreb and Karlovac. Table 2.1 shows that the compact karst covers 27,935 km^2, but it is still noticeable that areas with variable carbonate facies (breccias, conglomerates, limestone marls, etc. that are at least about 2% of karst) are not taken into account. If the area of Adriatic that belongs to Croatia

Table 2.1 Compact karst in Croatia covers 27.935 km^2, isolated karst in Croatia covers 652 km^2

County	County area in km^2	Compact karst area in km^2	Isolated karst area in km^2
Zagrebačka	3060	319	26
Krapinsko-zagorska	1229		45
Sisačko-moslavačka	4468		369
Karlovačka	3626	2911	
Varaždinska	1262		14
Koprivničko-križevačka	1748		65
Bjelovarsko-bilogorska	2640		8
Primorsko-goranska	3588	3588	
Ličko senjska	5353	5353	
Virovitičko-podravska	2024		12
Požeško-slavonska	1823		85
Brodsko-posavska	2030		0
Zadarska	3646	3646	
Osječko-baranjska	4155		0
Šibensko-kninska	2984	2984	
Vukovarsko-srijemska	2454		0
Splitsko-dalmatinska	4540	4540	
Istarska	2813	2813	
Dubrovačko-neretvanska	1781	1781	
Međimurska	729		0
Grad Zagreb	641		38
Total in km^2	56,594	27935	652
%	100	(49.4%)	(1.2%)
Total Karst		28597 (50.6%)	

without the volcanic rocks is included, then 73% of rocks in Croatia are susceptible to karstification (Garašić and Garašić 2016).

2.3 Short History of Karst Research in Croatia

In Croatia Karst research has a long tradition, which is understandable considering that more than half of Croatian territory has karst features. In addition to local experts, the Croatian Karst area was also studied by a number of the world's most famous geologists, hydrologists, speleologists and people from other related sciences. This is also the reason for the significant representation of Croatian terminology for karst phenomena in world literature. For example, the Croatian (Slavic) words like *ponor, polje, dolina, jama, kamenica* and some others related to karst phenomena and processes are found in almost all world languages as expert words of experts dealing with karst. Short history of karst research in Croatia was given by Malez (1984), Matas (2009) and Božić (2014).

Herak et al. (1976) in the Appendix to the *bibliography of karst of Yugoslavia from 1666 to 1974* had used determining the development stages of karst exploration in Croatia. It was based on hydrogeological and geomorphological considerations of individual authors in the study of karst. There are five stages, different in duration and results of the research. Shorter or narrower periods in which the corresponding changes have occurred are more often used to determine the beginnings and the endings of individual stages, rather than precise data on years and dates. There are also plenty of examples where the same authors worked through two stages.

The period until the mid-nineteenth century in the study of our karst areas is more in the sign of foreign than domestic researchers. Mainly, individual phenomena were discussed, with an emphasis on the underground world of the karst area, which was at the forefront because of the mystical perceptions that caused it. Among the local authors during this period they stand out; Gučetić from Dubrovnik, who published a work in 1584 with data obtained from surveys and meteorological measurements in the Šipun and Vjetrenica caves in the vicinity of Dubrovnik and Cavtat. This work is also considered one of the oldest speleological works in the world. Later, Getaldić also performed optical experiments in the Betina Cave in Dubrovnik.

The year 1689 is often taken as the beginning of scientific interest in the entire Dinaric Karst area, when Valvasor (1689) from Ljubljana published his famous work *Die Erhe des Herzogthums Crain–Slava vojvodine Kranjske*, which describes 70 caves and pits with maps and drawings that were largely filled with fiction, mysticism and perhaps vague ideas about the morphology of karst. However, this year often stands out as historical in terms of speleological research in the Dinaric Karst in general. In his books, Valvasor mentioned tens of caves in Croatia, in example Tounjčica Cave (Fig. 2.6).

An important role is also played by Marsiglia, who published his work in Paris in 1725, which describes, among other things, the subterranean river Lika in Croatia. The work of Hacquet (1787) contains, among other things, information about the caves of Velebit.

Travel writer Fortis in 1774 in his work *Viaggio in Dalmazia* describes the caves of Dalmatia. Ivan Lovrić, who explored the cave at the river Cetina spring during 1774 and 1776, is particularly prominent among domestic researchers and writers from this period.

In the nineteenth century, the area of exploration was expanded to include regional representations of karst, or particular topics such as vegetation cover of karst areas and karst hydrography. With dyeing and other means, underground connections between sinkholes and springs were explored.

Exploration in the second half of the nineteenth century was a sign of general geological ideas about karst. Examples and processes in our karst spaces have provided exceptional opportunities for analysis and discussion of karst in general; formation of karst, sinkholes, pits and unevenness in karst terrain. With water, which plays a major role in this process, the role of tectonics is increasingly emphasized.

It has already been pointed out that the Austrian authorities, due to their water supply needs and construction of roads, encouraged the exploration of karst areas and their specific features. Thus, the Zagreb Military Command organized a survey of water supply conditions in Vojna Krajina. In addition to Viennese geologist Titze, Croatian geologist Pilar participated in those surveys.

Pilar independently explored the area from Karlovac to the sea. In his addition to the book "*Die Wassernoth im Karste der kroatischen Militärgrenz*", which was also published in the Croatian language ("*Oskudica vode po krasu u hrvatskoj vojničkoj krajini*") in Zagreb in 1874, Pilar emphasized the geological basis on which special morphological and hydrological features were developed. He opposed the widespread notion that underground and above-ground forms in karst occur exclusively under the influence of erosion, and he believed in the connection of groundwater. He gave an important role to fracturing, faulting and folding, to which he attributed a significant role in the layout of groundwater. Its technical solution for the construction of wells for the use of water in karst is also worth mentioning.

Fig. 2.6 Tounjčica is one of the caves mentioned by Valvasor

Pilar, together with the Vienna geologists (Mojsisovic et al.1880), Mojsisovic, Titze and Bittner conducted geological surveys in Bosnia and Herzegovina. About that, in 1882 he published a special treatise entitled *Geological Observations in Western Bosnia*. Pilar is also known for investigating individual karst phenomena, such as caves in Gorski Kotar and Medvednica, and for his theories that the water sinking in Livanjsko Polje re-emerges along the edge of the Sinjsko Polje.

In 1835, Nikolajević described the caves in the vicinity of Dubrovnik in the first Croatian language newspaper, and Mlinarić in 1844 in *Danica Ilirska* described the cave near Sv. Ivana Zelina.

Klaić's "Natural Geography of Croatia" in 1878 describes 54 caves in Croatia.

E.A. Martel (the founder of French speleology), accompanied by Czech Putick, explores the Pazin Cave 1891–1893 (Martel 1896, 1897).

The increase of interest in karst research in Croatia at the end of the nineteenth century is also shown by the fact that on 20 February 1892, *the Committee for the Arrangement of Barač Caves near Rakovica* was established. This Committee is considered the first cave research association in Croatia. In the same year Barač's Cave was opened for tourist visits and thus became the first tourist cave in Croatia.

At the beginning of the twentieth century, the study of karst was carried out by the applying the cyclical theory of Devis and Penck. At this stage and after Pilar's death, Croatian Karst researchers were led by the geologist Dragutin Gorjanović-Kramberger. Gorjanović received his doctorate in 1879 in Tubingen. In 1896, he became a full professor at the University of Zagreb, and in 1909 he became a member of the Yugoslav Academy of Sciences and Arts (Gorjanović-Kramberger 1899, 1911, 1914).

Gorjanović published his first work about karst "Die Karsterscheinungen im westlichen Theile des Agramer Gebirges" in 1881 in the journal of Geological Institute in Vienna.

Gorjanović followed the works of Grund (1910), Cvijić (1895), Terzaghi (1913), Katzer (1909) and other leading experts in the field. After the founding of the Geological Commission within the Speleological Committee, he began a systematic exploration of caves in Lika. Gorjanović is also credited for making the first list of caves in Lika, which includes 156 caves. The list was published in the "News" of the Geological Commission in 1912. In the data collection for the list, Langhogger was particularly active. The list also shows a draft of the Samograd Cave near Perušić, which was made by county engineer Špiler from Gospić. This is also considered to be the first more thorough list of caves in Croatia and the first published geodetic survey of the design and profile of a Croatian cave. It is interesting to note that the famous Croatian naturalist Hirc, in his work The Geography of Croatia and Slovenia, indicates that there are about 500 caves in Croatia (recent data shows the number is much greater).

Gorjanović accepted Katzer's terms "shallow" and "deep" karst, which is evident from the title of his work Shallow Karst around Generalski Stol in Croatia, which was published in 1912. In Gorjanović's description of karst relief forms, great importance is attributed to tectonic changes, mainly folding, faulting, fracturing and erosion. He pointed out that the lowering of the ground water level occurs parallel with the lowering of "karstification" level. He claimed the sinkholes were the result of both surface corrosion and collapse. He also mentioned fractural sinkholes, fractural valleys, star sinkholes, etc. and pointed out that tectonics were involved in their formation. Blind valleys are created by the lowering of the base water, and at their bottom there may be a series of ordinary sinkholes. Gorjanović considered the Cretaceous flat and dry Ogulin plate to be a result of the above-ground and underground dissolution of carbonate deposits and the removal of that solution through the underground. He also thought and that the Upper Dobra River (Fig. 2.7) had two branches; one flowing the surface and other sinking at the foot of the adjacent hill. In his view, a fractured plate filled with sinkholes could not sustain the surface flow, so the flow moved to the underground. Gorjanović's third work was entitled "The question of flattening of karst areas and the creation of solitary stone pillars" (*K pitanju izravnavanja krševitih krajeva te postanak osamljenih kamenitih stupova)*, and was published in 1916.

In 1910, within the framework of the Geological Commission of the Kingdom of Croatia and Slavonia in Zagreb (the first scientific association to study underground karst phenomena in Croatia), a Committee for cave exploration

Fig. 2.7 Upper Dobra River in Gorski Kotar

Fig. 2.8 Plitvice Lakes in Lika

was established with the task of systematic work on exploration of the Dinaric Karst caves. Dr. Dragutin Gorjanović Kramberger was particularly prominent in the founding of the committee and its work.

Geologist Ferdo Koch in his work "*Plitvice Lakes—A Contribution to the Knowledge of Tectonics and Hydrography of the Karst*" published in the "News" of the Geological Survey (successor of the Geological Commission) in 1926 emphasizes the role of tectonics in the formation of lakes. Koch lacked a more reliable stratigraphic interpretation of the carbonate deposits, so he regarded the dolomites to be of Cretaceous age. That means that underneath the dolomites, limestones of the Lower Cretaceous, Jurassic and Upper Triassic would have to be found. Of course, such an assumption required reliable folds under the lake that would secure the bottom from losing water underground. Related to this were the ideas of stronger tectonic disturbances (or even earthquakes) as potential causes for the possible creation of sinkholes that are dangerous to the water status of lakes. It was not until much later that the Triassic dolomites were discovered, so the Cretaceous and Jurassic limestones were actually missing, which significantly reduced (probably even eliminated) any risk of lake water loss through the bottom. Koch's emphasis on the role of tectonics in shaping the Plitvice Lakes (Fig. 2.8), however, retained its significance. Along with the mentioned work on Plitvice, Koch wrote about Velebit and the tectonics of karst, Korčula and Pelješac, etc.

Among karst research geologists Josip Poljak (1922, 1934, 1935) played a significant role. He was particularly interested in the study of geomorphology and hydrography of the Dinaric Karst. Thanks to his frequent fieldwork, he stood out as an expert in karst. His concept, according to which for the qualitative research and explanation of karst phenomena it is necessary to take into account the diverse geological and tectonic parameters, is nowadays receiving more and more recognition. Poljak's most complex work is the *Geomorphology and Hydrography of the Ogulin and Ogulin Zagorje Environment*, published in the "*Memorial*" dedicated to Gorjanović in 1925/26 (Sakač 1984).

In the Academy's journal "Natural History Research", three extensions of his work to the *Cave of the Croatian Karst* were published, I (1913), II (1914), III (1924).

Writing about J. Poljak, Božičević (2012) states: "The advent of geologist J. Poljak has led to great advances in Croatian speleology in the field of systematic speleological research. His report On the caves of the Croatian Karst, as well as all his numerous speleological and geological works, shed light on the problems of the Croatian Karst one by one. The number of his works in the karst area is considerable and he has processed around 450 speleological objects". In the same paper, Božičević states "In the area of Dalmatia, Lika, Gorski Kotar and Istria, a number of our researchers is increasing. Among them, to name but a few M. Margetić, M. Kusijanović (published a brochure on the Močijska Cave near Dubrovnik in 1918), S. Vuksan, S. Vuković, I. Krajač, A. Premužić, V. Horvat and others" (Božić 2014).

In the decade before World War II, cave lists were completed, and during the war caves were increasingly used for war purposes (warehouses, shelters, etc.). In 1942, the Czech speleologist Karel Absolon published the results of his prewar research on the island of Brač.

Gjurašin (1942) stands out among Croatian Karst researchers of that period. In 1942/43 he investigated the mixing of salt and fresh water and other hydraulic regularities of karst. Gjurašin studied fractures through which the water in the karst underground moves and the speed of the underground flow. The results he obtained varied from 0.009 to 0.300 m per second. Variability has also been confirmed by recent measurements. The water flow between Bosiljevo stream and the Dobra River has a mean velocity of 0.00186 m / sec, and the flow between Mrežnica and Bistrac has the velocity of 0.110 m/s. This variability indicates that there must be considerable differences in the size of the fractures and in their dip. Ivan Kuščer (studied vruljas near Senj) did similar research.

The second half of the twentieth century represents one of the most fruitful stages in karst exploration in Croatia. At this stage, very significant research results were achieved, which is understandable considering that during this period there was an accelerated development of science in general. Geosciences got more intensively engaged in exploring karst areas and their peculiarities. Special attention was paid to exploration of Croatian Karst areas due to the design and implementation of numerous hydroaccumulations, tunnels, etc. The new constructions required additional structural, morphological and hydrological studies that brought much more light into the geomorphological interpretation of the terrain explored and conclusions about the much wider areas. This has been aided by intensified speleological exploration, groundwater tracing, geoelectric measurement, as well as feedback based on actual findings in the construction sites and behaviour of the structures in karst.

The remarkable results of the research were also contributed by the new organization of scientific and teaching work. It may be mentioned that in the Yugoslav Academy of Sciences and Arts (JAZU), in 1948, a Karst Research Commission was established, also known as the Karst Commission. On 12 January 1955, the Karst Expert Council for Research of Karst was established in Zagreb. A periodical publication "Krš Jugoslavije—Carsus Iugoslaviae" was launched, intended for domestic and foreign experts, to monitor the overall problem of karst. It published original scientific contributions, news, papers, material and literature reviews. The first issue of the publication was published in the autumn of 1957 and the last one in 1991. From 2012 in Croatian Academy of Sciences and Arts (HAZU) exist Karst Committee for Karst in which are well-known scientists about karst and speleology.

In Split, it is organized by the Inter-Academic Committee for Nature Conservation at JAZU. On 18–20 October 1976, a symposium on the Ecological Valorisation of the Coastal Karst was held.

Emphasizing the role of geologists in the fifth stage of the karst exploration of Croatia, described Herak et al. (1976, 2002). By applying new research procedures and new geological elements to the process of better understanding the intricate Dinaric Karst, many geomorphological phenomena in the karst have been explained, of which only formal assumptions existed in the past. Sinkholes can be mentioned as an example; more recent research has found that their formation, shape and arrangement depend on the complex orogenetic changes that cause the collapse and formation of cavities within the orogenesis of the carbonate complex involved. After the Neotectonic uplift, erosion occurred, and the former internal cavities reached the surface and their walls were exposed to corrosion. Explaining the formation of symmetrical sinks does not exclude the mechanical and corrosive action of water from the surface.

Herak and Stringfield (1972), an extensive and high-quality Karst monography, *Important karst regions of the Northern Hemisphere*, was published. Particularly noteworthy is the chapter on Karst Yugoslavia (with Croatia), which is very often cited in our and world literature on karst issues. In this paper, Herak also presented some new views on the general classification of karst with respect to temporally and spatially determined geological measures. He proposed to distinguish between "geosynclinal karst" and "epicontinental karst". In 1977, Herak presented a new complex classification of karst terrains based on tectogenesis. It takes into account all levels of tectonic changes and karst types associated with faults, folds and overthrusts. But as the term geosyncline had already lost its original meaning at the time, two basic types of karst were given new names. Instead of the geosynclinal karst, the term "orogenic karst" was introduced, which emphasizes that it is an integral part of the orogen (hence subsequently consolidated chain of highlands), and consequently the epicontinental karst was renamed "epiorogenic karst". Subsequently (1984) he considered the interdependence of caves and their geotectonic framework, and in 1986 he pointed out the importance of allochthony as the primary basis of karst plateaus that have been given the modern surface shape by the processes of neotectonics, erosion, denudation and corrosion.

On this basis, geologist Stjepan Bahun proposes a model according to which, in the formation of karst plateaus, orogenetic tectonics and neotectonics are stages of the same complex process. Bahun (1970, 1974, 1985, 1991) studied the sinkholes with great care, advocating a more comprehensive interpretation of their origin. He emphasizes that the process begins with complex orogenetic changes that cause the collapse and formation of cavities within the orogenesis of the affected carbonate complex.

Bahun and Herak have designed a map showing the geological basis of the Dinaric Karst in the international hydrological decade programme. It was created in cooperation with experts from Slovenia, Bosnia and Herzegovina and Montenegro. The map has been developed in accordance with new approaches on the importance of tectonics in the formation of karst terrains. It was completed in 1974 and presented at a gathering of international experts in Rome. However, its international publication was banned by the Yugoslav People's Army. It was only allowed for internal use.

Hydrogeological studies of karst have become increasingly important in the recent period. It has already been pointed out that three karst hydrological regions with specific geological and hydrological characteristics (Periadriatic and island karst, High karst belt and Shallow karst or Fluvio karst) can be distinguished in the Croatian Karst area. The first detailed account of Croatian Karst hydrogeology and its development was given by Herak et al. (1969), published a complete review of the karst hydrogeology based on the contemporary rock classification, differentiation of aquifers, complete and incomplete water stops caused by either impermeable rocks or tectonically hardened terrain, especially undisturbed anticlines.

The geological basis of hydrogeological relations in the Slunj and Vrbovsko areas was studied by Bahun (1968, 1970) who wrote about the relationship between the karst process and fluvial erosion in the Lika area, and with Pavičić in 1976 presented the hydrogeology of the Lower Lika flow. 1975 and 1976 were written about Lika and Gacko by Božidar Biondić and Goatti, and in 1976 by Pavičić and Fritz.

Garašić wrote about the influence of water on the geomorphology of karst areas in 1986.

Borelli and Pavlin wrote in 1967 about the problems of groundwater losses from karst reservoirs. Stepinac and Jurak also wrote about similar problems. Biondić (1988) points to the problems associated with the abstraction and protection of karst waters, and in Bahun warns of the need to know the geological basis for the protection of karst waters.

More frequent study of karst in the Adriatic coast and on the islands began after 1950 when the Laboratory for Geology of the Sea was established at the Institute of Oceanography and Fisheries in Split. The research covered the morphology of the bottom, including the morphology of vrulja, sediments, mixing of salt and fresh water in the coastal area, etc. Other institutions gradually became included in these studies. Using various geophysical methods, drilling, laboratory processing of sediments and water, etc. The need for drinking and industrial water has also prompted more complete hydrogeological research. Numerous expert and scientific papers have been written on the results of the aforementioned research. Thus, for example, Petrik 1957 wrote about the hydrological regime of Lake Vrana in Cres. From 1977 to 1981, Alfirević reports on the hydrogeological aspect of groundwater circulation on the eastern Adriatic coast, then on groundwater at the seafloor, and on bathymetric and sedimentary features of the Adriatic (Alfirević 1969a,1969b). Fritz and Pavičić (1982, 1987), Fritz and Bahun (1997) discussed the hydrogeological zoning of the Croatian coastal karst, and also they focused specifically on the hinterland of Split.

Speleological research has greatly contributed to a better knowledge of karst forms and phenomena in our national and world proportions. Speleology is most commonly defined as the science that investigates the origin, morphology, physical (microclimate, water) and biological properties of caves and other natural cavities in the Earth's crust, which can be directly observed and explored. It is possible to record their shapes and measure dimensions, to process the composition of rocks, to study river subways and underground lakes, to study the physical and chemical processes that have conditioned their formation and development, and residues of fossils, using the achievements of geology, mineralogy and morphology and taking into account considering the climatic, hydrological and biological conditions, especially the living world in them, and archaeological findings, evaluating the social significance of the caves, especially with regard to tourism (about 12 500 speleological objects have been registered in Croatia).

Studying the origin of caves in Croatian Karst on a number of examples, it can be concluded that they generally follow the fault areas—fault planes and fault zones. Earlier, speleologists had explored only a certain number of caves that have a natural entrance on the surface of the terrain. The construction of motorways, bridges, tunnels, viaducts and other larger structures in karst areas, revealed a number of caverns (caves without natural entrance from the terrain surface). Cavern research has led to some new insights. It has been detected that, as a rule, caves occur within the

Fig. 2.9 Upper flow of Kupa River in Gorski Kotar

groups on individual faults, depending on the type of rocks, the severity of the damage, the neotectonic activities, etc.

More recent cave diving exploration of submerged karst springs (Majerovo vrelo, vrelo Gacka River, vrelo Una River, Kupa River (Fig. 2.9), Ličanka, Glavaš na Cetini, Krnjeza, etc.) as well as caves that are nowadays below the sea surface (Vrulja, Zečica, Duba, Podstažišće, Modra špilja on Biševo itd) also contributed to the knowledge of differences in directions, strength, tectonic conditions and other factors of complex karstification processes. This is supported by the finding of speleothems in a cavity found in a borehole in the Adriatic Sea at a depth of 3125 m.

Among the Croatian authors who recently dealt with the problems of origin, spatial distribution and other characteristics of caves, Garašić, Božičević, Pavičić stand out.

In 1986, Garašić emphasizes the importance of neotectonics in the formation and morphological design of caves. As an example he gives Muškinja Cave (today is named as Varićakova Cave) that belongs to one of the largest cave systems in Croatia. Subsequently, in 1989 in Budapest, he presented a new understanding of the morphogenesis and hydrogeology of the Croatian Karst cave (Garašić 1989).

Garašić (2006) points out in the paper "Recent insights on karst speleogenesis" that karst research in the last ten years is related to speleo-hydrogeological studies, seismotectonic measurements, and especially to the intensive construction of roads, dams and other large structures. These investigations provided new insights into the processes taking place in karst areas. In one of his works, Garašić states that about 10% of speleological objects belong to the regressive type of speleogenesis (Maucci 1952). In this respect he particularly emphasizes the example of Red Lake near Imotski (Fig. 2.10). Referring to the results of cave exploration at the end of the twentieth century, Garašić (2001) points out that

Fig. 2.10 Crveno jezero (Red Lake) near Imotski

the process of creating this lake was not exclusively gravitational, as previously thought, but that it "moved from the underground to the surface".

In recent works that have been processed by data obtained from measurements and studies of over 1000 caverns (caves without natural entrance from the surface), which were found during the construction of motorways through Croatian Karst areas in the last fifteen years, Garašić (2005a) states that gravity and backward (regressive) karstification or exclusive regressive karstification occurring between 1500 and 1800 or 15–20% of the total number of known caves.

By measurements of absolute displacements (neotectonic activity), caves were found to be most numerous on the same fault (or fault zone) in the vertical arrangement in the case of neotectonic uplift and a smaller deviation from the direction of the neotectonic impact relative to the predisposed fault is important for speleogenesis. On the other hand, faults that create horizontal "extensions" of the caverns are almost as a rule in the area of neotectonic descent or in the direction of changing the direction of neotectonic impacts relative to the predisposed fault.

Comparing the neotectonic (present) direction and displacement strength, points out that it would be possible to compare speleological objects in similar rocks with great accuracy, and when comparing their karstification intensity, the data would be comparatively very significant.

In exploration of underground spaces, speleologists are increasingly using the structural (genetic) approach, hydraulic observation of underground reservoirs, branching and swallow-hole capacity, experimental verification of natural salinization conditions of springs in the coastal zone, etc. This is evidenced by the increasing number of hydrological works with more accurate data and analyses of the morphology of aquifers, including the morphology of vrulja,

Fig. 2.11 Karrens (škrape) on north Velebit

the structure of porosity, the quantitative degree of corrosion, etc.

Speleological research verifies and confirms tectonic parameters essential for the formation of karst morphological systems. For example, let us mention that Božičević (1964, 1965a, b) confirmed the significance of faults in the formation of the speleological system.

The above and numerous other indicators and results of the research indicate the high reputation of Croatian speleology. As an example, let us mention that Croatia is one of the leading countries in the world by number of caves explored. It also belongs to a small number of countries with explored caves deeper than 1000 m. Two of the three most famous subterranean rivers in the world are located in Croatian Karst. It also contains the largest karst field in the world (Ličko Polje), speleothems found in borehole at depth of more than 3125 m and groundwater reservoirs that exceed all known reservoirs in the karst world (e.g. Crveno Jezero with at least 16 million m^3 of drinking water, Kupa Spring—154 m. Una Spring—248 m diving depth with approx. min 150,000 m^3, etc.), and numerous previously unknown, endemic cave animals were found and described in our karst area.

2.4 Surface Karst (Geo)morphology

Ever since the initial works on karst geomorphology in the nineteenth century, surface karst forms were observed, processed and classified according to their dimensions, intensity and mode of occurrence, etc. Divisions were rarely based on geological factors, which, in addition to atmospherics, are a major factor in the formation of karst surface forms. Some have divided them into corrosion (i.e. those resulting from the chemical wear of the rocks) and erosion (i.e. those resulting from mechanical wear of the rocks). There are also those formed by the processes of denudation, accumulation, abrasion, Aeolian erosion, i.e. corrosion, etc. The smallest surface forms are karrens (scallops) (Figs. 2.11 and 2.12), followed by sinkholes (Fig. 2.13), karst bays, dolinas, uvalas (Čalić 2011) and the largest surface forms—karst polje (Figs. 2.14 and 2.15). The largest karst polje in the world is Ličko Polje (Fig. 2.16) between the Velebit Mountains (Sokač 1973, Perica 1998, Faivre 2000, Faivre and Reiffsteck 2002) and the Lika highlands. The second one is Gacko Polje also in Lika Karst region.

Fig. 2.12 Karrens (škrape) on south Velebit

Fig. 2.13 Sinkholes on Biokovo

Fig. 2.14 Karst polje near Vrlika

Fig. 2.15 Polje and Gacka River (Lika)

Fig. 2.16 The largest karst polje in the world is Ličko Polje (during flood)

3 Underground Karst Morphology (Speleology)

Abstract

The peculiarity of karst is evident in the appearance of underground karst forms—caves. Caves are natural cavities in the rock mass large enough for a person to "enter". The structure of karst can be compared to "Swiss cheese"; it is full of voids, but only some are visible on the surface. The quantity of the voids (caves) depends on the thickness and type of rock. Caves that do not have a natural entrance on the surface are called "caverns". So far, around 12,500 caves with natural entrances have been discovered and explored in Croatia. It is estimated that there are at least as many caves yet to be discovered. There are 3 times more caverns at depths up to 100 m, i.e. about 75,000, and at depths up to 500 m at least 10 times more or 250,000. These numbers are calculated from the statistical data about the caverns discovered while tunnelling (Garašić and Garašić 2015).

3.1 Short Review of History of Speleology in Croatia

In the Palaeolithic, about a million years ago, a prehistoric men lived in the Šandalji Cave near Pula in Istria, leaving traces of their residence. This was discovered by the famous Croatian speleologist, academician Mirko Malez. Short descriptions on history of speleology in Croatia were given by Malez (1984), Matas (2009) and Božić (2014).

Much younger, but equally as famous are the finds of Neanderthals in the Hušnjakovo Cave in Krapina (Hrvatsko Zagorje), who lived there 120,000–40,000 years ago. The findings of the youngest Neanderthal in Europe in the caves of Dalmatia (Mujina Cave) near Trogir are also well known.

The ancestors of the present-day men (Homo sapiens) resided in caves such as Vindija Cave near Varaždin, Mačkova Cave near Lepoglava, Bukovac Cave near Lokve, while other caves such as Veternica Cave near Zagreb were used occasionally.

Findings of academician Grga Novak from the caves on the island of Hvar are also very significant. In Grapčeva Cave, Markova Cave and the cave on Pokrivenik, a Neolithic culture that flourished from 5000 to 4000 years ago was found. The oldest drawing of a ship found in Croatia was discovered on a piece of ceramics.

In Bezdanjača Cave near Vatinovac (Horvatova Cave) near Vrhovine in Lika, in the Middle Bronze Age, about 70–120 m from the cave entrance, around two hundred people were buried 3500 years ago (Figs. 3.1 and 3.2). In order to reach the funeral site, the olden people of Lika had to master a vertical 30 m rock followed by a few smaller verticals of 15 and 5 m. According to available data from other well-known caves in the world, people first descended to the depths of 125 m in the eighteenth century.

The first historical record of a cave in Croatia dates back to 1096. It is a church file about a cave in the bay of Željina on the island of Ugljan. The file describes the boundary of the church property located in the immediate vicinity of the cave. The text is written in Latin on parchment (the original is kept in the National Archives in Zadar), with the word "pechica" (picture). This cave is mentioned in several other documents from 1166 and later.

Poet Petar Zoranić Sinjanin wrote the oldest book about mountains in the world and the first Croatian novel, *Mountains*, in 1536. In *Mountains,* he described the underground of the Velebit and Dinara mountains.

The first description of a particular cave in Croatia was given by the Dubrovnik nobleman Jakov Sorkočević in 1580 in a book on fish and shells of the Dubrovnik region. It was a description of the Šipun Cave in Cavtat. Four years later, in 1584, the Dubrovnik philosopher Nikola Gučetić published a treatise "On Aristotle's Meteors" (Dadić 1984), which is considered to be one of the oldest scientific discussions about cave phenomena in the world. The cave is also

M. Garasic, *The Dinaric Karst System of Croatia*, Cave and Karst Systems of the World,
https://doi.org/10.1007/978-3-030-80587-6_3

Fig. 3.1 Bezdanjača Cave on Vatinovac Hill or Horvat's Cave near Vrhovine in Lika

mentioned by the third Dubrovnik scientist Serafin Razzi in 1595.

In 1689, the Slovenian scientist Ivan Vajkard Valvasor, in the book "Slava vojvodine Kranjske" (only complete original copy is preserved at the Metropolitani Library in Zagreb, since the author himself sold it to the church in Zagreb), among other caves, describes the Druška Peć Cave at Učka and shows a drawing of the entrance of Tounjčica Cave (Fig. 3.3) near Ogulin.

Italian scientist Louigi Fernando Marsiglia also pointed out the interesting features of Croatian underground in his work "The physical nature of the sea", printed in Paris in 1725. He described the phenomena of the subterranean rivers in Lika. Italian abbot Alberto Fortis travelled the entire Adriatic coast of Croatia, from Istria to Dalmatia. In the book "Viaggio in Dalmazia" published in 1774, he described several caves in Dalmatia (Rudelić Cave near the Cetina River spring), Istria and on the island of Cres.

In 1776, Ivan Lovrić published his research in the work "Notes on the Path of Alberto Fortis and the Life of Stanislav Sočivica". Lovrić gave a very realistic description of his exploration of Gospodska Cave near the Cetina River spring. With a rope, he overcame a vertical jump of 17 m and explored about 400 m of cave channels to the lake. Because of this endeavour, Ivan Lovrić may be mentioned as the first Croatian caver—speleologist. Although there were those who entered the underground before him, it is not known whether they used a rope in these activities. For example, we know for a fact that people were descending into some vertical caves about thirty m in depth with the help of broken trees (e.g. about 3300 years ago in Bezdanjača under Vatinovec or Horvatova Cave in Lika, where human remains, and artefacts were found at depths of over 100 m).

The first topographic sign for cave, in public use, was given by the French cave researcher Belsazar Hacquet in 1787 on a topographic map of Croatia. Over the next half century, there has not been much speleological exploration due to political events in Croatia. The first description of research in the nineteenth century was given by Austrian botanist Franz Petter. In 1832, he described Zelena Cave on Ravnik Island. In 1835, Julije Fras published a book "Topography of Karlovac's Military Land" which describes eighty caves. He was the first person in Croatia known to use wooden ladders for mastering vertical sections of Barić Cave near Ličko Petrovo selo (near Plitvice Lakes in Lika). In 1842, in "*Danica Ilirska*", priest Stjepan Mlinarić described the cave near Sv. Ivana Zelina. It was the first cave description in Croatian language. The descriptions of some caves in the vicinity of Dubrovnik were given by G. Nikolajević and F. Carrara. In 1865 and 1866, J. Sapetz described the Ozalj Cave. In 1866, Vjekoslav Sabljar published a "*Mjestopisni rječnik*" (Dictionary of places) with the names of some caves in Croatia, and in 1868, he described some of them separately.

With the founding of the Croatian Mountaineering Society (HPD) in 1874, more systematic exploration of the Croatian mountains and caves began. In the same year, E. Tietz described some caves in Kordun. In 1875 and 1888, Johannes Frischauf described caves in the islands of Rab and Cres and in Kordun. In 1878, Vjekoslav Klaić described 48 caves in his work "Natural Geography of Croatia", and in 1878, Ljudevit Rossi mentioned Božić Cave near Slunj. In 1882, Geologist Dr. Mijo Kišpatić published the results of his paleontological exploration of Barač caves near Rakovica (Garašić 1992, 1997). In 1883, geologist Dr. Đuro Pilar (Rector of the University of Zagreb and the first professor of geology) published the results of his research from Kupički vrh Cave. In 1895, Ivan Devčić described several caves in Lika, and Hitzthaler described caves in the vicinity of Ogulin.

Significant cave research was also done by the members of the Club Alpino Fiumano (CAF), which was founded in Rijeka on January 12, 1885. The club members were of

Fig. 3.2 Bezdanjača Cave on Vatinovac Hill

Fig. 3.3 Tounjčica Cave entrance

Italian and Croatian nationality. In 1919, the club was renamed Club Alpino Italiano—Sezione di Fiume (CAI-SF). They explored the Croatian coast, Gorski Kotar and Istria between 1887 and 1930. During this period, they explored more than 200 caves. From 1902 to 1930, they published their journal *Liburnia*, with articles on speleology. The club's long-time leader was Guido Depoli (Božić 2014).

During 1855, cholera reigned in Dalmatia, so the wealthy citizens of Split took refuge in Vrlika, visiting the cave near the spring of the Cetina River, which was named Gospodska Cave. For these visits, the entrance to the cave was adapted (stone stairs). At the end of the nineteenth century, the first attempts of graphical representation of caves, i.e. drafts of caves appeared. The first sketch of a cave was made for Mala Pećina Cave near Muć Gornji in Dalmatia. It was created and published in 1882 by the priest and archaeologist Mihovil Granić. In 1887, geologist Đuro Pilar made the first draft of the cave on millimetre paper in (Velika Peć on Medvednica near Zagreb).

In 1893, Alfred Eduard Martel came to Istria to explore the swallow-hole of Pazinčica and some other caves in the Croatian Karst area (Martel 1896, 1897).

In 1903, Luigi Mioto drew up a geodetic draft of the Varnjača Cave near Kotlenica in Dugopolje. It was published in 1905 by Fritz von Kerner in Vienna. The geodetic draft of the Samograd Cave was made in 1899 and published in 1910 in the "*News of the Geological Commission of the Kingdom of Croatia and Slavonia*". (Gorjanović-Kramberger 1911).

In his book on Gorski Kotar in 1898, Dragutin Hirc published the first painting showing the exploration of a cave in Croatia (Hajdove hiže Cave near Delnice). The painting was created by a painter Vaclav Anderle in 1875.

In the first issue of the journal "Hrvatski Planinar" in 1898, geologist Milan Šenoa gave instructions on how to collect cave data. In 1899, a cave with numerous remains of a prehistoric man was discovered in the Hušnjakovo hill near Krapina, which was studied by Dragutin Gorjanović-Kramberger for the next ten years.

Until the end of the nineteenth century, only individuals with a few helpers explored the caves in Croatia. The first association to systematically start exploring the caves was the Mountain Tourism Association *Liburnija*, founded in Zadar in 1899. The members of the society engaged in all mountaineering and tourism activities, including (as early as 1900) exploring the caves. In 1903, they organized the first Croatian speleological expedition in Kornati Islands. Although members of Liburnija explored the islands, Velebit and Mosor mountains, not until 1908 have they established a special Cave Research Committee within their society.

Dragutin Hirc made a great contribution to the knowledge of caves in Croatia. In addition to several hundred different travel articles, in which he also described caves in the period from 1875 to 1909, his most significant work was "Natural Geography of Croatia", (published in 1905), in which he described about 300 caves.

The first professional speleological association was founded in Croatia at the Geological Commission for Croatia and Slavonia in Zagreb in 1910 under the name of the Cave Research Committee. The first chairman of the committee was geologist Dragutin Gorjanović-Kramberger. In addition to Gorjanović, there were several co-workers in the Committee (Koch, Schubert, Kerner and Josip Poljak). In 1921, Poljak presented his doctoral dissertation on caves in Croatia, becoming the first Ph.D. of speleology in Croatia.

Since 1909, the caves of Dalmatia have been explored by Umberto Girometta with high school students from Split, particularly after the founding of Cave Department in high school in 1911 with Ramir Bujas.

Since there are more vertical caves (pits) in Dalmatia than horizontal caves, the members of this association were called "jamari" (pit explorers) and were very well equipped and trained (Božić 2014).

In central Dalmatia, on land and islands, they explored many caves such as the knee-pit at Koliština near Prgomet, 138 m deep. There were more independent studies. In 1910, A. Waagen researched and described caves in the islands of Krk, Cres and Rab. In the same year, Ivan Franić described the caves in the vicinity of the Plitvice Lakes in the book "Plitvice Lakes". In 1911, A. Langhoffer published the work "Caves of the County of Lika-Krbavska" and, together with the list of caves, gave the first draft of a tourist cave in Croatia (Samograd Cave near Perušić). In 1912, A. Kormos explored the Bukovac Cave near Lokava in Gorski Kotar where he found the remains of a prehistoric man.

In 1918, Mijo Kusijanović published the brochure "The Močilj Cave near Dubrovnik". In 1922, Umberto Girometta published a major work, "The Caves and Caves of Central Dalmatia", in which he described 472 speleological objects. Geologist Marko Margetic published the work "New Caves and Pits of Central Dalmatia" in 1925. The caves on the island of Hvar were explored by M. Schneider from 1926 to 1927. Subsequently, in the Markova Cave, Grapčeva Cave and in the Pokrivenik Cave, the scientist Grga Novak, with his assistant Vladimir Mirosavljevic, discovered the remains of civilization from 2000 to 3000 BC. In 1925, Italian cavers descended to Ponor near Rašpor on Ćićarija in Istria and measured the depth of 450 m, which, at that time, was the deepest in the world. With accurate measurements of the cave in 2019, Croatian cavers measured a depth of 351 m.

Italian cavers Bertarelli and Boegan (1926) published the work "Two Thousand Caves" (Duemilla grotte) and described caves and pits in Istria, Gorski Kotar and the Croatian coast. Z. Rosandic wrote about Lika caves and their protection, as well as Ilija Sarirnic, Josip Pasaric, Josip Roglic, Zvonimir Dugagsky, Zora Karaman, Karel Absolon, V. Apfelbeck, J. Brait, A. Hofman and others. Also, noteworthy is the work of Ivan Krajač, who, with Antun Premužić, explored the caves near Plitvice Lakes (1928) and on Velebit (1930–1933). In 1930, he descended to the 120 m deep Varnjača Cave on Velebit with his wife Hela, using mountaineering equipment, making Hela Krajač the first woman in Croatia known to have descended into a cave. In the same year, Krajač descended 180 m into the Vrtlina Cave, while his assistant Mile Sjauš reached 195 m in depth. In 1927, at the instigation of Umberto Girometta, in Croatian Mountaineering Society Mosor, a "Section for the investigation of karst phenomena" was established, and members explored caves in the vicinity of Split. In 1929, they made a significant contribution to the popularization of speleology by arranging and opening Vranjača Cave for tourist visits: led by Mayor Rado Mikačić. Between the two world wars, there were several speleological explorations conducted by individuals, for example, Zagreb mountaineer and journalist Vladimir Horvat and Vladimir Redenšek. In 1934, the exploration of the Veternica Cave on Medvednica near Zagreb began, with numerous groups of various curious people participating. Ph.D. Josip Poljak also scientifically researched Veternica in 1934 (Fig. 3.4).

World War II interrupted almost all speleological research. Vlado Redenšek, Ivan Pancer and Frano Baučić

Fig. 3.4 Ph.D. Josip Poljak at the entrance of Veternica Cave (1934)

acted individually. In the midst of the war (1943), Frano Baučić founded *Spiljsko-pecinsko-ponorski odsjek* (caving section) of Mountaineering Society in Zagreb, which quickly dissolved due to poor activity. However, in 1945, Frano Baučić published the work "Podatci o pećinama", in which he processed 612 caves and pits in Croatia.

The first scientific speleological work after the Second World War was "The Age and Systematic Significance of the Croatian Cave Bear" by geologist Milan Herak, printed in 1947. The excavated material from Cerovac caves (Figs. 3.5 and 3.6), which was discovered in paleontological studies by geologists Crnolatac, Ogulinac and Malez, was processed there (Malez 1965).

Some Important Names of Croatian Speleology

Ph.D. Josip Poljak (born in Orahovica on 15 November 1882, died on 20 August 1962 in Zagreb) was curator at the Croatian Museum of Natural History for over 50 years, and he has published over 200 articles in various professional and popular editions, such as "Priroda", "Hrvatski Planinar", "Vienac", "Napredak" and " Lički calendar"—describing Croatian caves and karst and geology.

He affirmed his affection, interest and need for speleological research with the founding of the Cave Research Committee in 1909.

The result and proof of the systematic nature of his speleological research were the acquisition of a doctorate in 1922 with the title: "Caves of Croatian Karst"—the first doctorate in Croatia in the field of speleology.

From 1945 to 1955, Poljak collaborated with other geologists as an experienced hydrogeologist and expert on the underground of our karst in the preparation of projects for the construction of hydroelectric power plants and artificial reservoirs in the Dinarides.

His last speleological article was published in 1950 in magazine *Priroda* with three photographs - with title "New Cave near Fužine in Gorski kotar" which was discovered during the construction of the Vinodol Hydro Power Plant. Today, it is a tourist cave named "Vrelo" with artificial reservoir near Fužine (Božićević 2012).

Academician Ph.D. Mijo Kišpatić (born in Osijek on 22 September 1851–died in Zagreb on 17 May 1926) was a geologist, petrologist, mineralogist who published a paper in 1881 about his findings of the bones of a cave bear in the Barač Cave near Kršlja in Croatia. Less known is that Kispatic's first visit to Barač's caves was in 1877, and it is impressive that on 20 February 1892 he established the "Committee for Research and Arrangement of Barac Caves". The Committee numbered 12 people, and under its operation, on 15 August 1892, Barac caves were prepared and opened for tourists, which was a rarity at the time. This fact is nowadays considered the documented foundation and start-up of speleological organizations in Croatia (Garašić 1992, 1997).

Academician PhD. Mirko Malez, (born in Ivanec on November 5, 1924, died in Zagreb on August 23, 1990) was a world-famous Croatian speleologist, geologist, palaeontologist and archaeologist (Fig. 3.7). He started practicing speleology as a student. Malez performed palaeontological and archaeological research in the caves. He was an exceptional palaeontologist and a head of the Department of Quaternary Palaeontology and Geology of the Yugoslav Academy of Sciences and Arts—JAZU (today the Croatian Academy of Sciences and Arts—HAZU). He was one of only four doctors of speleology from Croatia (Poljak 1922; Malez 1963; Božičević 1985; Garašić 1986). He was the president of the Speleological Society of Croatia for 24 years and one of the representatives of the Federation of Speleologists of Yugoslavia at the founding of UIS—Union Internationale de Speleologie, in 1965 at the 4th World Speleological Congress in Postojna. He was a member of the Organizing Committee of the 2nd Yugoslav Speleological Congress (Split 1958) and president of the Organizing Committee of the 9th Yugoslav Speleological Congress (Karlovac 1984). He explored more than 1000 caves in Istria and Lika in Croatia, and many more in Bosnia and Herzegovina, Montenegro, Macedonia and Serbia. He was the editor of the scientific journals "Carsus Iugoslaviae", "Acta Paleontologica" and "Acta Geologica". He was one of the founders and editors of the magazine "Speleolog" (1953), author of a monograph on the Veternica cave (1963), a book on the Cerovac caves (1965) and editor of the Proceedings of the 9th Yugoslav Speleological Congress (1984) of 900 pages. He published over 400 professional and scientific papers. He was a full member of JAZU, and an external member of the Austrian Academy of Sciences.

Ph.D. Srećko Božičević (born in Velika Trnovitica on 16 January 1935, died on 15 March 2015 in Zagreb) is a Croatian geologist and speleologist. He was general secretary of Croatian Speleological Society more than 40 years and explored 2100 caves in Istria for his doctoral dissertation. He was author for more than 200 scientific and professional papers (Fig. 3.8).

Here, we will mention some speleologists in the past who research caves and published data about this caves in Croatia: Dragutin Gorjanović Kramberger (1912) about 156 caves, Josip Poljak (1914–1928) about 150 caves, Umberto Girometta and Margetić (1923, 1925) about 472 caves in Dalmatia, Boegan and Bertarelli (1926) about 1800 caves in Istria, Frano Baučić (1945) about 600 caves, Mirko Malez (1950–1960) about 550 caves from Lika, Učka and Istria,

Fig. 3.5 Lower Cerovac Cave

Fig. 3.6 Upper Cerovac Cave

Fig. 3.7 Academician Mirko Malez in Bezdanjača Cave (1965)

Srećko Božičević (1985) about 2100 caves from Istria, Mladen Garašić (1986) about 5263 caves from Croatia, etc.

3.2 Types of Speleological Objects in the Dinaric Karst of Croatia

Caves in Dinaric Karst of Croatia are formed as a result of tectonic, lithostratigraphic and hydrogeological conditions. The most common carbonate rocks in the Dinaric Karst are of Mesozoic and Paleogene ages. Therefore, caves are most commonly found in those rocks. Due to intense tectonics, fractures and faults play the main role in formation of underground karst phenomena (tectogenesis). Vertical and subvertical paleotectonics and neotectonics are dominant, so vertical or very steep speleological objects are the most common. Speleogenesis in Dinaric Karst is all-inclusive and qualitatively and quantitatively special due to the intensive influence of water (groundwater, surface water and sea) on the rocks in the geological past (but also today). In Dinaric Karst, the most frequent division of caves is by slope and arrangement of underground channels, namely *caves, pits and combined speleological objects*. Another division is based on morphology such as *simple, knee, branched, multi-storey or multi-level and systematic*. Caves can also be divided according to their hydrogeological functions (*spring, swallow-hole, estavela, vrulja and underground flow*). In Croatian language, the vertical speleological objects differ from the horizontal speleological objects according to their entrance openings. There are many synonyms for vertical underground karst forms (pits) in Croatian language, namely jama, jamina, jamurka, propast, perast, golubinka, golubnjača, zvekača, zvekara, đot, etc., and many synonyms for horizontal underground karst objects (caves) like spilja, špilja, špilj, spela, pećina, šupljara, peć, pećinica, etc. (Garašić 1991).

3.2.1 Caves

Caves are horizontal or slightly inclined speleological objects (Culver and White 2005). Vertical channels can sometimes appear in them, but they are negligibly small in relation to the total length of the horizontal cave channels (Fig. 3.9). Due to the dominant vertical tectonics, the total number of horizontal speleological objects in the Dinaric Karst of Croatia is 4–5 times smaller than the number of vertical ones. However, their layout and distribution depend on many factors (altitude, location in karst, rock type, karstification phase, etc.).

3.2.2 Pits

Pits are vertical or intensely inclined speleological objects (Gunn 2004). Sometimes completely horizontal or slightly inclined channels appear in them, but they are negligibly small in relation to the total length of the vertical pit channels. The number of pits in the Dinaric Karst is several times greater than the number of caves, which is explained by the large thickness of carbonate rock complexes, paleo and neotectonic uplift of local geotectonic plates, depth to the base of erosion, altitude, etc. (Fig. 3.10).

3.2.3 Combined Speleological Objects

When it is difficult to determine whether the speleological object is a cave or a pit based on the total known length or depth of the channels, then we are talking about the so-called combined speleological objects. For instance pits can have a horizontal entrance and afterwards continue vertically. However, their total number in the Dinaric Karst area is very small. Only a small percentage of the total number of known speleological objects are combined speleological objects. The photos (Figs. 3.11 and 3.12) below show Mijatova Cave (Garašić 1976). The vertical entrance to this combined cave is 30 × 30 m wide and fifty metres deep. This cave is longer

Fig. 3.8 Ph.D. Srećko Božičević in Veternica (1966)

Fig. 3.9 Horizontal entrance of Pozijatva Cave

than 1000 m. The vertical entrance was created by the later processes of speleogenesis. Another example of a pit with a cave entrance is Klanski Ponor (Boegan 1930). This combined speleological object was created by water that penetrated through the horizontal and slightly inclined bedding planes and fractures about 60 m after which it reached the vertical fault zone and continued vertically downwards. According to some new measurements, this combined speleological object is assumed to be about 320 m deep.

3.3 Speleogenesis in the Dinaric Karst of Croatia

Relations between the initial, major and fossil stages of speleogenesis, types of karstification (gravitational, regression, complex) together with their origin (inner, central and Outer Karst Belt) provide new data and scientific insights, suggesting that the study of caverns is the basis of karst exploration. There are far more caverns than caves with natural entrances. Cavern speleogenesis is directly related to lithostratigraphic, hydrogeological and tectonic predispositions. These are mainly vertical speleological objects (pits), of a simple or knee morphological type, with a hydrogeological function of intermittent or permanent springs, which are in the main stage of speleogenesis.

Fourteen main types and six semitypes of speleogenesis were recognized in Croatian speleogeology. There are data for about 12,500 caves from different regions of Dinaric Karst in Croatia. Their speleogenesis has been affected by different geological conditions (Garašić et al. 1997b, 2010, 2017a).

The knowledge about different types of speleogenesis proves that speleological objects in neotectonic active zones in Croatian Karst always play an important hydrogeological role such as springs or swallow-holes. Young caves that originated between Miocene and the begging of Quaternary are not in the last phase of speleologenesis.

Fig. 3.10 Vertical entrance of Jezeranka pit

Studies about speleogenesis on a large number of examples in Croatian Karst led to the conclusion that caves are located directly along the fault plains or along the fault zones. Speleological research was, until recently, limited only to caves with natural entrances on the terrain surface. Numerous caverns, discovered during the construction works in karst, provided the better insight on speleogenesis along the fault plains or in the fault zones. It has been detected that speleological objects always occur in several groups along the single fault (vertical or horizontal sequence) (Figs. 3.13 and 3.14), depending on the type of rocks, intensity of karstification, neotectonic activity, etc.

The greatest number of speleological objects occurs along the same fault (or fault zone) in the vertical arrangement in case of neotectonic uplift and in the horizontal arrangement in case of neotectonic sinking.

The probability of occurrence of vertical or horizontal caves was calculated based on the data from 3877 faults and 925 caverns formed along their fault planes or fault zones. These values range from 30 to 35%. When neotectonic predispositions of faults are taken into account, the probability of a presumed layout of the speleological object ranges between 55 and 65%. Neotectonic activity was identified by direct measurement of absolute displacements in the surrounding speleological objects.

Fig. 3.11 Mijatova Cave is combined speleological object located in Kordun

Fig. 3.12 Large chamber in horizontal part of Mijatova Jama

During the study of caverns in all stages of speleogenesis, a large number of caves developing from the underground towards the surface were detected. This phenomena is explained by neotectonics and the influence of groundwater. In the first two development stages, the cavern expands along the fault planes from the underground towards the surface. In the third phase, a cavern becomes a cave with an entrance on the surface of the terrain. In the fourth phase, due to the impact from the surface, intensive rock wear and collapse of the cave begin.

Examples of regressive karstification in the Dinaric Karst of Croatia can be seen in numerous "relics" on the surface in the vicinity of Imotski in Dalmatian Zagora. A large number of mega sinkholes have been genetically transformed from "valleys" into vertical caves. In some of them groundwater is present constantly, in some occasionally, and in some almost never. The presence of water depends on the degree of speleogenesis acting from the interior towards the surface of the terrain. Crveno Jezero (Red Lake) (Figs. 3.15 and 3.16) and Modro Jezero (Blue Lake) near Imotski are directly connected to the groundwater level. However, although they are only a few hundred metres apart from each other, the collected data cannot confirm the correlation between the changes in groundwater levels measured in them. Water is always present in Red Lake with fluctuation of up to 50–60 m per year. Blue Lake often dries out during the dry months.

Speleogenesis is different in different karst areas along the Croatian motorways. Therefore, karst can be separated in different zones (Herak et al. 1969). Zonation is proved by detailed speleological research on sites along the motorway performed from 1991 till today.

The zone between Zagreb and Karlovac is a lowland. It is a non-karst terrain that is predominantly built of Quaternary and Plioquaternary deposits. No speleological phenomena are found in this area.

Fig. 3.13 Fault as an indicator of tectonics in Varićakova Cave

Fig. 3.14 Fault in Varićakova Cave

The zone between Karlovac and Bosiljevo is so-called Inner Karst Belt with speleological phenomena predominantly related to Plioquaternary and Triassic deposits.

Rijeka Direction (A-6 Croatian Highway):

The zone between Bosiljevo and Fužine is so-called Central Karst Belt, with numerous speleological phenomena in Mesozoic carbonates.

The zone between Fužine and Rijeka is so-called outer karst zone, where vertical speleological phenomena with signs of strong fault tectonics are exclusively present.

The zone between Rijeka and Rupa is a transition area of Ćićarija Mountain with predominantly vertical and in smaller extent horizontal speleological objects.

Istria (A-8, A.9, Istria "Y", Croatian Highway):

In hydrogeological sense, Istrian area is made of almost horizontal layers of so-called Vrsar Anticline. The layers are fractured by local tectonics, and they transport water vertically to rivers Raša and Mirna or to the main Istrian Karst springs.

The area of Ćićarija Mountain features reverse faults and overthrusts with flysch watertight barriers. The Učka Mountain area features contacts of flysch and carbonate sediments. Groundwater from all the caverns of the Istrian area springs in Kvarner Bay through a multitude of vruljas.

Split Direction (A-1, Croatian Highway):

The area between Bosiljevo and Mala Kapela Tunnel and along the edges of karst fields through Lika features horizontal caves with a great influence of groundwater.

The zone between Mala Kapela Tunnel and Sveti Rok Tunnel is so-called Central Karst Belt. It features the abundance of groundwater with dominantly horizontal or oblique and sometimes vertical groundwater movements. The area is rich in estavelas (Fig. 3.17).

The zone between Sveti Rok Tunnel and Maslenica junction features extremely fast vertical groundwater circulation and slow horizontal movements.

The area of Ravni Kotari features mostly horizontal and to a lesser extent vertical movements of groundwater.

Fig. 3.15 Diving platform on Crveno Jezero (Red Lake)

Sometimes a waterproof flysch substrate is located in the immediate vicinity of the roads.

The area of Dalmatian Zagora is similar to the area of Lika. The groundwater flow is more intense, since it features the groundwater from Herzegovina flowing to the Adriatic basin. In this area, particular attention should be paid to the vertical circulation of groundwater. The groundwater level fluctuation of 236 m, which is among the highest recorded in the world, was recorded in a cave between Zagvozd and Vrgorac.

The area south of Neretva River towards Dubrovnik has the characteristics of more often vertical (rarely slanted) caves with direct discharge of groundwater from Herzegovina through vruljas and springs into the sea.

The analysis of simplified speleogenetic features of the main road sections in Croatian Karst of Croatia shows that caves in the main phase of speleogenesis have great potential for both vertically and horizontal groundwater transport. Groundwater additionally changes the engineering geological and geomechanical properties of the rock through which it flows. This fact needs to be taken into account when constructing in karst of Croatia.

Croatian Karst is one of the most famous "locus tipicus" areas of classical karst in the world. Underground karst forms are represented, among other things, by over 12,500 speleological objects (Garašić 1991, 2001), with different morphological, morphogenetic and hydrogeological features (Garašić 1986).

Despite the large number of caverns found and explored in our karst so far, there is a relatively small number of publications in this area of speleology. Caverns found along the Croatian highways were systematically investigated since 1991. Their genesis, morphology, depths and lengths are different, but they confirm the scientific assumptions about the differentiation in the degree of karstification. The degree of karstification depends on the level, type, amount and aggressiveness of the groundwater, as well as tectonic predispositions and stratigraphic properties of the environments in which they occurred.

All major road tunnels (Učka, Tuhobić, Sveti Rok and Mala Kapela tunnels) had at least one major speleological object that needed sanation. In some tunnels, more than 70 caverns were remediated (Brinje Tunnel). Bridges had to be built over the caverns in some tunnels (Veliki Gložac, Plasina and Vrata tunnels). Today, some caverns are in use as a water supply (Učka Tunnel). Some tunnels pass above the water pumping ground (Mravinci, Katarina, Grič and Vrata tunnels). In some caverns, the presence of seawater was detected (Pećine Tunnel). All of the mentioned examples represented hydrogeological problems that had to be solved.

The following is a list of some viaducts and bridges whose pillars or abutments are placed directly above the caverns: Anđeli, Vela Draga, Delnice, Golubinjak, Hreljin, Severinske Drage, Osojnik, Zečeve drage, Bijakuše, Maslenica Bridge, Krka River bridge, Dobra River bridge, Bistrac, Kamačnik, Rječina, etc.

No problem (such as changes in the movement, quantity and quality of groundwater) caused by the unprofessional handling of hydrogeological challenges in karst occurred in Croatia.

In the area between Rijeka and Karlovac, all three stages of speleogenesis are present. The indicative ratio of the main to the fossil phase of speleogenesis is different for each of the mentioned zones. The average ratio for the whole area of Croatian Karst is 3:1 (3.11:1). In the area between Karlovac and Bosiljevo, the ratio is 1:7 (1:6.89). The cavities are usually filled with collapsed material, calcite or, in most cases, clay. In the central karst zone, i.e. in the area between Bosiljevo and Fužine, the ratio of main to fossil phase of speleogenesis ranges from 2:1 (2.36:1) to 3:1 (3.05:1). In the

Fig. 3.16 Cave diving expedition to Crveno Jezero (Red Lake) in 1998

outer karst zone, i.e. in the area between Fužine and Rijeka, the ratio is 4:1 (4.20:1).

In the area between Bosiljevo and the Sveti Rok Tunnel, the ratio of main to fossil speleogenesis phases is 2:1 (2.21:1) to 3:1 (3.22:1). In the outer karst zone, i.e. in the area between Sveti Rok Tunnel and Maslenica junction, this ratio is 4:1 (4.44:1). All known and investigated caves along the motorway route, as well as those in its immediate vicinity, were taken into account in the calculation of the relationship between the various stages of speleogenesis.

The initial stage of speleogenesis is present at all sites where karstification takes place. However, due to the narrow width of initial fractures and fissures, it is not possible to directly observe this stage of speleogenesis. In the inner karst zone, i.e. in the area between Bosiljevo and Fužine, in fifty speleological objects, a parallel presence of initial, main and fossil phase of speleogenesis was detected in the ratio 1:1:1. This is a relatively rare occurrence. Only about 1.58% (183 caves) with all stages of speleogenesis present were found in Croatian Karst.

Based on the analysis of the data from a large number of studies, it can be concluded that most caves in the Croatian Karst have not yet endured all three stages of speleogenesis. This is due to geological, tectonic, hydrogeological and climatic factors (Palmer 2007; Pamić et al. 1998). For example, caves in high mountain ranges, with the process of gravitational karstification, have not yet "reached" the less permeable or impermeable substrate, and there is no fossil phase of speleogenesis.

Comparisons between caves with natural entrances on the surface and caverns are particularly interesting.

All three phases of speleogenesis are not present in the same speleological object in any recorded case of caverns (925 samples). Their statistical distribution is as follows: 599 (64.75%) of caverns are in the main phase of speleogenesis, 223 (24.10%) are in the fossil phase of speleogenesis, and 103 (11.15%) are in the same proportion of initial and major phases of speleogenesis. The ratio of main to fossil phase of speleogenesis in caverns is 599: 223 (2.68: 1). This is evident in the area of Central Karst Belt, where the percentage

Fig. 3.17 Estavela Tović near Fužine

of the fossil phase is increased. Intense tectonic activity and the large amount of precipitation enriching the groundwater of the Central Karst Belt (Gorski Kotar, Lika) are the main factor for this occurrence. The fossil phase of speleogenesis is especially present in caverns found in tunnels with small overburden (Brinje, Hrasten and Sopač).

Taking into account the position of caverns (overburden) in relation to the surface, it can be concluded that in the inner karst zone (Karlovac–Bosiljevo), most caves in the fossil phase of speleogenesis are present at depths up to ten metres, followed by the main phase of speleogenesis underneath. In the central karst area (Bosiljevo–Fužine, Bosiljevo–Sveti Rok Tunnel and Dalmatian Zagora), the main phase of speleogenesis appears at depths of several metres, and the fossil phase occurs between 100 and 300 m. In the outer karst area (Fužine–Rijeka, Sveti Rok—the Maslenica junction and Ravni Kotari), it is a rare occurrence. If the fossil phase is present, it occurs at depths between 200 and 500 m.

If we add types of karstification processes in particular areas the afore mentioned results, we could conclude the following:

With the appearance of additional regression karstification, the Central Karst Belt (Bosiljevo–Fužine; Bosiljevo–Sveti Rok Tunnel; Dalmatian Zagora) has, in addition to the main speleogenesis phase, an increased number of objects that are in the fossil speleogenesis phase. In civil engineering sense, the largest number of speleological phenomena that can interfere with construction is expected in this area. Caverns are sometimes unstable and can be of different dimensions.

Dominant gravitational karstification in the outer karst zone (Fužine–Rijeka; Tunnel Mala Kapela-node Maslenica–Ravni Kotari) causes the predominant appearance of vertical speleological objects that are in the main stage of speleogenesis. In civil engineering sense, a large number of speleological objects that can be very large in size and pose problems during remediation are expected in this area.

The occurrence of both gravitational and regressive karstification in the inner karst zone (Karlovac–Bosiljevo) significantly increases the number of speleological objects in the fossil phase compared to the number of speleological objects in the main phase of speleogenesis. In civil engineering sense, such speleological objects cause problems due to their different engineering-geological and geomechanical properties. Their number is relatively small compared to other karst areas.

The major role in speleogenesis is played by the tectonic predisposition of the area where speleological objects were formed. Karstification within horizontal or slightly slanted layers is known to progress more slowly than within steep or vertical layers. Tectonics played a crucial role in speleogenesis of caverns found in some of the tunnels (Tuhobić, Vrata, Sopač and Veliki Gložac). These relatively rare speleological phenomena are predominantly of tectonic origin and are exclusively related to the proximity of fault plains.

So far, less than 3% of speleological objects created solely by tectogenesis have been registered in Croatian Karst. This indicates they were formed without significant erosion and corrosion effects of groundwater. Aside the fault zones, the phenomena are also found in the scaffolds of local anticlines (Sopač Tunnel), where groundwater played a secondary role in speleogenesis.

Hydrospeleogenesis is always a complex process in which water plays a crucial role.

Erosion and corrosion (mechanical and chemical speleogenesis) occur together in the formation of speleological objects. So far, 93% of speleological objects created by erosion and corrosion have been registered in Croatian Karst.

Whirlpool pots, so-called hieroglyphics and leopard skins, were detected in some of the caverns there. At the sites of heavily flattened rocks, the strength of the chemical or mechanical action of the groundwater (the area of the Sljeme Tunnel, the foundations of the Dobra Bridge and the foundations of the Zečeve Drage viaduct) is undoubtedly proven.

On the terrain surface, carbonate rocks (limestones, limestone dolomites, dolomitic limestones and dolomites) of different textures and age respond differently to erosion and corrosion. In the underground, a different reaction has been confirmed. The most intensive karstification was detected within the Jurassic and Lias limestones, followed by the Dogger and Malm sediments, and the Upper Cretaceous limestones.

The carbonate clasts of Lika, Velebit, Mala Kapela and the Croatian coast were first described as facies of Promina deposits. However, their imprecise relationships and differences in composition were the reason the name Jelar deposits was introduced. Their formation was later associated with more significant tectonic disturbances (reverse faults and overlaps) followed by vertical faults. Promina deposits in the area of Ravni Kotari and Promina Mountain, as well as areas with Jelar deposits, are always in transgressive contact with older deposits. These are carbonate breccias of Paleogene–Neogene age. Karstification is extremely intensive in these deposits. The third longest subterranean river in the world—the Gacka River, springs in Jelar deposits. Zrmanja and Krupa rivers partly flow through the Promina deposits (Figs. 3.18 and 3.19). The largest underground cavities in Croatian Karst with halls larger than 100 m in diameter were found in Jelar deposits in the area of southern Velebit. Even the deepest speleological objects in this area, partly or completely, pass through the Jelar deposits. This indicates a later intensive karstification (depths of more than 450 m). Better solubility of breccia enabled the creation of a large number of karst phenomena. The number of cave occurrences per km^2 in Jelar deposits exceeds the number of cave occurrences in some Jurassic and Cretaceous limestones.

Anhydrite and gypsum are sedimentary rocks of purely chemical genesis, unlike carbonates that can also be of organic genesis, and conglomerates and breccias that can also be of mechanical (clastic) genesis. Because of its specific gravity, gypsum is lighter than carbonate rocks. It is a so-called diapiric elevation. Specifically, gypsum rocks rise from carbonate or other massifs, creating diapiric folds, with numerous fractures. Sometimes they intertwine with surfaces of faults. These movements are measured in hundreds of metres and in some places several kilometres in length. In Croatia, diapir gypsum beds and the only speleological objects in gypsum have been found in the area of Kninsko Polje. These are mostly smaller speleological objects that have been degraded by their subsequent orogenetic movements and the gypsum solubility to their present dimensions (20–50 m on the surface of the terrain). Raises are estimated to be over 1500 m. Therefore, the existence of larger speleological objects in gypsum is possible, but at greater depths (caverns).

Neotectonic activity was detected in some speleological objects in the outer karst zone, for example, in Hrasten, Sveti Rok, Tuhobić and Učka tunnels. There, some speleothems are torn apart by the neotectonic movement of the blocks. These are millimetre movements of local type.

All morphological characteristics of speleological objects (shape, length, depth, orientation, slope, etc.) are strongly related to the factors of speleogenesis, its type, phase, intensity, as well as location (different karst belts).

It is well known today that the most complex processes of karstification (corrosion, erosion, abrasion, denudation, accumulation, secondary and tertiary sedimentation, etc.) advance in different directions depending on the hypsometric position, geological (lithostratigraphic) predispositions and hydrogeological conditions.

The majority of speleological phenomena are formed by gravitational karstification. Only about twenty per cent are formed by so-called reverse or regression karstification. Examples of a so-called complex karstification are caverns in Vrata and Učka tunnels (influence of watertight rocks).

Fig. 3.18 Krupa River canyon

Fig. 3.19 Krupa and Zrmanja rivers

The gravitational type of karstification progresses intensively downwards, the reverse or regressive type of karstification progresses upwards, i.e. towards the surface, and the complex type of karstification progresses equally in all directions. Gravitational or surface karstification is always influenced by gravitational water (precipitation) and is very easily visible. It is very pronounced in Jurassic and Cretaceous carbonate sediments. It is favoured by tectonic fracturing (Garašić 1986). Reverse or regressive karstification is visible only in the underground (in caverns) and is a consequence of hydrogeological conditions (changes of groundwater level, constant presence of groundwater during the first and second stages of speleogenesis, etc.), lithostratigraphic factors (intensively developed in more soluble deposits, in Jurassic and Cretaceous fine breccias, in Plio-quaternary breccias, etc.) and in blocks with neotectonic elevation. Reverse karstification is present in the area of neotectonic uplifts (e.g. the area Jablan–Sušica or southern Velebit), and its intensity is proportional to the intensity of the uplift. Indicators that show reversible or regression karstification with safety are, for example, the so-called airbags (Figs. 3.20 and 3.21) on the cave ceilings (Fužine area, Josipdol and Brinje). The intensity of reverse karstification, unlike gravitational karstification, depends less on tectonic fracturing and more on specific hydrogeological conditions. A complex type of karstification was found in a cavern that is in contact with Paleozoic clastic watertight deposits and the Mesozoic dolomitic limestone complex in Vrata Tunnel and in contact with flysch in Učka Tunnel. Progression of karstification with equal intensity in all directions is a rare occurrence.

Roads and/or tunnels passing above the water springs represent the greatest possible engineering geological, hydrogeological and civil engineering problem, for example, above the Jadro spring (Mravinci) in flysch materials or above the Zvir spring in Rijeka (Katarina Tunnel) in karstified Mesozoic carbonates.

Of the 925 caverns explored in detail, 791 or 85.50% are vertical speleological objects, and 134 or 14.50% are horizontal speleological objects. No combined or complex speleological objects were recorded. Considering the construction of individual sections of the road, which mostly pass through the outer and Central Karst Belt, a minor change from the general ratio of the relationships in Croatia is somewhat explanatory—vertical: horizontal: complex (V: H: C = 78: 21: 1).

Caves located on the surface of Učka Mountain were connected to the cavern found in the tunnel by the underground stream. The length of this system in August 2021 was 6596 m long with the height difference of −430 m. The cavern was entered by diving.

The large number of newly opened and explored speleological objects on the Zagreb–Rijeka and Zagreb—Split—Dubrovnik motorway has contributed to a better understanding of the processes of karstification, speleogenesis and speleomorphology of karst areas. It provided valuable scientific knowledge that will allow easier and safer rehabilitation of caverns encountered during the construction of karst roads.

Fig. 3.20 Effects of regressive karstification in Vrelo Cave

Fig. 3.21 Airbags in Vrelo Cave

The volumes of the investigated caverns vary from several tens of cubic metres to tens of thousands of m^3. Their spatial orientation is generally parallel to the so-called Dinaric direction of provision (northwest–southeast) with about 67% of the caverns. About 29% of caverns are oriented almost perpendicularly to this direction of provision (northeast–southwest), and about 4% of the caverns surveyed do not have a dominant direction of provision. During detailed speleological research in Tuhobić Tunnel (Garašić and Kovačević 1992; Garašić and Vivoda 2003a), a certain match was detected in the systematic appearance of three vertical caverns together at a maximum distance of up to 50 m between them. They were created in Jurassic limestones. Similar phenomena were later detected in "Sveti Rok" Tunnel, also in Jurassic limestones (Garašić, 2004).

The depth of karstification in the inner karst zone was estimated to be 50–100 m deep, in the central karst zone to a depth of 150–600 m and in the outer karst zone to a depth of 500–1000 m or more. This is followed by zones of inclined or even horizontal circulation, which divert the groundwater to different basins.

Speleological research has greatly facilitated the design of cavern remediation projects. Some regularity in their appearance with respect to types of karstification and speleogenesis was perceived. In several places, the original routes of the roads were moved in order to protect the groundwater streams discovered by speleological studies.

3.3.1 Hydrogeological Factors of Speleogenesis in the Dinaric Karst

Hydrogeological factors affecting the speleogenesis of the Dinaric Karst area are related to the amount and types of precipitation that fall on the rock surface, their annual distribution, chemistry, changes in groundwater levels that indirectly participate in the karstification index, i.e. the extension of secondary porosity of rocks to dissolution porosity. It is a surface water and groundwater that is abundant in the Dinaric Karst area as in its past and today.

Sometimes, after seismic activity, permanent or occasional changes in groundwater runoff (groundwater flow, flow direction, quantity, quality and temperature) occur in karst due to insufficient knowledge of hydrogeological relationships.

After the 3.7 Richter scale magnitude earthquake occurred on 7 September 2018, the springs of the Vrljika River (Opačac, Utopišće and Jaul) near Imotski completely dried out. The average flow during the year in Opačac is between 1 and 16 m^3/s. A similar situation occurred after earthquakes in 1923, 1942 and 2014. No changes were detected in Red Lake and Blue Lake, which are located about 1.5 km away from the springs. After a slight tremor of the soil on 11 September 2018, the water began to flow again (Marijanović 2018) due to the hydrogeological and tectonic factors that are certainly related to recent speleogenesis and paleospeleogenesis.

Hydrogeological characteristics of rocks are evaluated according to the lithological composition of the rocks, the genesis of the areas where they were found, the degree of deformation of these rocks, the apparent velocities of underground flows, the locations of springs and swallow-holes in the observed (catchment) areas and other elements that may affect the formation and movement of surface and groundwater.

The name carbonate rock includes a complex of different lithological members from pure limestone to dolomite of different lamination. Although the entire lithologic complex of carbonate rocks enters the range of water-permeable rocks with the possibility of forming aquifers, hydrogeological features are evaluated ranging from well water-permeable (mainly limestone) to poorly permeable rock (mostly dolomite). Consequently, the higher the limestone content, the more permeable the water permeability, and with the higher dolomite content, the permeability decreases. The name clastic rock also encompasses a complex of rocks of different lithological composition and different degree of water permeability; however, the differences are related to the range between poorly permeable and watertight rock masses.

Carbonate and clastic rocks form the basic rock mass of Dinarides. Their water permeability and their position in the structure of the Dinarides also depend on their hydrogeological function in the formation of basins and the network of underground flows. The third group of deposits was formed by erosion processes during the youngest Quaternary geological period. They have variable lithological composition and relatively small thickness relative to the rocks of the basic structures of the Dinarides. It is predominantly unbound and semibonded deposits whose permeability depends on the content of the clay component which reduces the permeability. Considering groundwater flows in the deep karst underground and barriers built of clastic rocks, Quaternary age deposits have an indirect influence on the karst underground, but as cover deposits affect the rate of infiltration of precipitation into the underground.

Carbonate rocks can generally be divided into two groups of varying degrees of water permeability. The first group contains well-permeable carbonate rocks that make up the bulk of aquifers with extremely rapid subsurface flows, especially in zones of strong faults with deep penetrations of the karstification processes (Fig. 3.22). The large portion of the Mesozoic rock complex, ranging from Lias to Upper Cretaceous, has all the characteristics of a highly water-permeable medium. Carbonate rocks from the beginning of the tertiary (foraminiferal limestones) are of the same hydrogeological character. Rock porosity is fractural or secondary. The total average porosity of limestones is about 3%, which is very low relative to the intergranular porosity deposits. The large mass of limestones and the connectivity of underground channels in karst allow rapid flows of large quantities of groundwater during rainy periods. Subsurface flows are concentrated mainly to tectonically damaged zones within the limestones. The second group of weakly water-permeable carbonate rocks includes the rocks dominated by dolomites over limestones. These are primarily Upper Triassic dolomites that extend along the entire Dinarides and form part of the underlying upper mountainous area formed after the formation of the Dinaric carbonate

Fig. 3.22 Fractures accelerate the processes of karstification (Segestin, South Velebit)

platform. The sedimentation conditions on this platform were quite different, so regardless of the shallow sea environment in general, conditions suitable for the processes of dolomitization and formation of clastic sediments existed. Higher occurrences of dolomite are associated with the beginning of Jurassic (Lias) and the Malm, when there was considerable differentiation of the Dinaric Platform. Dolomites also have fractural porosity, but due to their lower solubility in water than limestones, dolomites generally do not have more developed karst forms. The caves are predominantly of fractural type. The so-called Jelar deposits also belong to carbonate rock group. Lithologically, they are dominated by coarse limestone breccias (Velebit), locally with a considerable amount of clastic components. Therefore, these rocks are usually given the hydrogeological status of the whole variable permeability (Fig. 3.23).

Clastic rocks are estimated to be generally watertight because clay sediments prevail in the lithological composition. It is a lateral and vertical alteration of shales, siltites, sandstones, marl limestones, conglomerates and breccias that give this rock mass variable water permeability with the possibility of forming smaller aquifers with the appearance of small springs. However, as a whole, depending on the depth range, in the predominantly karst carbonate medium of the Dinarides, they represent barriers to groundwater movement. There are generally two stratigraphic types of clastic rocks in the Dinarides. These are mostly clastic rocks of the Paleozoic and Triassic ages that form the cores of anticlines in Gorski Kotar, Lika and Dalmatia. They mainly form the basis of the carbonate mass of the Dinaric Platform and mark the beginning of sedimentation on that platform. Another stratigraphic type consists of clasts of Paleogene age (flysch) whose occurrences in the coastal Adriatic area, in Istria and on the islands have an important hydrogeological function of barriers to groundwater movement. There are numerous large karst springs formed on the fault contacts of carbonate mass of the Dinarides (as aquifers) and clastic flysch deposits (as a barrier). The equivalent of Jelar deposits in the Velebit Mountain area are Promina deposits in Ravni Kotari. Due to their high content of fine clastic components, they have similar hydrogeological characteristics as the older flysch. Deposits of variable water permeability and relatively small thickness cover the complex of Quaternary age deposits that were formed during the terrestrial phase of the Dinaric Platform development. These are lake, fluvioglaclinal and alluvial deposits of karst fields, followed by terra rossa, as the most common cover deposits on carbonate rocks, such as rock-creeps and rock-creep breccias, talus and eluvial deposits. The porosity of these deposits is intergranular. Their thickness varies from place to place, but not to the extent that could significantly affect the formation and movement of groundwater in the karst medium, except in the marginal areas of karst fields where lake deposits may have a function of a barrier.

3.3.2 Lithostratigraphic Conditions of Speleogenesis

Lithostratigraphic factors are related to the type and geological age of the rocks in which speleological objects are created or can occur. There is no universal rule as to when, where and why speleogenesis is more intense or even missing in the same or very similar rocks of the same geological age. However, the specific solubility of rocks in aqueous solutions, mild acids from the atmosphere, porosity, primary stratification, exposure to chemical and geological processes through geological past, orientation and thickness of layers, etc., lead to certain conclusions regarding the Dinaric Karst.

For example, speleogenesis is more intensive in limestones or dolomitic limestones of the Upper Cretaceous than speleogenesis of the Upper Triassic dolomites. It is very pronounced in Paleogene limestone breccias and the most pronounced in Lower Jurassic limestones. These conclusions are derived from the number of known speleological phenomena on particular surfaces of karst terrains in the Dinaric Karst. So by changing the degree of exploration of the terrain, such conclusions will change, but only in percentage terms. Namely, the great tendency for speleogenesis is very noticeable in Jurassic limestones. There is also an abundance of speleological phenomena in Paleogene breccias, Eocene limestones or Senonian (Upper Cretaceous) limestones. This is a big difference compared to all other lithostratigraphic members found in the Dinaric Karst in Croatia, where speleogenesis is less intense.

3.3.3 Tectogenesis

Tectonics is one of the basic conditions for the formation of caves and pits in the Dinaric Karst. In rare cases, tectonics is absolutely dominant in speleogenesis. It occurs most often in poorly permeable or even watertight tectonically active rocks (such as massive dolomites) whose primary fissures have not yet been expanded to speleological dimensions by groundwater or atmospheric water (Herak 1984; Garašić 1985).

The present-day rock mass movements and/or tectonic movements that have emerged from the Miocene age to the present day are called neotectonics. They are very noticeable in caves. Absolute movements and their result (direction, intensity, slope, etc.) can be monitored on selected neotectonically active faults and fractures. Precision instruments can record movements in microns. This sort of measurement was performed at several locations in caves of the Dinaric Karst in Croatia (Garašić 1980, 1986).

Fig. 3.23 Adriatic and Dinaric tectonic plates border area (Jelar Breccias)

3.3.4 Sea Level Uplift in the Dinaric Karst Area of Croatia

Within the caves along the Croatian part of the Adriatic Sea, there are undisputed indications of sea level changes, mostly rising, in geological past. Specifically, speleothems that could have arisen exclusively in the dry phase of speleogenesis have been found within speleological objects that are today partially or completely submerged in the sea. Analysis and detailed measurement of the age of some speleothems from those caves shows that in the last 18,000–20,000 years the level of the Adriatic Sea has risen between 86 and 106 m (Faivre et al. 2010).

With a geological interpretation, Surić et al. (2002, 2005, 2010, 2014, 2015) conclude that in spite of very favourable coastal features, late Pleistocene–Holocene relative sea level changes along the eastern Adriatic coast are still not completely resolved mostly due to the intensive and complicated regional and local neotectonics (Ćosović et al. 2008). The north Adriatic area as subsiding one, and proposed a reconstruction of possible very slow (local) uplift which is supported by well-dated submerged speleothem sand tectonic reconstruction. In order to reconstruct the late Pleistocene–Holocene sea level rise along the Eastern Adriatic Coast, eight speleothems were collected from three submerged caves along the Croatian coast from depths of −38.5 to −17 m.

3.3.5 Water in Speleological Objects in the Dinaric Karst of Croatia

Groundwater is always present in the caves of the Dinaric Karst in Croatia - as vapour inside the air, as running water, or as ice and snow in solid form. The state of matter depends on the external and internal temperature, the season and atmospheric conditions and the position of the cave, i.e. the altitude of its entrance. About 35% of caves in the Dinaric Karst of Croatia always have water in liquid state in greater or lesser amount, from the dripping water to the significant watercourses with waterfalls, lakes or permanently submerged cave channels (Fig. 3.24). Substantial changes in groundwater levels have been recorded during different seasons, sometimes more than 200 m. For example, in

Fig. 3.24 Underground lake (Cave near Aržan)

Fig. 3.25 Đula (Đulin ponor)—Medvjedica Cave system—ground water streams

Fig. 3.26 Đula—Medvjedica Cave system—underground lake

Pavlinovići, in 1977, a 236 m level fluctuation was recorded. Some caves such as springs, swallow-holes and underground river flows have groundwater present in almost all their parts, for example, Đula–Medvjedica Cave system (Figs. 3.25 and 3.26), which gets filled with water during the rainy season (over 16 kms of cave channels). In the Dinaric Karst of Croatia, groundwater is always present, but during the year, the amount can change significantly.

3.3.5.1 Springs

Thousands of springs are known in the Dinaric Karst area of Croatia, some of which are occasional and some permanent, i.e. those from which water flows throughout the year, regardless of possible low groundwater levels. Their flow varies throughout the year, and the ratio of minimum to maximum flow can range from 1: 10 to 1: 500 or more. Following is a list of some well-known springs such as Bijela and Crna River springs near Plitvice Lakes (Figs. 3.27 and 3.28), Crno Vrelo near Kordun Ljeskovac, Čabranka River spring in Gorski Kotar, Cetina River springs (Fig. 3.29), Čikola River spring, Dobra River spring, Dretulja River spring, Gacka River springs, Krčić spring, Krupa River spring, Kupa River spring (Fig. 3.30), Kupica River spring, Ličanka River spring, Lika River spring, Miljacka River spring, Mrežnica River spring, Ombla River spring, Rječina River spring, Rudnica spring, Rumin spring (Fig. 3.31), Slunčica River spring, Tounjčica River spring, Una River spring, Vrljica spring, Zagorska Mrežnica River spring and Zrmanja River spring.

All rivers and streams in the Dinaric Karst area in Croatia originate from caves or pits. For example, the Una River flows from a pit (Figs. 3.32 and 3.33) that has been explored

Fig. 3.27 Plitvice Lakes—National Park with many caves

by diving up to the depth of 248 m (entrance at 395 m alt, Black Sea basin) (Casati and Garašić 2016). Kupa River (entrance at 321 m alt, Black Sea basin) springs from a pit deeper than 154 m. Dobra River springs from the Gojak Cave, which was explored in the length of 2160 m. Gacka River (Adriatic basin) springs from a cave longer than 1100 and 105 m deep (Figs. 3.34 and 3.35), and Krnjeza River (Adriatic basin) springs from a pit deeper than 106 m. Sinjac creek (Black Sea basin) springs from the cave lake 203 m deep (Kovačević, 2013). Cetina River (Adriatic basin) springs from a cave system over 1550 m long and 109 m deep (Figs. 3.36 and 3.37). Rječina River spring (Adriatic basin) is 35 m deep and 350 m long. Kamačnik River (Black Sea basin) springs from a pit over 118 m deep, etc. Syphons in caves can sometimes be more than 100 m deep. For example, a syphon in the underground lake of Gospodska Pećina Cave (Adriatic basin) with a depth of more than 107 m (Figs. 3.38 and 3.39) or a syphon in an underground lake in the Zagorska Pećina Cave (Black Sea basin) that is more than 113 m deep, etc. There are about 9000 springs known in the Dinaric Karst of Croatia. Many of them have small, narrow fractural entrances so they are not considered to be caves. It is noticeable that the karst springs of the Black Sea basin are deeper than the springs of the Adriatic basin, which can be explained by the subduction of the Adriatic below the Dinaric. The Dinaric area (mainly the Black Sea basin) rises faster in the neotectonic time than the Adriatic (Adriatic basin), so that the active zone of karstification with the syphon groundwater movement is deeper in hypogenic area (Garašić and Garašić 2017a).

Fig. 3.28 Plitvice Lakes—UNESCO world natural heritage

3.3.5.2 Swallow-Holes (Ponors)

Swallow-holes are speleological objects in which, during the year, water constantly or occasionally sinks underground. They are usually developed along contacts between watertight and permeable rocks regardless of their stratigraphic affiliation. Examples of vertical swallow-holes in Croatian Karst are *Ponor na Bunovcu* 534 m deep on Velebit in Triassic deposits, *swallow-hole near Rašpor* 351 deep and 6,080 m long. Examples of mostly horizontal swallow-hole channels are *Đulin Ponor–Medvedica* Cave system (Figs. 3.40 and 3.41), 16,396 m long, *Panjkov Ponor–Varićakova* Cave system (Figs. 3.42 and 3.43), 13,218 m long (Kovačević 2010).

Swallow-holes appear systematically in all karst fields (polje), for example, swallow-holes in Gračačko Polje (Jelar ponor, Tučić ponor), in Lipovo Polje (Markov Ponor), in Gacko Polje (Fig. 3.44) (ponor Perinka—Fig. 3.45) (Božičević 1966), in Drežnica Polje–ponor Sušik (Fig. 3.46). Swallow-holes also appear in the high mountains (Velebit) on the contact of watertight and permeable deposits (Ponor Crne Vode, Ponor Pepelarica (Fig. 3.47), swallow-holes on Štirovača, etc.). Some are always or constantly active, i.e. rivers, creeks or streams are entering them all year round, while some consume water only in the rainy part of the year or occasionally. There are swallow-holes that are completely filled with clay materials and no longer perform the

Fig. 3.29 Glavaš—one of Cetina River springs

hydrogeological function of water immersion, since significant changes in the amount, intensity and position of surface flows have occurred. These are fossil or paleo swallow-holes. Sometimes ponors are filled with alluvial clastic sediments, so their intake is reduced, but still exists. These are the so-called sieve swallow-holes, for example, swallow-holes of the River Dretulja, Lička Jesenica, Jaruga in Jezerane (Figs. 3.48 and 3.49), etc.

The swallow-hole of Pepelarica streams (there are a few streams named Pepelarica) is the deepest active swallow-hole of the Middle Velebit (Fig. 3.50), which was first investigated in 1981 and in 1983 explored to syphons (Fig. 3.51), without the use of diving equipment, to a depth of 2 m. The depth of swallow-hole was measured to be −358 m but was partially corrected to be −348 m in 2017 (up to the water level in the syphon), although it is not certain that the water level was the same as in 1983. Some additional cave channels were also explored on that occasion.

Đulin Ponor and Medvedica Cave (under Ogulin town) (Malez 1955; Čepelak 1986) represent the natural drainage channels for the water of the Ogulin Dobra River and its tributaries. Geologically, the present situation is only one moment in the process of which, knowing the laws of the karst, we can see the past and foresee the future. However, human intervention in this natural process may change the

Fig. 3.30 Kupa River spring by night

picture of the expected event in the development of this hydrological system. Some of the consequences of the diversion of the River Dobra from its natural underground flow to the artificial are already evident just thirty years after this operation.

Tracing of watercourses proved the existence of underground circulation from Ogulin swallow-holes towards the springs: Bistrac, Kromari and Gojak. The Bistrac spring receives water partly from Zagorska Mrežnica and enriches it with precipitation water, collected between the swallow-hole and the spring. From the geological map, it is evident that the main tectonic lines spread perpendicularly to the underground water flow and hindering it. There is also a syncline along this path, which additionally decreases the water quantity. The reason why water flows in this direction, however, is to be sought in the hypsometric relationships of the relief. The water filling the intersected area can only be overflowed at the lowest points, i.e. the mentioned springs.

Perinka Swallow-hole is ponor on Gacka River near Švica in Lika (Božičević 1984).

Lika River swallow-holes are located near Lipovo Polje in Lika (Turner 1960). The longest ponor is Markov Ponor (Božičević 1968, Bakšić 2001), explored up to 2405 m in length, with a vertical difference of 84 m. Diving research still continues to this day.

3.3.5.3 Percolating Speleological Objects

Caves from which water never springs or sinks into from the surface, but have a permanent stream of water, are called percolating caves. These are usually caves with vertical entrances, for example, Rokina Bezdana (Figs. 3.52 and 3.53) near Jezerane, Babina Jama near Lovinac, Mijatova Jama (Fig. 3.54)/or horizontal entrances (Veternica Cave near Zagreb, cave in the Tounj quarry, Tamnica Cave). Sometimes one of the entrances is connected to the surface with a short syphon (Jopićeva Cave (Fig. 3.55) near Krnjak). Sometimes we are talking about *potential flow speleological objects* such as cave system Crnopac (South Velebit, Crnopac), 54,709 m long and 797 m deep (Barišić 2019). Research has not yet succeeded in reaching a steady stream

Fig. 3.31 Rumin River spring

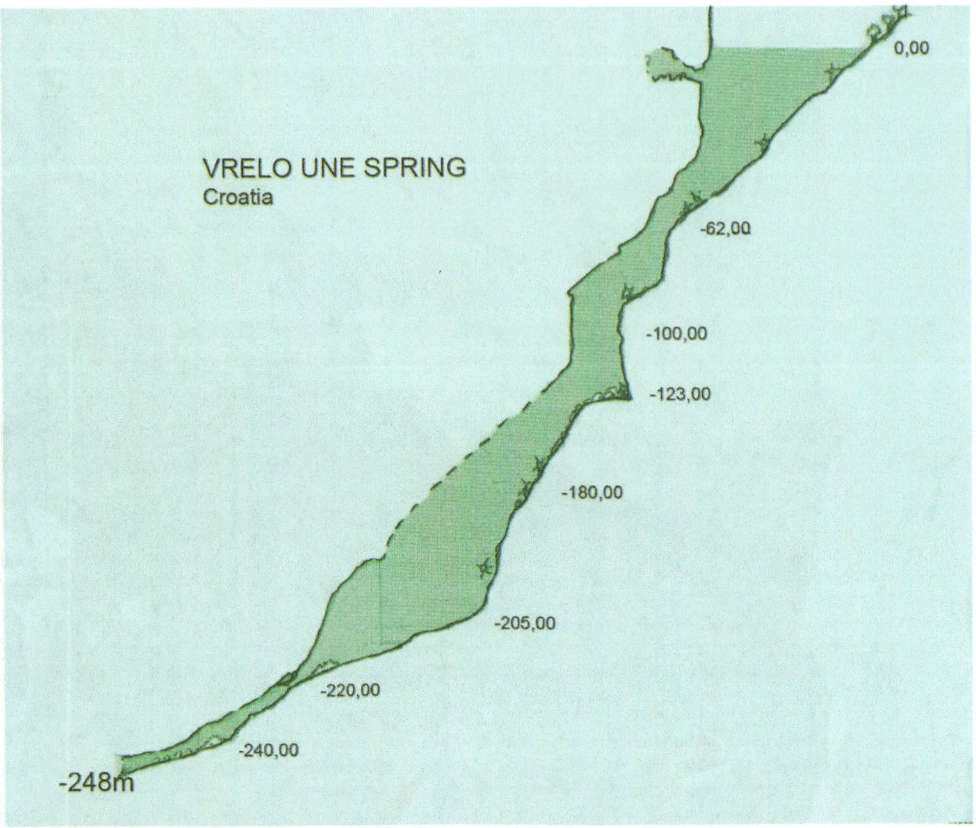

Fig. 3.32 Profile of Una River spring (Vrelo Une) by Luigi Casati

Fig. 3.33 Cave diving in Una River spring

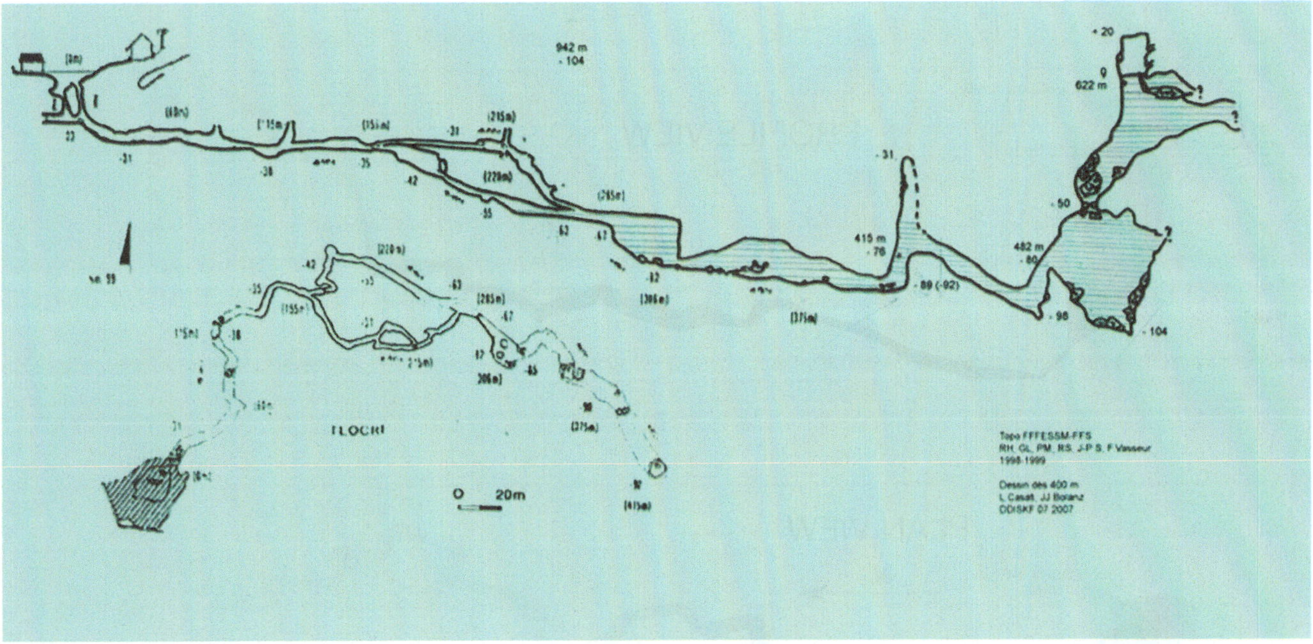

Fig. 3.34 Majerovo vrilo (spring of Gacka River)—profile and plan by Frank Vasseur and Luigi Casati

Fig. 3.35 Cave diver JP Bresser in Majerovo vrilo

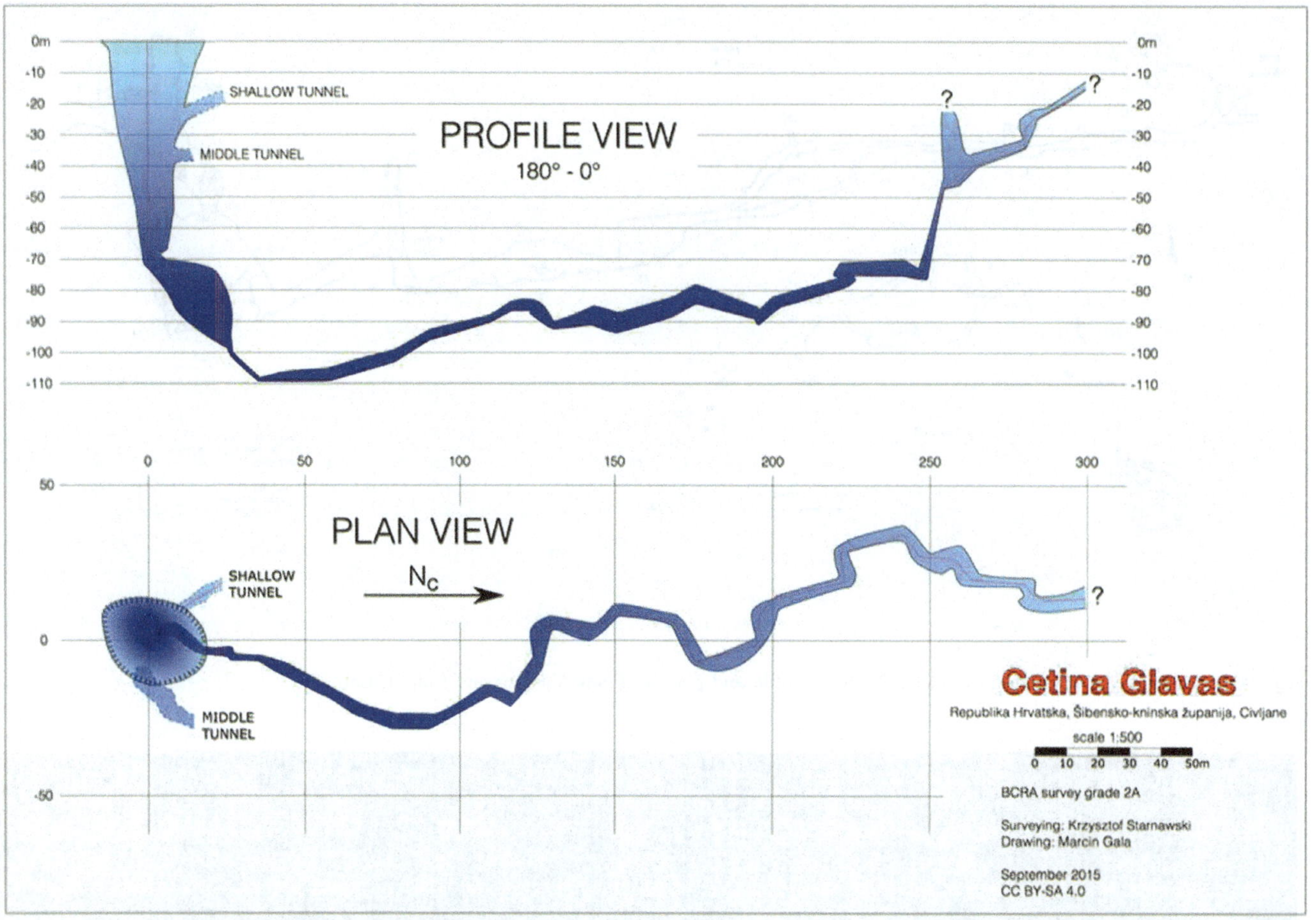

Fig. 3.36 Profile and plan of Cetina River spring Glavaš by Krzysztof Starnawski

Fig. 3.37 Cave diving in Glavaš spring

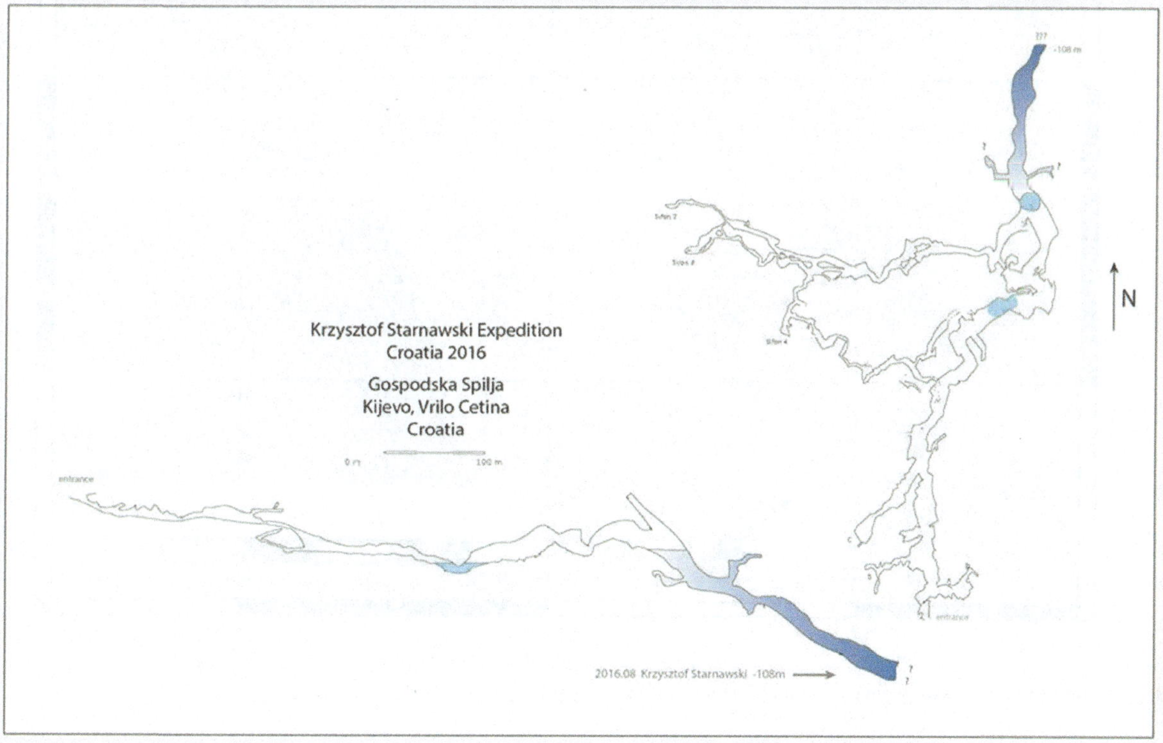

Fig. 3.38 Profile and plan of syphon in the underground lake of Gospodska Pećina Cave by Starnawski

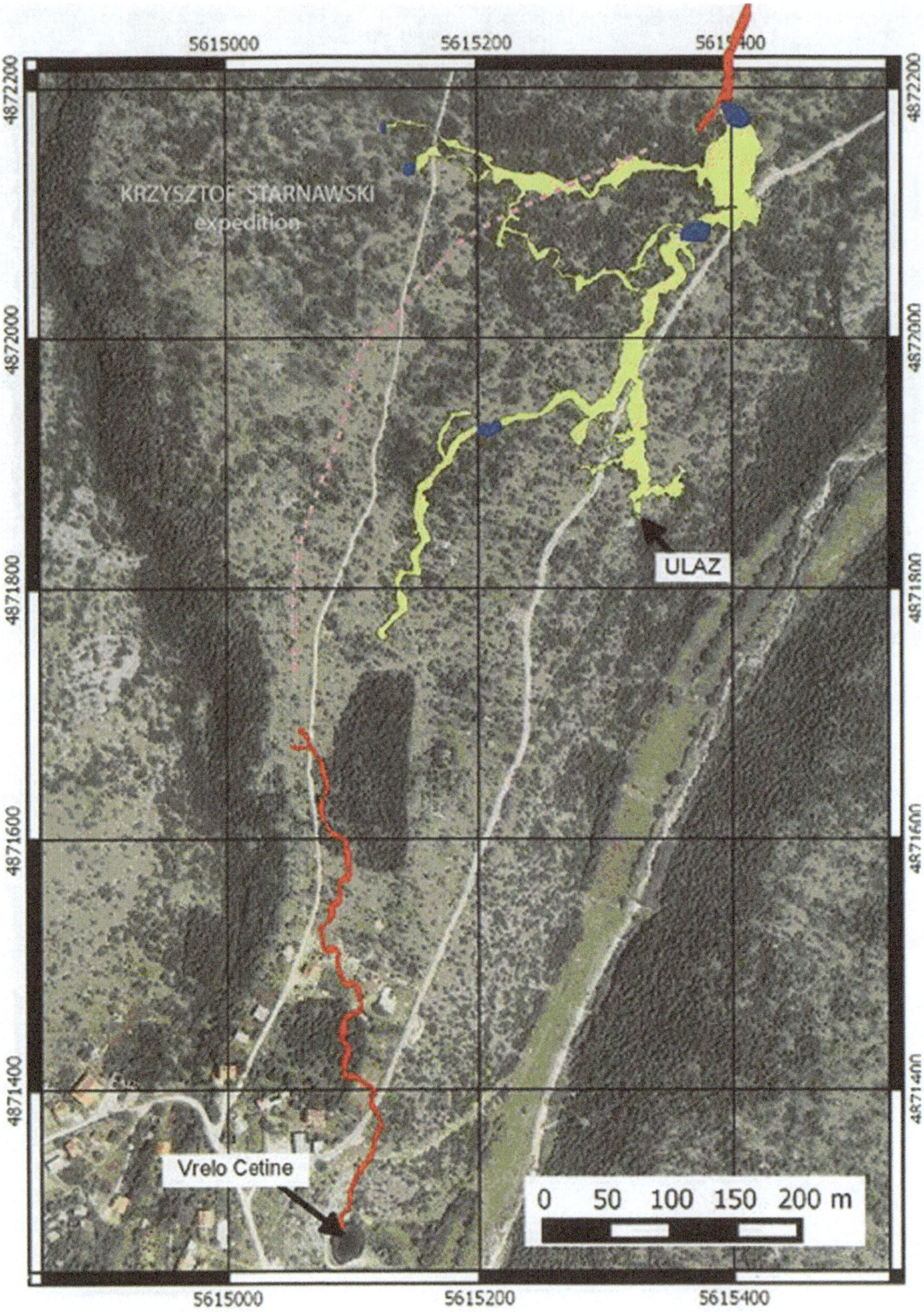

Fig. 3.39 Position of Gospodska pećina Cave and Glavaš spring

Fig. 3.40 Swallow-holes (ponors) of Đulin ponor–Edvedica Cave system

Fig. 3.41 Underground lake in Đulin ponor–Medvedica Cave system

Fig. 3.42 Underground lake in Panjkov ponor—Varićakova Cave system

Fig. 3.43 Panjkov ponor—Varićakova Cave system

Fig. 3.44 Gacko Polje in Lika

Fig. 3.45 Perinka swallow-hole in Gacko Polje

Fig. 3.46 Sušik swallow-hole in Drežnica Polje in Gorski Kotar

Fig. 3.47 Pepelarica swallow-hole on Velebit

Fig. 3.48 Jaruga swallow-hole in Jezerane (summer)

Fig. 3.49 Jaruga swallow-hole in Jezerane (winter)

Fig. 3.50 Middle part of Velebit Mountain

Fig. 3.51 Expedition to Pepelarica swallow-hole in 1983

Fig. 3.52 Waterfalls in Rokina bezdana near Jezerane in Lika

Fig. 3.53 Strong underground water streams in Rokina bezdana

Fig. 3.54 Mijatova Jama near Mateško selo in Kordun

of groundwater flowing from Gračac Field towards the Zrmanja, Krupa and Krnjeza Canyon. For the time being, just like in the case of Munižaba (9911 m in length and 510 m in depth), we are talking about so-called suspended water, i.e. groundwater that occasionally appears and flows but is in no way connected to the steady water flows at lower levels (Bajo et al. 2009). If Crnopac Cave system (56 km), Munižaba (10 km) and Cerovac caves (8 km) were connected, the system would be approximately 70 kms long. This is likely to happen in the near future as some caves are only about ten metres away from each other.

Future research in the south-eastern part of Velebit Mountain, which is predominantly built of limestone breccias, will reveal different levels of individual speleological

Fig. 3.55 Bent spring syphon is one of entrances to Jopićeva Cave system

objects. This has already been noticed in, for example, the Crnopac Cave system, Cerovac caves, etc. The karstification level of the former karst field is dominant here. The uplift of the Velebit and the relative lowering of the present-day Gračac Karst field is the cause of the appearance of different levels within the speleological objects. The study on their genesis could provide the conclusions on the speed and time of the Neotectonic uplift of that part of the Velebit Mountain.

Caves in Croatia—Caves in the Central Part of Dinaric Karst 4

Abstract

Due to different speleogenesis, geology and morphology, caves in the Dinaric Karst of Croatia can be divided in three areas: The Outer Karst Belt, the Central Karst Belt and the Inner Karst Belt. The Outer Karst Belt includes caves along the coast and below the surface of the Adriatic Sea, on islands and in high mountains along the coast (Učka, Velebit, Mosor, Biokovo, etc.). It also includes the areas of the Istria and Pelješac peninsulas and the plains of Ravni kotari and Dalmatian Zagora. Hydrogeologically, this area belongs to the Adriatic basin. The Central Karst Belt of the Dinaric Karst of Croatia is in the area of Gorski Kotar and Lika with numerous karst fields (Polje) and mountains around these karst fields. There are mostly subterranean rivers belonging to the Adriatic or Black Sea basin in this area. The Inner Karst Belt of the Dinaric Karst in Croatia includes the areas of Banija and Kordun that belongs to the Black Sea basin. In addition to the mentioned karst belts, there are separate karst areas in Croatia (Hrvatsko Zagorje, Samoborsko and Žumberačko gorje, Medvednica, Slavonske planine) featuring a lot of caves, pits and all surface karst forms. These karst areas belong to the Black Sea basin.

4.1 External (Outer) Dinaric Belt

The outer Dinaric belt includes the area along the Adriatic Sea (Korbar 2009), the Adriatic seabed, the Adriatic Islands, the high mountains along the coast, Istria and coastal Dalmatia. In this area, speleological objects were mostly formed by the action of intensive tectonics, with excellent lithostratigraphic conditions (limestones, dolomite limestones, sandstones, limestone breccias, etc.). In high mountain areas and high parts of the island, abundant precipitation and groundwater are also responsible for the creation of caves. Mostly vertical speleological objects were formed. (Fig. 4.1).

4.1.1 Caves of the Adriatic Sea in the Dinaric Karst of Croatia

Surić et al. 2007 describes the research of speleolothems from the Medvjeđa spilja on Mali Lošinj and Pit by the coast of Iški Mrtovnjak. It undoubtedly proves that they were formed in the land phase, i.e. while the sea level was about 86 to 116 m lower than today.

The coastline of the Adriatic Sea and the area subjected to karstification:

(a) During the maximum of the last glaciation with a sea level of approximately 120 m below present (Alfirević 1969a);
(b) Today sea level is higher.

Similar findings were given by Garašić (2005) in Podstažišće pit on the island of Brač, a lake with fresh, brackish and salt (sea) water exists. Cave is connected to the sea through unknown passages. Lake is deeper than 70 m, and speleothems are present at that depth (underwater). Podstražišće is a pit (Figs. 4.2 and 4.3) on the south side of the island of Brač, where groundwater is located at the depth of 48 m (Fig. 4.4). Even though the entrance to the pit is about 100 m away from the sea, only the first 12 m of groundwater is fresh, continued by over 70 m of salt (sea) water. Throughout the pit, in both dry and submerged areas,

M. Garasic, *The Dinaric Karst System of Croatia*, Cave and Karst Systems of the World,
https://doi.org/10.1007/978-3-030-80587-6_4

Fig. 4.1 External Dinaric belt—Omiš Dinara viewed from Biokovo Mountain

speleothems can be traced, which were apparently formed before sea level rose. At a depth of -54 m, stalagmite and stalactite samples were taken for analysis (Fig. 4.5).

In parts of the Adriatic Sea seabed that were mainland 18,000 to 30,000 years ago, there are many fully developed submerged caves. They are sometimes covered by sediment. Caves whose entrances are on the steep slopes of today's coast have the hydrogeological role of intermittent or permanent vruljas. It is the entire northern Adriatic, from under the Velebit part (Kuhta and Novosel 2000) of the Adriatic to the south of Zadar and the southern part of the Adriatic with almost all islands, the coast below Mosor, Biokovo, etc.

The submerged karst of the Adriatic Sea could also be considered a paleokarst, but only if the karst forms and relief were greatly altered by subsequent processes (filled, eroded, abraded, etc.). However, it is possible that the real paleokarst (Juračić et al. 2002), which is present on the mainland, is also under the sea, but no such locality has been confirmed.

4.1.2 Caves of the Adriatic Islands

There are 1244 islands in the Dinaric Karst of Croatia in the Adriatic Sea. A total of 78 of them are larger than 1 km^2, 524 of them are between 1 km^2 and 0.01 km^2, while 642 are smaller than 0.01 km^2. Over 1230 islands are built of carbonate sediments that are amenable to karstification. Only a few are of volcanic origin, and one is of Aeolian origin. There are numerous speleological objects on their surface or in their immediate vicinity below the sea. Most known caves and pits have been explored on inhabited islands; the island of Brač (about 260 caves), Korčula, Vis, Krk, Cres, Lošinj,

Fig. 4.2 Postražišće Pit on the island of Brač

Ugljan, Dugi otok, Mljet, Hvar, etc. Caves were also explored on some uninhabited islands (e.g., in the Kornati Islands, etc.)

Podgračišće II or Titina jama on island of Bračs is the deepest cave of all on the Adriatic Islands. With recent measurements, its depth is 329 m. It was explored in several stages during 1960, and the final research was conducted in 1971.

In Cave Pogračišće 2, the author of this book spent 47 h waiting for help at depth of −248 m due to an unexpected radio interruption (Fig. 4.6). A week later, the team explored Podgračišće to the bottom depth of −329 m. The idea about organizing a special cave rescue organisation in Croatia emerged from that accident in 1971 (Figs. 4.7 and 4.8).

Banićeva spilja or Čampari Pit (Fig. 4.9) is the deepest cave on the island of Cres. It was surveyed in 1973 to the depth of −101 m. Skeletons of a cave bear were found near the horizontal entrance, but also in the deepest parts of the cave. Most likely, several bears fell into the steep pit. The remains of the cave bears are kept at the *Institute of Paleontology and Geology of the Quaternary* of the Croatian Academy of Sciences and Arts in Zagreb. These findings also indicate the recent association of the present-day area of the island of Cres with the mainland (Istria) about 20,000 to 30,000 years ago when cave bears lived freely in the area and in the caves (Malez 1975).

Several sea caves (Figs. 4.10 and 4.11) were found on the island of Cres.

Fig. 4.3 Underground lake in Podstražišće Pit

Medvjeđa špilja (bear cave) on the island of Lošinj is today flooded by the sea. Its terrestrial origin is evidenced by the bones and teeth of the cave bear (Malez and Božičević 1964, 1965). The cave was first explored in the 1960s and later in 2005. The Bear Cave speleothems were analysed by Lončar et al. (2015).

4.1.2.1 Caves of the Pelješac Peninsula

About 70 caves are known on Pelješac, 40 of which are known only from literature. It is assumed that the number of caves could be much greater.

The deepest cave is Mladen Cave, deep -235 m, located near the highest point of Pelješac Peninsula—Sveti Ilija (Bočić and Baćurin, 2002).

For the purpose of construction of the Pelješac Bridge in 2008, hydrogeological and cave explorations were conducted by Mladen Garašić in 2008 and 2019 along the coast of the area around Brijesta. A considerable amount of fresh water was found in caves at depth of −36 m. Today, it is used as a public water supply.

4.1.3 High Mountain Caves

The high mountains in the Outer Karst Belt are Učka in the Kvarner Bay area, Velebit in Hrvatsko Primorje area, Mosor and Biokovo in Dalmatia, and Snježnica in the Dubrovnik Littoral area. These mountains rise steeply from the Adriatic Sea. They appear at the contact of two tectonic plates (African plate with Adriatic and Euro-Asian plates with Dinaric). They are the result of neotectonic uplift. Altitudes of these mountains range from about 1200 to 1750 m.a.s.l. The width of mountain ranges can be up to twenty kilometres in some places. These are areas with a lot of rainfall, built mostly of Mesozoic carbonate rocks that are heavily damaged by intense tectonics and favourable hydrogeological conditions. Because of this, the number of speleological objects in these areas is also large. Caves are mostly vertical (pits). The deepest pits and the longest caves in the Dinaric Karst of Croatia (northern and southern part of the Velebit Mountain and the Biokovo Mountain) were also found there.

Fig. 4.4 Cave diving in Podstažišće Pit

On Učka Mountain, more than 200 caves are known. Systematic cave exploring started with Malez (1960). The longest and the deepest one is the cave system Zračak nade —Cavern in tunnel Učka with depth of −430 m, and length of 6596 m.

On Velebit Mountain, more than a thousand, mainly vertical, caves are known.

On the Biokovo Mountain, there are about 400 known caves of which 135 were explored. About 180 species and subspecies of underground animals have been identified in caves of Biokovo of which many are endemic. The depth and potential of the pits also indicates the high degree of biodiversity of the underground fauna and the extreme need for further exploration of the underground of this mountain. High seismic activity in this area is the result of neotectonics and the vicinity from the contact between the African and Euro-Asian plates.

On Biokovo Mountain, the deepest known caves are "Jama Njemica" (depth −954 m, length 1512 m), "Mokre noge" (depth −831 m, length 1343 m), "Jama Amfora" (depth −788 m) (Bockovac 1999, Bakšić and Lacković 2002), cave system "A-1 (Vilimova jama)—A-2" (length 1732 m, depth −589 m), etc.

The Velebit Mountain is the longest mountain in the Dinaric Karst (Fig. 4.12). It is situated between north-eastern Adriatic coast and Lika region in Croatia. This mountain is about 150 km long and 10 to 30 km wide. The highest altitude is about 1750 m, and the lowest altitude is sea level

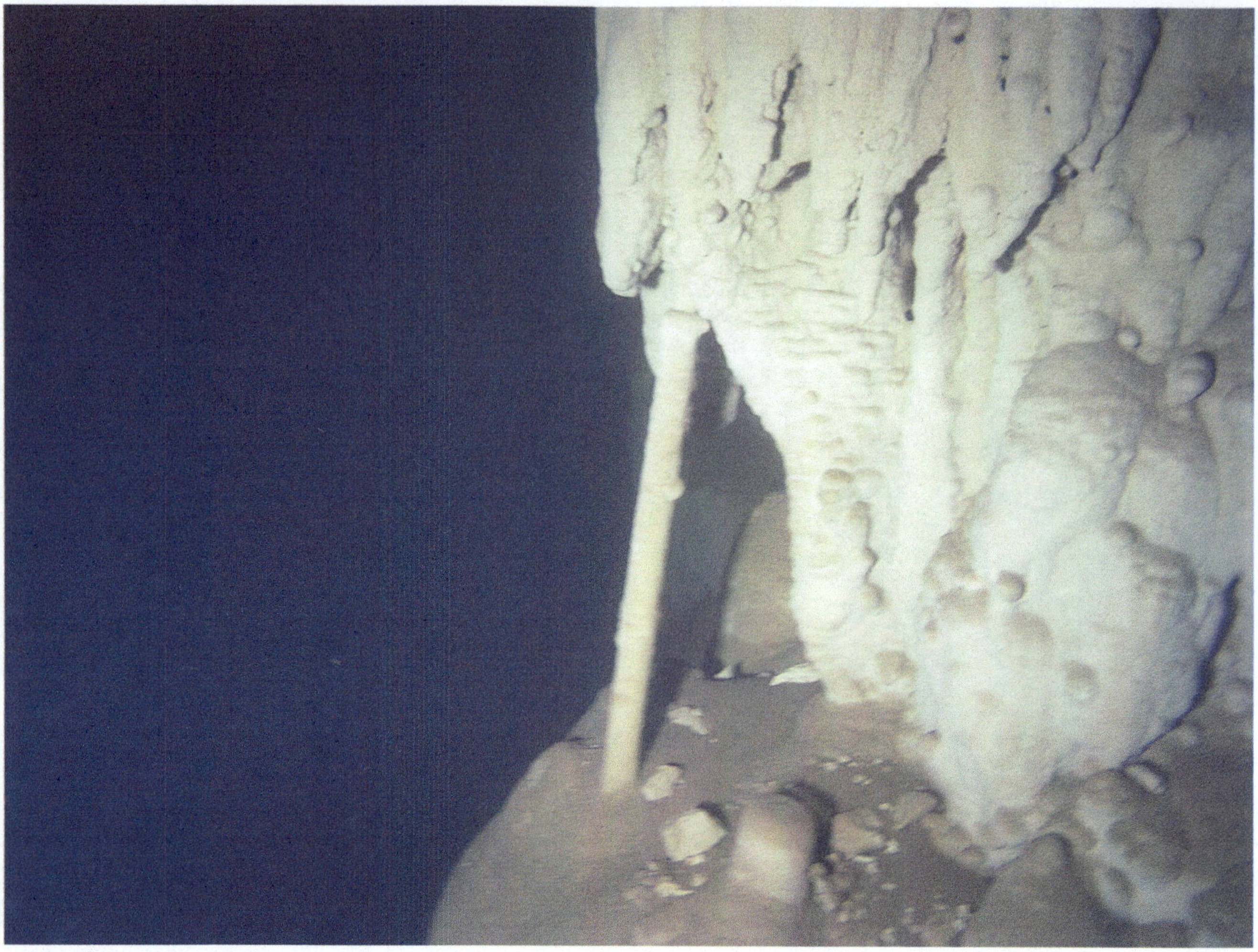

Fig. 4.5 Speleothem samples were taken for laboratory analysis from the depth of −54 m in Podstažišće cave

(0 m). This is the Classic Karst region with a few thousands of known caves. Velebit is a nature park, with two national parks: NP Paklenica and NP Sjeverni Velebit.

Four deepest caves in Croatia are located on Velebit Mountain (Bakšić and Paar 2006). They are: cave system Lukina jama (Fig. 4.13) – Trojama (depth −1431 m, length 3741 m) (Bakšić 1994, Jalžić 2007), Slovačka jama (depth −1324 m, length 6416 m) (Bakšić 1998, Lacković et al. 1999), cave system Velebita (depth −1026, length 3346 m) and Cave Nedam (depth −1226 m, length 2873 m). Some new observations were made in deep caves there by Paar et al. (2019) and Stroj and Paar (2018).

The value and beauty of nature on Velebit have long been acknowledged, and the mountain range was included in 1978 in the world network of biosphere reserves within the UNESCO programme of "Man and the Biosphere" (MAB) or the wealth of plant and animal species and mushrooms, the variety of karst forms and the landscape.

Fig. 4.6 Author waiting for help in Pogračišće 2 cave at depth of −248 m

Fig. 4.7 Author and the rescuer at the entrance of Podgračišće 2 cave

The Karst Mountain Velebit is geologicaly, hydrogeologicaly and tectonicaly divided into four (A, B, C, D) regions:

A. Northern part between the mountain passes Vratnik and Veliki Alan, with the highest peak Mali Rajinac 1699 m)—with four deepest caves (Lukina jama –1,431 m; Slovačka –1,324 m; Nedam –1,250m; Velebita –1,026) in Croatia.
B. Middle part (between Veliki Alan and Baške Oštarije with the highest peak Šatorina 1624 m)—with the deepest known sinking hole/Ponor Pepelarica −358 m (or 9 m less).
C. Southern part (between Oštarije and Mali Alan (peaks Vaganski vrh–1757, Sveto brdo–1753 m)—with the deepest known sinking hole (Ponor na Bunovcu −534 m). Here is located National Park Paklenica (Fig. 4.14) with about hundred known caves. Manita Peć (Figs. 4.15 and 4.16) Cave is on the first place from the year 1923.
D. South-eastern part (with Crnopac as the most remarkable peak)—with the longest cave in Croatia (cave system Crnopac –54,709 m long and −797 m deep).

Cavers first started exploring region C in 1976–1977, while exploration in region B started in 1982–1983, in region A from 1992 to 2020 and finally in region D from 1989 to 2020.

This cave is entirely formed in Triassic rocks (limestones, dolomites, shales, conglomerates, etc.). Because of

Fig. 4.8 On the surface after cave rescue

complexed hydrogeological conditions, groundwater can very quickly flood some parts of the cave. Therefore, organizing a new expedition in this dangerous cave took 35 years. In that occasion, the map was corrected and some new channels were discovered. However, the expedition held in 2017 performed no diving in the final syphon, which was found and entered on 23rd of July 1983. This challenging cave calls for further research.

4.1.4 Caves of Istria, Ravni kotari and Dalmatian Hinterland

Pit *Ponor kod Rašpora*, also called *Abisso Bertarelli*, was the deepest cave in Istria (Bertarelli and Boegan 1926). This cave was known as the deepest cave in the world in the year 1925 with (uncorrect) depth of −450 m. New explorations and surveys, conducted between 1970s and 2019, showed the depth of −351 m, and length of 6080 m (Kukuljan and Glavaš 2012, Korač and Barudžija 2017, Glavaš 2011, Rubinić et al. 2013)). Today, Ponor kod Rašpora is the second-longest and second-deepest cave in Istria.

On 20th and 21nd September 1924, a depth of 307 m was reached. Further descent was prevented by the lack of equipment. In November 1924, a depth of 365 m was reached. After the conversion in Trieste, it was adjusted to 381 m which was, at the time, a new world record in the depth.

The next, carefully planned and organized expedition was carried in August, 1925. The measured depth of 450 m was reached. The news of a new world record was reported by telephone, which was pulled to the bottom of the pit. In honour of distinguished member and president of the Touring club Italiano, Luigi Vittori Bertarelli, the pit was named Abisso Bertarelli. Unfortunately, due to heavy rainfall, otherwise dry streams that end up in the pit have swelled and surprised the researchers. The water appeared abruptly in Ponor, and soon, the stream became too strong and fatal for Karlo and Blaž Božić, who were among the five locals from Rašpor assisting the explorers in the upper parts of the pit. This was the first cave accident in any Croatian cave.

Caverns along Istrian Y highway (A-9). During exploration and mapping for the Dragonja—Vodnjan motorway 1989–1990, which later became part of the *Istrian Y* highway, many speleological objects were discovered. Those caves needed to be explored for the purpose of road construction. Four caves, located directly on the route, were discovered and explored near Rogošić in 1996. The deepest one was 96 m. The protection of these caves was

Fig. 4.9 Banićeva spilja or Čampari Pit is the deepest cave on the island of Cres

Fig. 4.10 Sea caves near Punta Križa on the island of Cres

recommended instead of burying them. During the construction of the second phase of that highway, in 2008 and 2010, another eight caverns were found along the route.

The Pit near Burići (Figs. 4.17 and 4.18) near Kanfanar was found in 1991 by Eugen Ujčić, a member of the Pazin Speleological Society from Pazin with the help of a local resident. This was followed by numerous researches by Pazin and Poreč speleologists. The description and draft of the cave was first published in 1998 in the book Kanfanar and Kanfanarština, and later in the newsletter of the Karlovac speleologists "Speleo'zin". The draft and description of the cave by author Silvijo Legović from SD "Proteus" was published in both publications. The pit is 127 m deep, and the horizontal length is approximately 90 m (Figs. 4.19 and 4.20). The entrance to the pit near Burići is at an elevation of 250 m above the sea level. Entrance vertical is 16 m deep.

Podublog pit near Marčana in Istria is 202 m deep and was the first cave in Croatia which was explored in 1969 only with ropes without ladders. Cavers used Gibbs and Hiebler asscenders on the dinamic ropes. Fig. 4.21 shows members of Podublog 1969 speleo expedition.

Baredine Cave is the only touristic pit (vertical cave) in the Dinaric Karst of Croatia. According to the report by Boegan (1928, 1930), the cave was explored during a speleological action carried out by the *Commissione grotta della Soccieta Alpina delle Giulie*, in 1924, when it was recorded under serial number 1807 and referred as the cave SE of Gedici or Baredine Cave. In the summer of 1973 in Poreč, at the instigation of S. Legović, a group of speleologists-enthusiasts was formed. Later, that group became the Speleological Society Proteus. In the same year in Baredine Cave, a new 36-m vertical was discovered, leading to the underground lakes. The measured water depth in the first lake was 6 m and 16 m in the second lake. This led to the realization of the total depth of the Baredine Cave of 132 m. When visiting the "E. Boegan" company in Trieste, speleologists from Poreč found information that Italian speleologists explored the pit up to 80 m in 1928. The cavers

Fig. 4.11 Sea cave near Punta Križa

of that time did not notice the passage that was about 70 cm wide, through which the pit extends further down to the very bottom.

Speleologists (SD Proteus) initiated protection of important speleological sites of the Poreč area. In addition to the Baredine Cave, Markova and Pincinova Caves were proclaimed protected natural monuments in 1986. The Baredine Cave is protected as a geomorphological monument of nature. In 1995, Baredine Cave was successfully opened to the public.

North-western Istria is made up of a wide Cretaceous zone, as a vast anticline above the Jurassic core in the area between Rovinj and Poreč. The geological structure around the Baredine Cave is completely different than the shell structure of the Alveolina and Numulitic limestones and the Paleogene basin of clastic sediments around the Italian Grotte Gigante area. Although the tectonic structure is autochthonous from the western Istrian Jurassic–Cretaceous anticline to the Pazin syncline, the epirogenetic movements on it were very strong in the younger orogenetic phase (Durn et al. 2003). Because of this, intensive processes of karstification of the existing carbonate deposits have occurred, that is, the appearance of numerous speleological objects along the tectonic and fault lines.

According to geological surveys and the analysis of the sediment in which the Baredine Cave was formed, the existence of Lower Cretaceous limestones with dolomite layers in the substrate was established. Some measurements

Fig. 4.12 Velebit Mountain is the longest mountain in the Dinaric Karst

of rock corrosion on karst surface near entrance of Baredine Cave have done by Cucchi et al. (1985, 1994).

The importance of Kobiljak Pit near Buzet in Istria was noticed by the speleologists from Trieste (Societa Alpina delle Giulie) at the beginning of the last century. The research was conducted in 1927 and 1929. Cavers reached the bottom of the second vertical shaft. Thanks to a low water level, they crossed the first syphon lake using an aluminium boat and reached the depth of −157.50 m (Hlaj and Poropat 2008).

In 1976, Drago Opašić–Billy with his friends from Pazin explored this cave and also reached a syphon which, on that occasion, was flooded. They assumed that further advance was impossible without the use of diving equipment. That research is documented on film made by Pazin speleologist Dano Brečević.

In February 2002, members of SD Proteus held a new survey. In the middle of the pit, the syphon was passable, so they descend to the bottom. At the end point of the lower lake, they widened a narrow crack and entered a new

Fig. 4.13 Lukina jama is the deepest cave system in Croatia (−1431 m)

Fig. 4.14 Paklenica Canyon of in Paklenica National Park

Fig. 4.15 Manita Peć cave in Paklenica National Park

Fig. 4.16 Manita Peć cave in Paklenica National Park

Fig. 4.17 Burići Pit near Kanfanar in Istria

submerged channel. After about twenty meters, the channel closes with a syphon and can only be continued with the use of diving equipment. The entrance is located at an altitude of 380 m. The length of the explored part of the pit is 460 m, and the total depth is -285 m. This cave still represents a challenge for further research.

The role of the Ponor of Pazinčica cave on underground hydrological system in central and southern Istria has described Otte (1968).

Cave Čude is located near river Zrmanja in Ravni kotari (Dalmatia). This cave is explored more than 980 m in length, containing several syphons (Fig. 4.22). In spring, cave is totally filled with water. In some parts of this caves, relicts of transported rocks (paleozoic clastites) from Lika region are found. This proves that this periodical spring cave is connected with Lika Caves via channels under the Velebit Mountain. The potential length of this cave might be more than 15 km (Garašić, 1986).

4.2 Central (Middle) Dinaric Belt

The central Dinaric Karst Belt in Croatia extends over the Gorski Kotar and Lika areas. However, the area of Ćićarija in Istria, as well as the Dalmatian Zagora, in its north-eastern parts belongs to the Central Karst Zone. In the area of Ćićarija, it crosses into Slovenia, and in the area of Dalmatian Zagora, it crosses into Bosnia and Herzegovina. In this area, the processes of karstification go deeper than in the Inner Karst Belt. They range to an average of 200 to 800 m

Fig. 4.18 Burići Pit is 127 m deep and approximately 90 m long

Fig. 4.19 Speleothems in Burići Pit

Fig. 4.20 Burići Pit in Istria

in thickness. In this karst area, there is an irregular border between the Adriatic and Black Sea basins. The occurrence of speleological objects is common. The interesting feature of this area is the appearance of swallow-holes, karst fields and estavelas. In this area, there are much more horizontal caves than in the Outer Karst Belt, but less than in the Inner Karst Belt (Herak et al.1968). Mostly vertical and horizontal speleological entrances occur on the surface. In the interior of the rocks, that is, in cavernous form, vertical speleological forms are predominantly present. The reason for this is the thickness of the watertight layers, but also the neotectonic activity. In the case of neotectonic uplift of rock complexes, pits (vertical cracks) are predominantly formed, while in the case of neotectonic lowering of rock complexes approaching watertight layers, horizontal speleological objects (caves) are mostly formed.

4.2.1 Caves of Lika

Estavela is a speleological object with the role of spring and the role of swallow-hole. Estavela Velika Pećina is one of the most affluent in the role of spring and creation of the occasional surface lake (approx. 1 million m^2) with a volume of about 5–7 million m^3 in only 5 days. However, in the role of swallow-hole, it takes approximately 30 days for the complete water withdrawal to underground. The difference is understandable because the underground is completely

Fig. 4.21 Members of Podublog Pit speleological expedition in Istria 1969

saturated with water in the rainy months. During the periods of the highest water levels, the volume of Lake Blata (Fig. 4.23), Begovac or Blaćansko Lake is estimated to be above 10 million m^3 of water, sometimes with depths exceeding 20 m. The main swallow-hole (except Velika Pećina located in the far south of the field) of this lake is Mala Pećina, which is in the northeast part of the field. It is a sieved swallow-hole that also draws water from Lake Blata towards the Slunjčica River spring.

As a natural rarity, it attracted the attention of Austrian natural scientists back in the nineteenth century. Since 1835, the lake has been studied by independent teams. The first team was led by hydrologist Julius Fras, followed by General Palfy from Krajina. Palfi's team sought to determine the directions of underground runoff by throwing coloured pieces of wood into Velika Pećina (at that time, it was a common method of underground flow tracing). Appearance of the coloured wood on the lake surface in the spring caused great confusion in some villages (the Ogrizovići reportedly recognized them as "dragon ribs"). A large flood hit the lake on 6th July 1885. According to Hirc, the entire Bijela draga was flooded.

Hirc further states that the water was extremely cold and swirling and that it had blackened all the sunken trees. The lake is known to dry at the end of August; however, from 1888 to 1893, it never dried up. In 1895, Šenoa explains that the flooding took away the mowed hay that clogged the caves (swallow-holes) and thus caused the permanent flood. In the early twentieth century, the lake was explored by Cvijic and especially by Poljak.

Fig. 4.22 Čude cave near Obrovac

When the river bed (tunnel) becomes a syphon, the amount of water that appears after snowmelt or heavy rainfall cannot flow through it. This excess water, which exceeds the flow capacity of the syphon, is discharged through an auxiliary ascending tunnel connected to Velika pećina (Grand Cave). So in the humid part of the year, Velika Pećina is the spring that recharges the lake. In the dry part of the year, when the groundwater is almost dry, water from the lake begins to flow through Velika Pećina (Fig. 4.24) in the opposite direction and replenishes the groundwater level. Therefore, in the dry part of the year Velika Pećina becomes a swallow-hole and the lake quickly dries.

Mala Pećina has no dual function, and in it, the water only sinks. From the part of the lake that is lower than the level of the overflow, the water fails to return to Velika Pećina, which makes this lower part of the lake persist longer. Some believe that the water that sinks into Lake Blata as spring is re-occurring in Sinjac (Fig. 4.25) (a lake in the village of Jezero in Plavča Draga), but this is not true. It was never proved by colour water tracing. The latest diving research points to a deep karst spring (−203 m) that is not geologically linked to Blata but to the Kapela Mountains. From 1949 to 1991, the water level of the lake was continuously monitored. The maximum depth of the lake was measured in the fifth month of 1984 and amounted to

Fig. 4.23 Lake Blata or Begovac in Lika

Fig. 4.24 Velika Pećina cave near Lička Jasenica

Fig. 4.25 Sinjac spring near Plaški in Lika

19.40 m. The last cave survey was conducted by speleologists from Zagreb (Garašić and Kovačević 1990).

4.2.2 Caves of Gorski Kotar

In speleological terms, Dragutin Hirtz was among the first people to start exploring the Gorski Kotar area. He mentioned several caves and pits in his work, *Gorski Kotar* from 1898 in which he also presented the first drawings showing the cave exploration in Croatia, like the drawings of Hajdova hiža, the entrance to Ponor Vele vode and cave Malenica near Crni Lug. The drawings were made by painter Vaclav Anderle. The book specifically describes Pilar's Cave (*Pilarova ledenica*), known for its icebergs, *Hajdova hiža* with the magnificent size of the entrance hall and Muževa hiža. In 1905, Dragutin Hirc published the monumental work Natural Geography of Croatia, where, among other things, he mentions *Pilarova ledenica* near Mrkopalj in Gorski Kotar. He also mentions that there are about 300 known caves in Croatia, but predicts as many as 500 in the future. Mirko Malez explored tens of caves near Lokve, Čabar, Skrad, etc., in Gorski Kotar (Malez 1956a).

Up to mid-2017, 124 speleological objects have been explored in Šverda area, 3 of which are deeper than 250 m (Kukuljan et al. 2018).

About 30 speleological objects have been explored in the area of Risnjak National Park (Buzjak 2016).

In the area of Bitoraj Mountain, about 50 pits were explored. The highest ratio of occurring cave entrances per square kilometre was found in the Jurassic limestones of the area west of Samarske and Bijele stijene (between 150 and 160 per km^2).

Ličanka river spring, also known as Veliko vrelo (Figs. 4.26 and 4.27), was explored from 1993 till today by Croatian and international cave divers. Several very complicated syphons were dived in this cave. Total length is 1816 m, and the deepest point is −48 m in sumps. Explorations are ongoing.

Stupina jama (Fig. 4.28) is the deepest cave in Gorski Kotar, near Ravno area. Its depth is −413 m, and its length is 625 m. The final large chamber is filled with speleothems (Figs. 4.29 and 4.30). The cave was explored from 1992 to1995 when tens of vertical caves were found near Lič and Fužine in Gorski Kotar (Garašić 1997a).

Sopot (Figs. 4.31 and 4.32) is cave near Severin na Kupi in Gorski Kotar. This is periodical cave spring, entirely filled with water during spring. Cave was first measured to be 259 m long, but with new cave diving techniques, several hundred meters more were explored (Figs. 4.33 and 4.34).

4.3 Inner Dinaric Karst Belt

Caves or horizontal speleological objects in the inner karst zone are more numerous than pits, i.e. vertical speleological objects. This is due to the small thickness of the rocks with the possible vertical circulation of water. Caves of about a hundred meters depth were found in only a few places. Their average depth is between 30 and 60 m which is also the

Fig. 4.26 Syphon in Veliko Vrelo or Ličanka River spring near Fužine

thickness of the most common limestones present on the surface. Below them are Mesozoic dolomites or dolomitic limestones or limestone dolomites that have a lower water permeability. Sometimes, when they are poorly fractured, they perform the function of hydrogeological insulators. A great number of long and complex speleological objects in the Inner Karst Belt have been developed at the sites of these contacts.

4.3.1 Kordun and Banija Caves

Cave in Šušnjar is geologically interesting because it is completely formed in marls. These are not Eocene marls, like the ones found in Istria or Dalmatia, but in the younger Neogene alterations of marls and marl limestones (Garašić and Kovačević 1991).

Jopićeva Cave was discovered in 1968. The length of all the cave passages makes 6710 m, and the depth is 57 m, thus being the longest cave in Croatia was in 1974 (Fig. 4.35). Jopićeva Cave is located near Brebornica Village in Kordun. The waterproof base is Triassic dolomite. It is covered with a rather thin sediment of Cretaceous limestone, where all the cave passages of Jopićeva are situated (Fig. 4.36). Dolomite base is uncovered on many places due to surface erosion, what makes basis for several bigger and smaller streams (Figs. 4.37 and 4.38), the strongest one being the Durlić creek (Garašić 1975; Čepelak 1981).

Tamnica Cave is one of the conductors of water in this area recharging river Mrežnica. This is a large cave forming

Fig. 4.27 International cave diving team in Ličanka cave spring

a cave system with Zalja Cave and Šutina jama. The system also includes relic caves Pećinica, Vrt and a few others. Tamnica Cave was again explored in 1978, 1980 and 1981, and the most recent surveys were conducted in 2006 and 2011. The cave is made up of mostly muddy channels (Fig. 4.39) that are completely or partially filled with water, so diving equipment was used for the exploration (Figs. 4.40 and 4.41). The total length of the Tamnica, Zalja and Šutina Cave system today is about 2 km. Some parts have not yet been physically connected, but the channels were proven to be very close. The level of karstification at this site is extremely low due to the appearance of Neogene marls near the surface (Garašić 1986).

The longest cave in Kordun is the cave system Panjkov ponor–Varićakova Cave (Fig. 4.42) near Rakovica. It is mainly filled with groundwater (Figs. 4.43, 4.44 and 4.45). The research of this cave system started in 1983, and many cave diving expeditions were carried out through the years. Today, the length of this cave system is 13,218 m.

4.4 Cave Occurrences in Separate Parts of the Dinaric Karst Area in Croatia

4.4.1 Caves of Žumberak and Samobor Mountains

In the area of the Žumberak Nature Park, geological material determines the formation of karst, which covers as much as 90% of the Park's territory. Numerous surface karst forms can be found in this area. Swallow-holes and short subterranean rivers are common, especially in the western part of the park. In addition to surface phenomena, the karst characteristics of the Žumberak area and part of the Samobor highlands are confirmed by numerous underground phenomena, namely caves and pits. Over 137 speleological objects have been explored in the area so far. Currently, the deepest pit is Bedara Cave (more than 1020 m long and about 113 m deep) and the longest cave is Provala, currently

Fig. 4.28 Stupina jama near Ravno is the deepest pit (−413 m) in Gorski Kotar

Fig. 4.29 Final chamber of Stupina jama

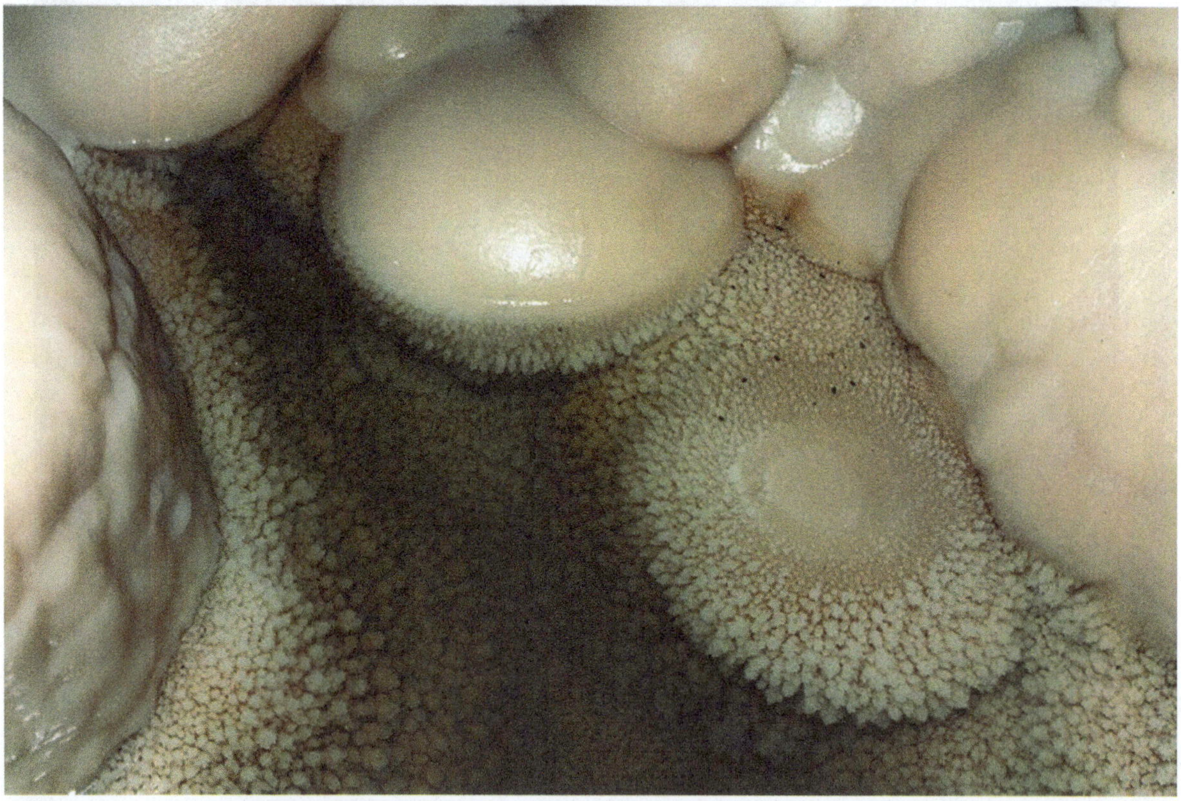

Fig. 4.30 Speleothems in Stupina jama

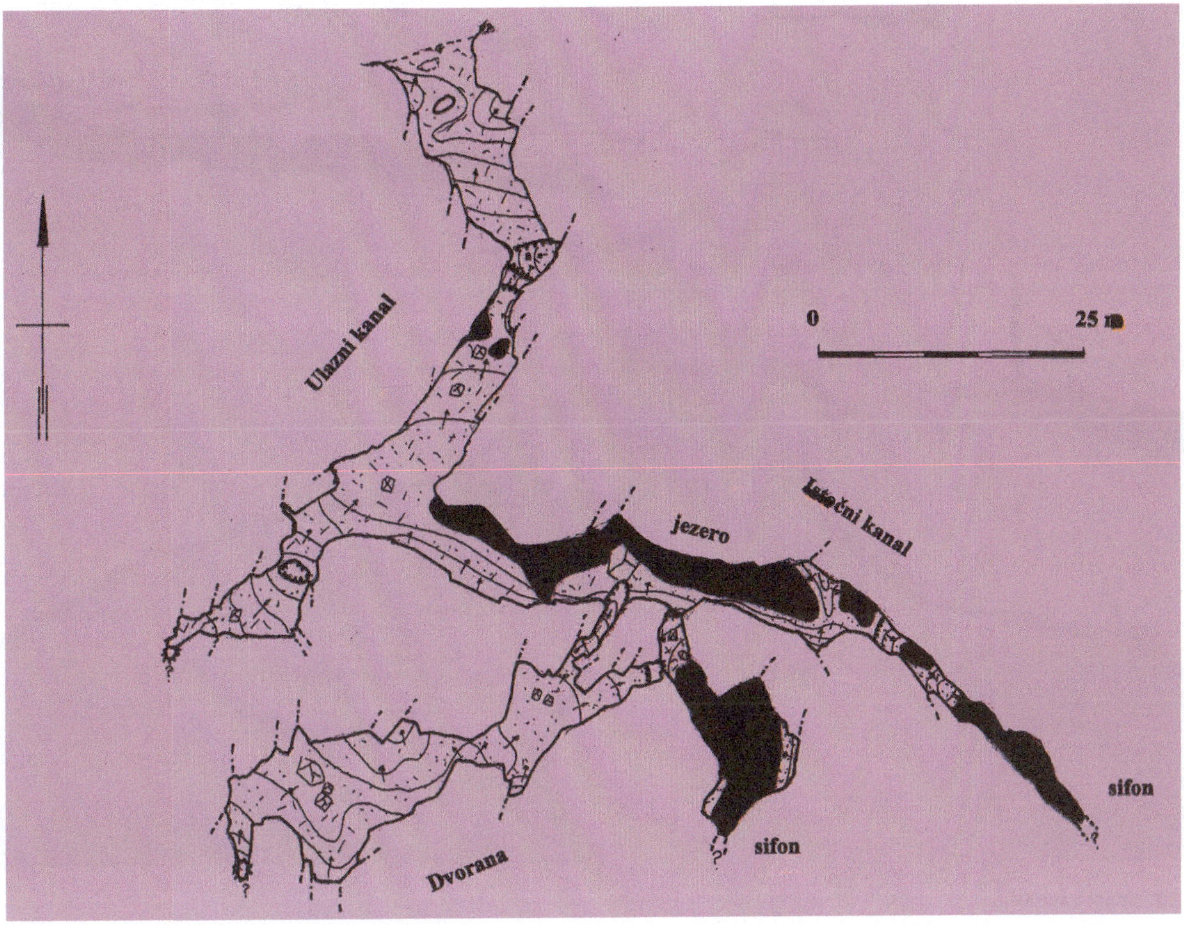

Fig. 4.31 Plan of Sopot cave near Severin in Gorski Kotar

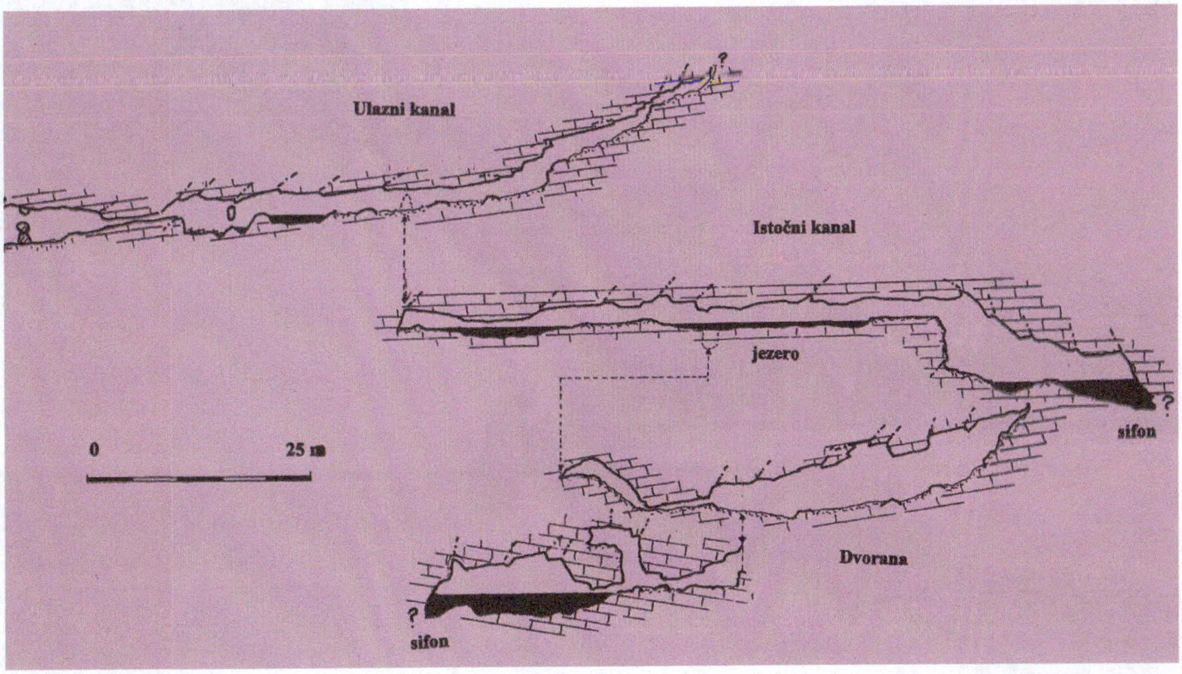

Fig. 4.32 Profile of Sopot cave

Fig. 4.33 Sopot cave lake

Fig. 4.34 Final lake with syphon in Sopot cave

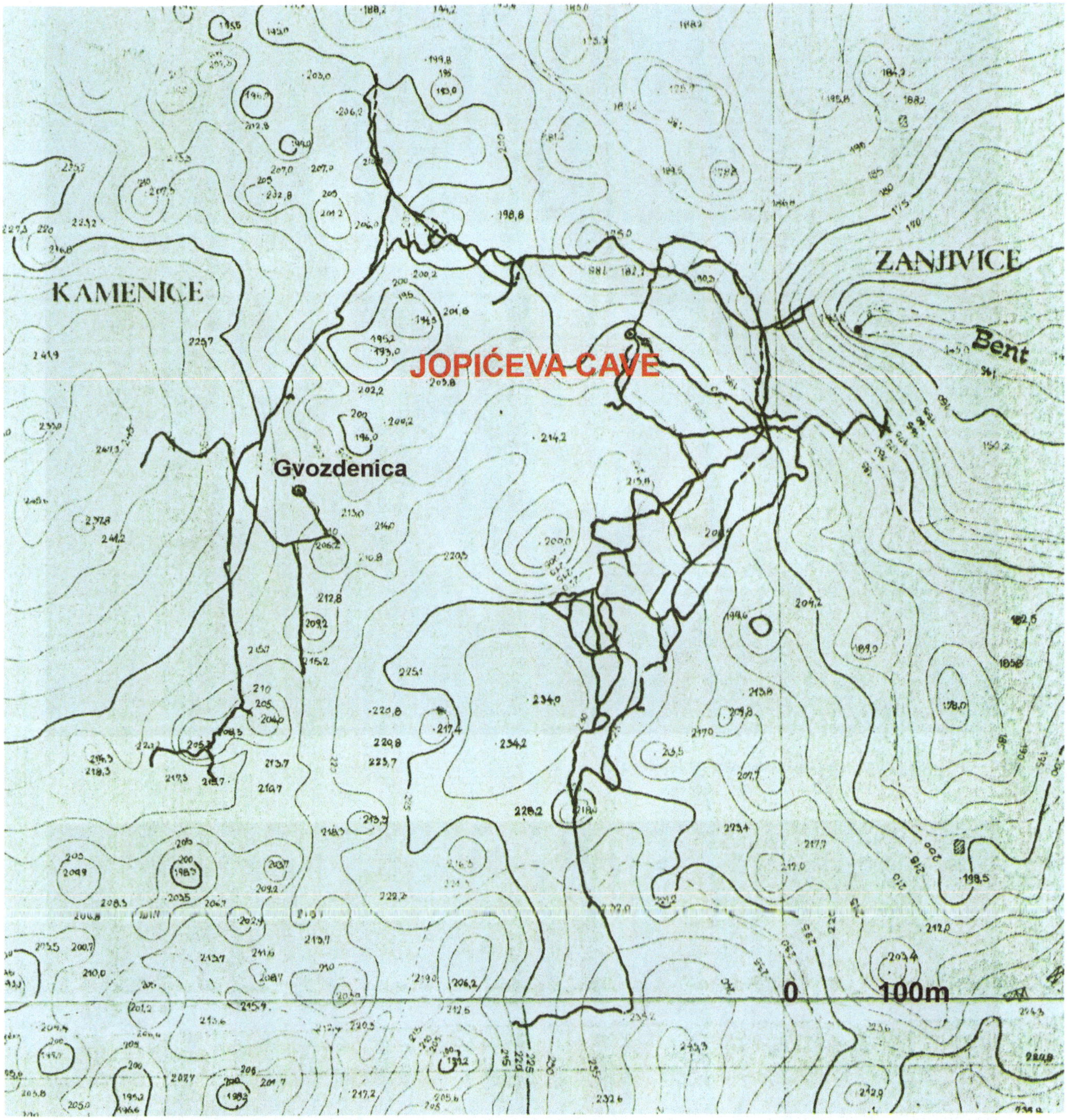

Fig. 4.35 Map of Jopićeva and Gvozdenica caves in Kordun

Fig. 4.36 Cretaceous limestone cave passages of Jopićeva cave

Fig. 4.37 Different water streams in Jopićeva cave

Fig. 4.38 The lower part of Jopićeva cave (1974)

Fig. 4.39 Tamnica cave in Kordun

Fig. 4.40 Tamnica cave water stream

Fig. 4.41 Total length of Tamnica, Zalja and Šutina cave system is about 2 km

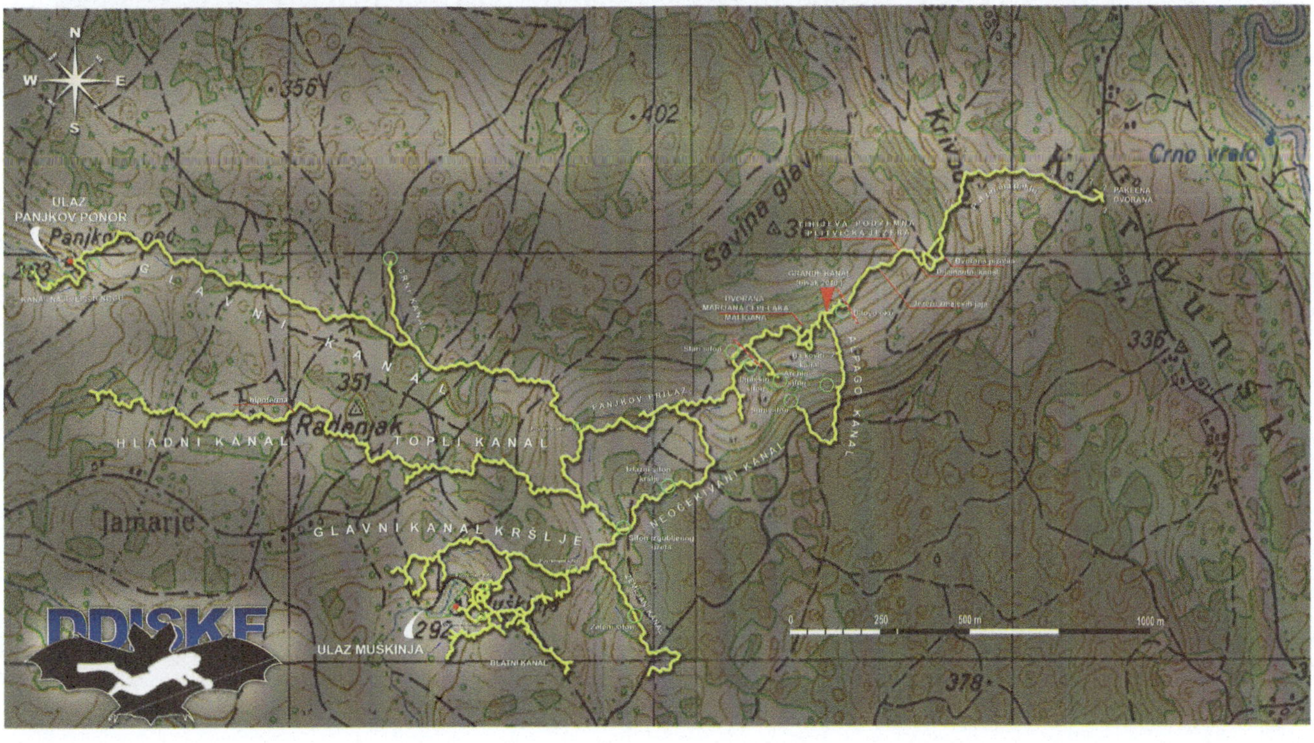

Fig. 4.42 Plan of Panjkov ponor—Varićakova cave system

Fig. 4.43 Channels of Panjkov ponor—Varićakova cave system

Fig. 4.44 Underground lake in Panjkov ponor—Varićakova cave system

Fig. 4.45 Cave channels with water in Varićakova cave

Fig. 4.46 Confluence of coloured and uncoloured streams in Veternica cave near Zagreb (1968)

Fig. 4.47 Groundwater flow tracing in Veternica

Fig. 4.48 Veternica—contact between lithotamnium limestones and Triassic dolomites (Dark Chamber)

Fig. 4.49 Fault in the upper part of Veternica cave

Fig. 4.50 Narrow horizontal cave channel formed along the fault plane (Veternica, Ramzes)

Fig. 4.51 Alpinistic channel in Veternica

with 1862 m of topographically recorded channels (Buzjak 2002).

4.4.2 The caves of Mount Medvednica

Veternica Cave is located on Medvednica Mountain near Zagreb, the capital of the Republic of Croatia. Its entrance is located 8,100 m from the central town square (Ban Josip Jelačić Square). The entrance was first mentioned at the end of the nineteenth century, but the first speleological exploration began in 1934 (Poljak 1934). Individuals from many generations of Zagreb cavers began their caving activity in Veternica (Malez 1961b, 1963). It is a complicated, multi-storey cave with several watercourses (Figs. 4.46 and 4.47) and several syphons (Čepelak 1977). Research is still ongoing (Lacković et al. 2011).

Veternica is located in the Medvednica Nature Park right on the outskirts of the city of Zagreb. In 1979, it was protected by the Law on Nature Protection as a geomorphological monument of nature.

It is 6128 m long, with a height difference of 151 m (Fig. 4.48). The first 380 m of the cave is adapted for visitors (track and electric lighting). It was named after the airflow (Veter = wind) at the entrance. The entrance to the cave lies at the altitude of 320 m above sea level. Veternica was formed on contact between lithotamnium limestones and dolomites (Figs. 4.49 and 4.50). It consists of main channels (2622 m) and an intricate system of horizontal corridors (Figs. 4.51 and 4.52), channels and halls. The initial part of the cave was continuously inhabited from the Palaeolithic to the Iron Age. In Veternica, the remains of some 70 species of animals (cave bear, cave lion and leopard, cave hyena, rhino, prairie, huge deer, wild cat, etc.), flint

Fig. 4.52 Veternica is a complexed cave with several water streams (Limun)

artefacts (spikes, scrapers, daggers, etc.) were found. The earliest findings of man (Neanderthal) belong to the culture of the developed Mousterian. Although they were originally believed to be 43,200 years old, the recent studies show they that are between 50,000 and 100,000 years old.

In the area of the Medvednica Mountain, 66 speleological objects are known, of which 64 are within the nature park.

4.4.3 Caves of Zagorje and Slavonian Mountains

In the area of Hrvatsko Zagorje (Varaždin and Krapina Zagorje counties), 165 speleological objects (caves) are known, and in the area of Nature Park Papuk, 41 caves are known.

5 Caverns Found and Explored During Construction Work in the Dinaric Karst of Croatia

Abstract

The uniqueness of Croatian Karst is in the large number of explored caverns discovered subsequently during construction works. During the construction of hydraulic structures, dams, tunnels, roads, railways, bridges, viaducts, quarries, ports, warehouses, airports, etc. in Croatian Karst, over 1,200 caverns were explored in detail. Some of them are about 10 km long, and some are 450 m deep. All of them were rehabilitated, and special attention was paid to the protection of groundwater in their vicinity. In some caverns, special constructions were built, such as bridges or pumping stations for water supply. Engineers and construction workers showed that large water accumulation projects can be completed even in permeable conditions of Dinaric Karst.

Caverns are speleological objects (caves and pits) with no natural entrance on the surface of the terrain. They are discovered exclusively by geophysical exploration, exploratory drilling, construction works in tunnels, notches, cuttings, foundations, etc. In the Dinaric Karst in Croatia, caverns are explored since 1950 (Nikola Tesla Hydro Power Plant, Gojak Hydro Power Plant). So far, over 1,200 caverns in Croatia have been systematically documented and processed. Care has always been taken not to disturb groundwater movements during their rehabilitation and not to jeopardize the stability of structures. If at all possible, the caverns should be left in their primary form in relation to the construction site.

Particular attention was paid to cavern investigation during the construction of motorways in Croatia from 1990 to the present. A total of 23 tunnels were built along the A-1 (Zagreb-Split-Dubrovnik) motorway, 12 tunnels were built along the A-6 (Rijeka—Zagreb), 5 tunnels were built along the A7 (Rupa—Križišće), and one tunnel was built along the A8 and A9 (Istrian "Y") motorway. A total of about 80 road tunnels have been built. In addition, 11 tunnels were built on the access roads to A-1, A-6 and A-7 (through the mountain Biokovo, and in the bypass of the cities of Rijeka, Šibenik and Omiš). Caverns were discovered and explored in almost every one of them. They were also found on the construction of bridges and viaducts along the aforementioned motorways, as well as along the state road D-1 (from Macelj to Split). During the construction of the motorways, special quarries were opened for construction purposes. Caverns (Vrata, Grobnik, Vrbovsko, Tounj, Debeljaca, Strinjine, etc.) were also found and explored in them. Map of Croatian highways is shown in Fig. 5.1.

Since the caverns had no names prior to their discovery, names were subsequently assigned to be distinguished from one another. They are rarely named after people, but sometimes this is done in recognition of significant speleologists or geologists, such as Poljakova Spilja—cavern on the right bank of the Sklope Dam or the Horvatova Spilja—cavern on the left bank of the Lika Dam (Figs. 5.2 and 5.3). Caverns are sometimes named after locations such as the Cave in the Tounj Quarry, Drenovac cave (the cavern in the Drenovac—Gojak Tunnel) or the cavern below the residential building in Rijeka. However, caverns are most commonly named after the location mark on roads where they are found. These are markings on projects and are not changeable. For example, cavern at Km 207 + 425, Bristovac Tunnel, Sveti Rok Tunnel—Maslenica bridge, motorway A-1.

The existing caverns and those that are unfortunately no longer present or have been altered by the necessary construction work are presented in this book for the first time. Mainly, unpublished professional reports stored in libraries of institutes, faculties and academies were used for the purpose of this book.

M. Garasic, *The Dinaric Karst System of Croatia*, Cave and Karst Systems of the World,
https://doi.org/10.1007/978-3-030-80587-6_5

Fig. 5.1 Map of Croatian motorways

5.1 Caverns Found and Explored in the Foundations of Various Buildings

During the construction of building foundations, hotels, shopping centres, factories, oil pipelines, gas pipelines, etc. in the area of Dinaric Karst in Croatia, caverns are often discovered. For example, cavern appearing during the construction of family houses in Samobor (2007), in Novi Vinodolski (2001), Jusići (1991), on the island of Hvar (1972), on the island of Ugljan (2017) (Fig. 5.4), etc. Caverns were found in the foundations of smaller buildings. They were investigated and documented. Some of them still serve as basements and wine cellars because of the ideal constant temperature of about 12 °C. During the construction of a large settlement and a skyscraper in Rijeka, about ten caverns were found. They were remediated in the safest way. During the construction of numerous hotels in Karst, many caverns were found, which were explored and repaired in the most appropriate way. For example, in the hotel settlement Orašac (2010) near Dubrovnik, the hotels Belvedere (1983) and the Villa Argentina (2008) in Dubrovnik, the Epidaurus Hotel in Cavtat (1988). When constructing shopping malls "Rijeka Tower Centre" (2008) in Rijeka or "Lidl" in Delnice (2014). During the construction of the pipeline through the Dinaric karst area in Croatia, caverns were found in the area of Krnjak (1983) and in the area near Lokve (1994). At the alumina and aluminium plant in

Fig. 5.2 Horvatova Spilja—cavern on the left bank of the Lika River Dam (Sklope)

Fig. 5.3 Horvatova Spilja—speleothems

Obrovac (Velebit), two caverns were found in 1975 while constructing a pool for the disposal of the puddle, and their remediation was specifically proposed because of the risk of possible contamination of the nearby Zrmanja River.

During the construction of a cargo terminal for Rijeka seaport in Škrljevo, several caverns were found (Figs. 5.5 and 5.6). One of them opened (2016) after the majority of work was completed. The works were carried out above the cavern which was discovered accidentally when a roller for stabilizing the terrain collapsed in it. Fortunately, no one was hurt. The dimensions and location of the cavern were such that remediation required stability that would withstand the weight of a cargo container, but also to take care of the environmental conditions of the groundwater.

During the construction of the Amarin Hotel in 2015, four caverns were discovered. They considerably slowed down the construction works. All caverns were explored and repaired. Some were connected to the sea even though they were located hundred meters away from the coast.

During the construction of the Park Hotel (Figs. 5.7 and 5.8), the most elite hotel on the Croatian coast, designed on the site of the former old Park Hotel in Rovinj, in 2016 and 2017, nine caverns were found (Figs. 5.9 and 5.10). Some of them had contact with the sea. It was designed to use the sea temperature to heat and cool the air conditioners in the hotel with the help of heat pumps. This has proven to be economically viable.

5.2 Caverns Found and Explored in Constructing a Hydropower Plants

During the construction of all hydropower plants with accompanying facilities built after the Second World War in Croatia and located in the karst area, speleological research was carried out both in the preparatory and in the construction and exploitation phase. In the preparatory stages for the construction, surveys were conducted from the

Fig. 5.4 Cavern appearing during the construction of family houses on the Island of Ugljan

surface of the terrain to assist with geological mapping. This was done in the area of future reservoirs, for example Gusić Polje, Sabljaci, Bukovik, Kruščica, Buško jezero, Peručko jezero, Lokvarsko jezero, Bajer, Letinac, Lepenica, Dobra Lake, etc. These are speleological objects that mostly had natural entrances on the surface. The largest number of caverns, i.e. previously unknown speleological structures, was explored during the construction of dams and their hydrotechnical supply and drainage tunnels. Research began with the construction of the first dam and tunnels for Nikola Tesla Hydroelectric Power Plant (1950—caverns discovered in the Lokve Tunnel, Vrelo Cave), Gojak Hydroelectric Power Plant (1956, caverns discovered in the Drenovac Tunnel). For the needs of the Senj Hydroelectric Power Plant (1960–1965), 3 tunnels were built with a total length of 33 km through the mountains (Lika-Gacka Tunnel; Marasi-Gusić field tunnel), including the Velebit mountain (Gusić Polje-Hrmotine Tunnel). Today (2020), the construction of the Senj 2 Hydro Power Plant is being planned, including the construction of another tunnel through Velebit.

For the Sklope Hydroelectric Power Plant (Fig. 5.11), which uses water from the Lika subterranean river, two large caverns were explored at the barrier site. They are called the Poljak Cave (1964) (after Josip Poljak) on the right bank (Fig. 5.12) and the Horvat Cave (1967) (after Vladimir Horvat) on the left bank (Figs. 5.13, 5.14 and 5.15) of the Lika River. A special three-ring injection method resulted in the accumulation of heavily karstified rocks (Figs. 5.16 and 5.17), which at the time was a worldwide success for geotechnics (Božičević 1964, 1965a, 1965b, 1968; Malinar 2018).

The Zakučac hydroelectric plant on the Cetina River (Magdalenić 1971) with two tunnel tubes (Fig. 5.18) is the only large-diameter hydrotechnical tunnel built by the Tunnel Boring Machine. During the construction of the second tunnel tube, numerous caverns (1976—1978) were found (Figs. 5.19 and 5.20) which slowed down the speed of tunnel excavation by this method (Bojanić et al. 1980). It was concluded that in the Croatian part of the karst, this tunnelling technique will no longer be used.

"Robbins" TBM (Tunnel Boring Machine) was stopped by a big cave on approximately half of hydrotechnical tunnel "Zakučac 2" near Omiš in Croatia. After this incident in tunnel, cavers entered the cave through the entrance above the TBM/7.1 m diameter main beam (on the right side of photo). The cavers explored and made survey of cavern and made very detailed geological map. The photo (Fig. 5.21) was taken from the cave towards to TBM head on 17 May 1978. The author of this book is standing in front of TBM in the future 9894-m-long tunnel. TBM had problems in this karstic faults zones for several months. This was the first and the last time TBM was used in karstic rocks of Croatia. Tunnel "Zakučac 2" is still functional and in excellent condition.

The hydrotechnical tunnel Obrovac (Velebit) (1980–1982) is 8191 m long, and a deviation of the tunnel has been made (Fig. 5.22) because go round the large cavern that was on the tunnel route. Inside the tunnel, 44 caverns were found (Figs. 5.23 and 5.24).

For the needs of HPP Lešće (2006), sixty speleological objects were surveyed on the Dobra River in the canyon and in the vicinity. In the barrier site of the dam, two smaller caverns, which were later completely filled with concrete for the safety of the project, were found.

Fig. 5.5 Cavernson Škrljevo cargo terminal near Rijeka

The Dubrovnik HPP was built from 1960 to 1965, and water from the subterranean river Trebišnjica was used with the drop to the sea near the Plat area. Unfortunately, during the breakthrough of a drainage tunnel from an underground engine room, 8 workers were killed by the appearance of water in the caverns (1962). Later, speleological research (1999, 2014) showed the complexity of completely submerged cave channels at sea level, where an accident occurred due to insufficient knowledge of the hydrogeological parameters of the nearby Robinson spring. As a result of a fire at the HPP Plat engine room, three workers were killed in 2019 trying to reach the surface through a drainage tunnel.

The cavern is located approximately 375 m from the portal of the access tunnel to the engine room of the HPP Dubrovnik in Plat, in Dalmatia, on the south side of the tunnel, at an altitude of 7 m.a.s.l. The investigated part of the cavern belongs to the Upper Triassic dolomites, which sometimes pass into limestone dolomites and the upper parts of the cavern to dolomitic limestones. Morphologically, the cavern can be divided into several units: entrance, narrow passages and labyrinth, upper floor with branches, vertical jumps, water channels with siphons and wide and large channels and halls. The total length of the investigated dry part of the cavern is 428 m with a height difference of 23 m (Figs. 5.25 and 5.26). However, the fully submerged part of the cavern and the channels behind the siphon were still explored approximately 350 m in length by diving. Investigations have been discontinued because of the danger of a possible sudden and strong influx of groundwater, which was also the cause of the tragic events of 1962, when eight people were killed here during the breakthrough of the southern drainage tunnel. Part of the groundwater flowing through the cavern exits through an unfinished tunnel after approximately 350 m into the Adriatic Sea, and a part flows at the Robinson spring, which is about 500 m away from the access tunnel. According to the indicators, it is evident that groundwater rises up to 20 m.a.s.l. in the rainy season. I note that both the first and second drainage tunnels, with their half profile at an altitude of 0 m, are built approximately 3 to 5 m below sea level. The north drainage tunnel, with all its engine rooms, a drainage tunnel (520 m), a pressure pipeline (250 m) and a supply tunnel "Plat" (16,500 m) with access tunnels (300 m, 510 m) have been in operation since 1965, and the southern one has never been completed just because of the dangerous and under-explored caverns. The cavern from the access tunnel to the engine room is connected to the

Fig. 5.6 Cavern on Škrljevo cargo terminal near Rijeka

Fig. 5.7 Grand Park Hotel in Rovinj

drainage tunnel 2. It is unknown whether the drainage tunnel 1 contained caverns and whether they were investigated at all.

Later, I was informed that through a cavern found in the access tunnel at the end of February 1965, water had entered the engine room and that, after rehabilitation, that entrance to the cavern was walled up. In order to explore unknown parts of the cavern during October 1999 (Fig. 5.27), the entrance to the cavern was reopened. At that time we reached the drainage tunnel P-2, and shortly after, the vicinity of the Robinson Spring. The big storm on 11 December 2006 caused the access tunnel and engine room to flood again from that cavern. Due to the risk of sudden and unpredictable strong groundwater arrivals, research in this cavern has not yet been completed. In January 2019, three men died in the same tunnel.

For the purposes of the construction of the Ombla HPP, research has been carried out for about thirty years in the Vilinska Cave and at the Ombla River spring—the shortest karst river in the Dinaric Karst (Milanović 1996). Over three km of cave canals have been explored, but the most significant cavernous parts of this system have been discovered by the excavation of exploration tunnels. Diving research was conducted from the tunnel in the upstream portions of Omble River, about 650 m from the Ombla river spring. It is actually a cavern that extends to the spring and the sea, but the physical passage of the cave divers has not yet been accomplished.

Caverns were also found when digging smaller tunnels through which sewage pipes were drawn, for example, in the Prapratno Tunnel (1988) near Ston on the Peljesac peninsula, 890 m in length.

Fig. 5.8 Nine caverns were found during the construction of Grand Park Hotel (Rovinj)

Fig. 5.9 Caverns discovered during the construction of Grand Park Hotel

Fig. 5.10 Grand Park Hotel—Some caverns were connected with the sea

Fig. 5.11 Sklope Hydroelectric Power Plant

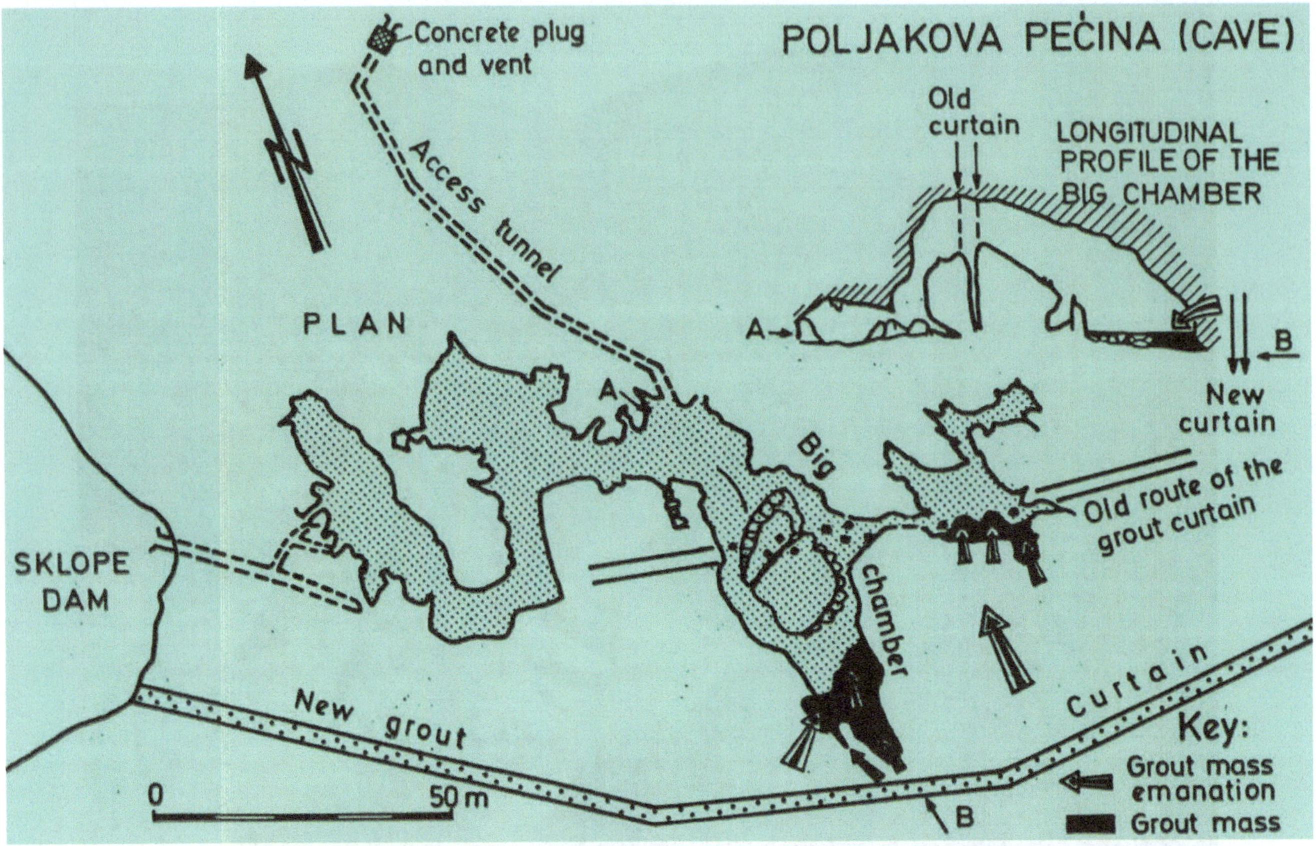

Fig. 5.12 Plan of Poljakova cavern on right bank of Lika River

The Blato Tunnel (2005) on the island of Korčula, 833 m long, through which a 400-mm-diameter collector pipe passes. Two caverns were found during tunnel construction work.

The Vidova gora Tunnel (1971) was dug on the island of Brač. It is a hydrotechnical tunnel with a diameter of 2.35 m and a total length of 8542 m. For the purpose of building the water supply of the island with water from the Cetina River, a TBM rotary drill set of 2.35 m in diameter was used. The works in the tunnel "Vidova Gora" were performed by the Yugoslav National Army from 1971 to 1976. Water passes through hydrotechnical tunnels (in full profile and under pressure), and there is no possibility of installing a "door" through which caverns can be reached. The entrances to the caverns were enclosed by concrete. An engineering geological map of the tunnel was made by the Army during the tunnelling. Only 29 out of 177 "large and dangerous" caverns were investigated. Unfortunately, due to lack of time, there was always pressure to hurry up with the explorations, so the deepest objects were investigated up to approx. 93 m, but the largest vertical difference in some objects was even 156 m. A large amount of water was detected in several caverns (Baučić 1984).

A dozen large caverns were found in the Konavosko Polje Tunnel, which is 2100 m long and was dug for fifty years. Some of them were later repaired. Recently, reparation of the tunnel is being planned.

The Konavle drainage tunnel was made in the form of a horseshoe profile with a semicircular calotte of 1.95 m radius and a total height of 3.90 m. The length of the tunnel

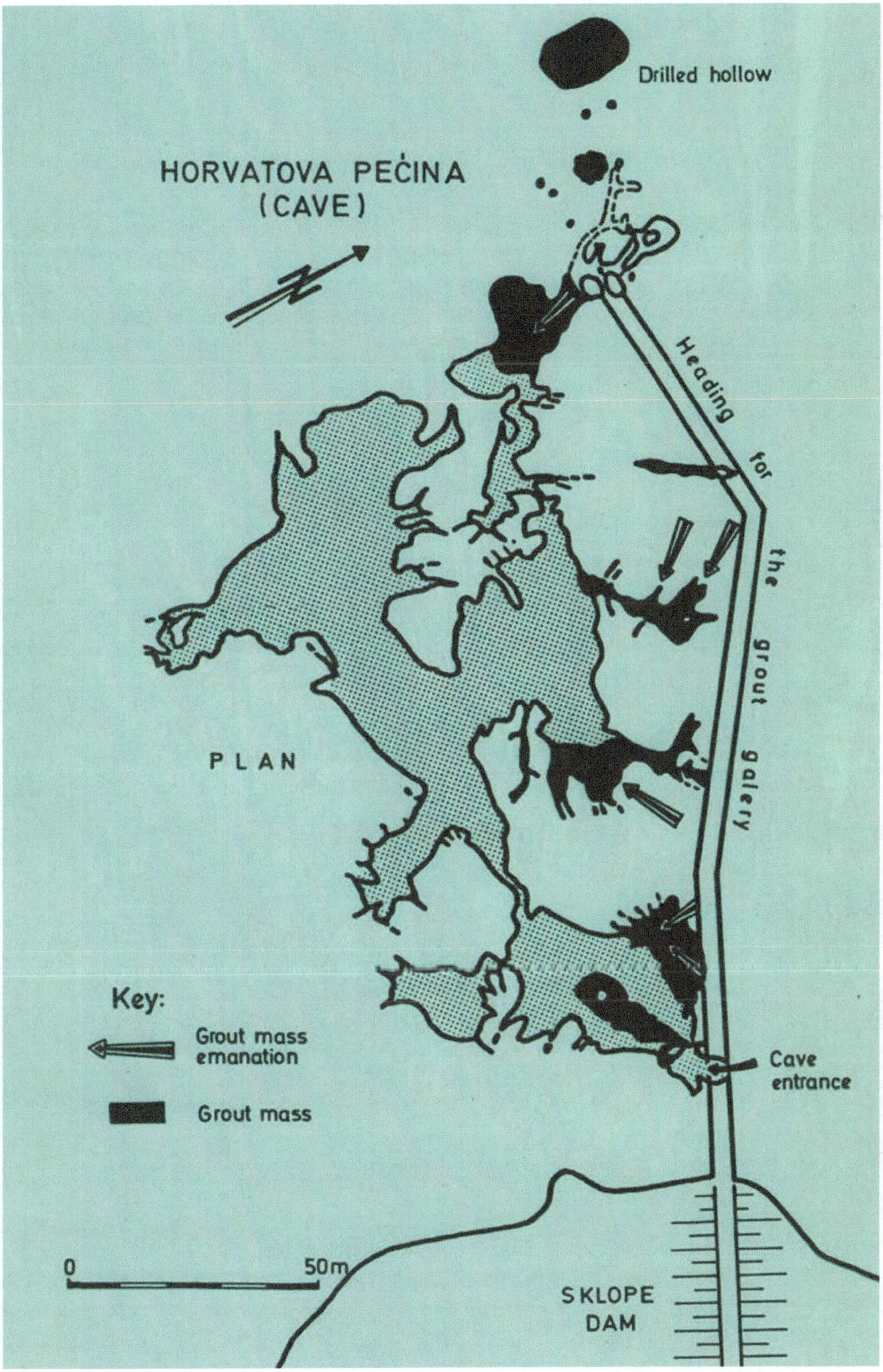

Fig. 5.13 Plan of Horvatova cavern on left bank of Lika River

Fig. 5.14 Horvatova cavern

Fig. 5.15 Horvatova cavern speleothems

Fig. 5.16 Horvatova cavern—evidence of neotectonic movements on speleothems

Fig. 5.17 Recent and neotectonic earthquake activity on speleothems in Horvatova cavern

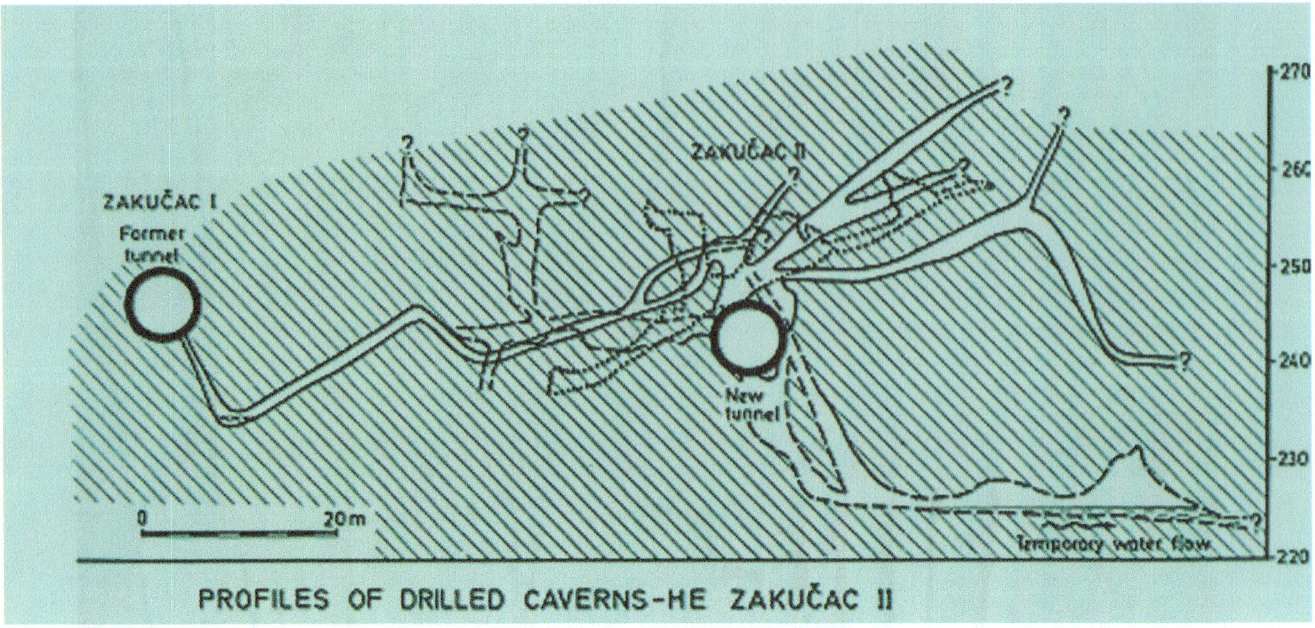

Fig. 5.18 Zakučac hydroelectric plant on Cetina River with cavers (profile map)

Fig. 5.19 Zakučac hydrotechnical tunnel built by the Tunnel Boring Machine (TBM)

Fig. 5.20 Numerous caverns were found in Zakučac hydrotechnical tunnel

Fig. 5.21 Tunnel Boring Machine in one of the caverns in Zakučac tunnel

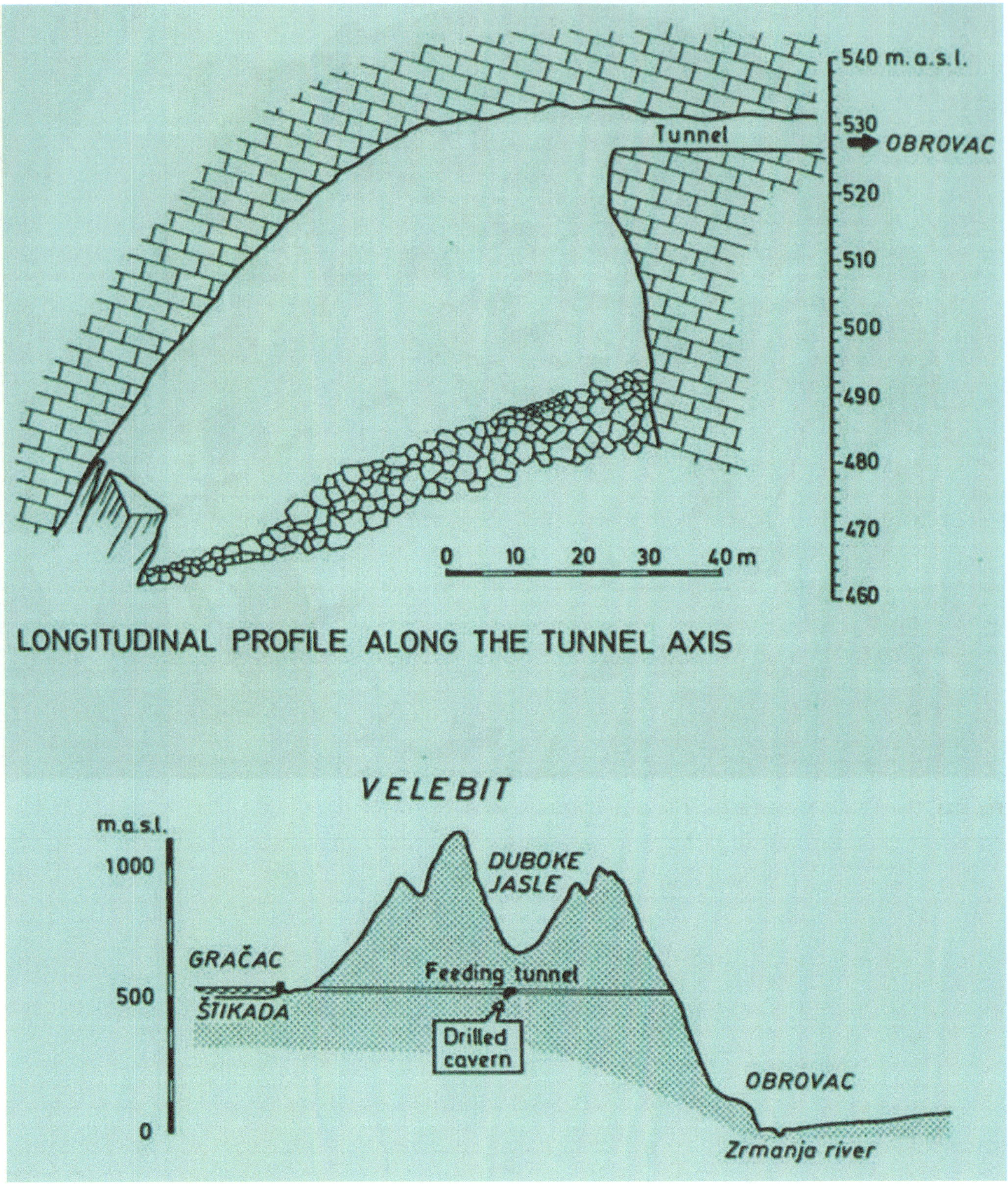

Fig. 5.22 A large cavern in Velebit (Obrovac) hydrotechnical tunnel

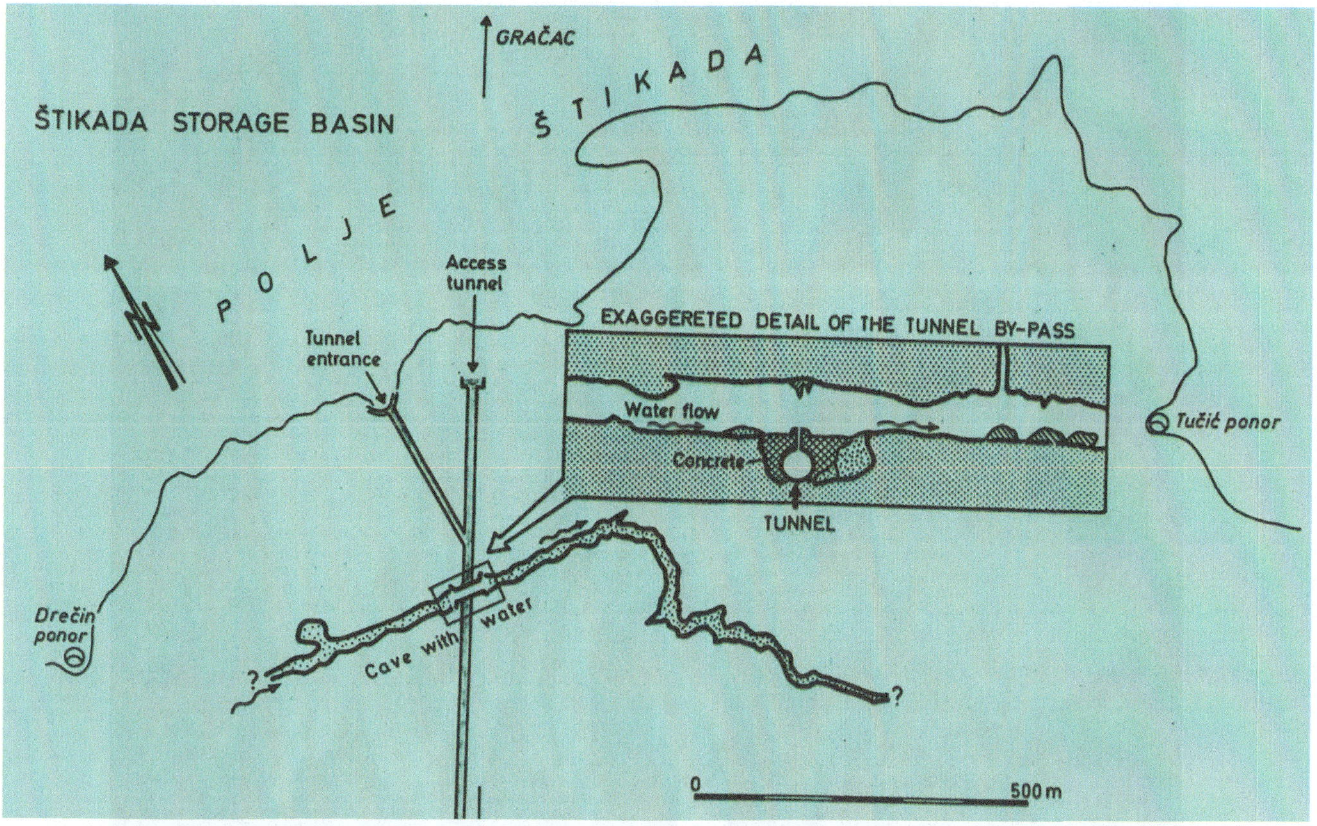

Fig. 5.23 Profile and plan of a cavern near the north entrance of Velebit Tunnel

was 1,957 m. The following works were then reconstructed: —deviation of the tunnel in the length of 160 m, thus avoiding the unstable zone—increasing the flow profile of the tunnel by lowering the level by 80 cm—coating the sections with a concrete lining in full profile where necessary. Thanks to the reconstruction, the tunnel is in good working order today, with an estimated throughput of 60 m^3/s for the full profile. At the entrance of the drainage tunnel, a limnigraphic station was installed for continuous measurement of the flow through the tunnel. Due to the influence of the slowdown, the flows are not an unambiguous function of the water level (Roglić and Baučić 1958).

In Vrgorac hydrotechnic tunnel, also called "Maričevac Tunnel" (between Baćina Lakes and Krotuša on Vrgoračko Polje) that is 2138 m long and constructed from 1934 to 1938, only several small caverns are known.

Fig. 5.24 Rare speleothem formation from the cavern in Velebit hydrotechnical tunnel

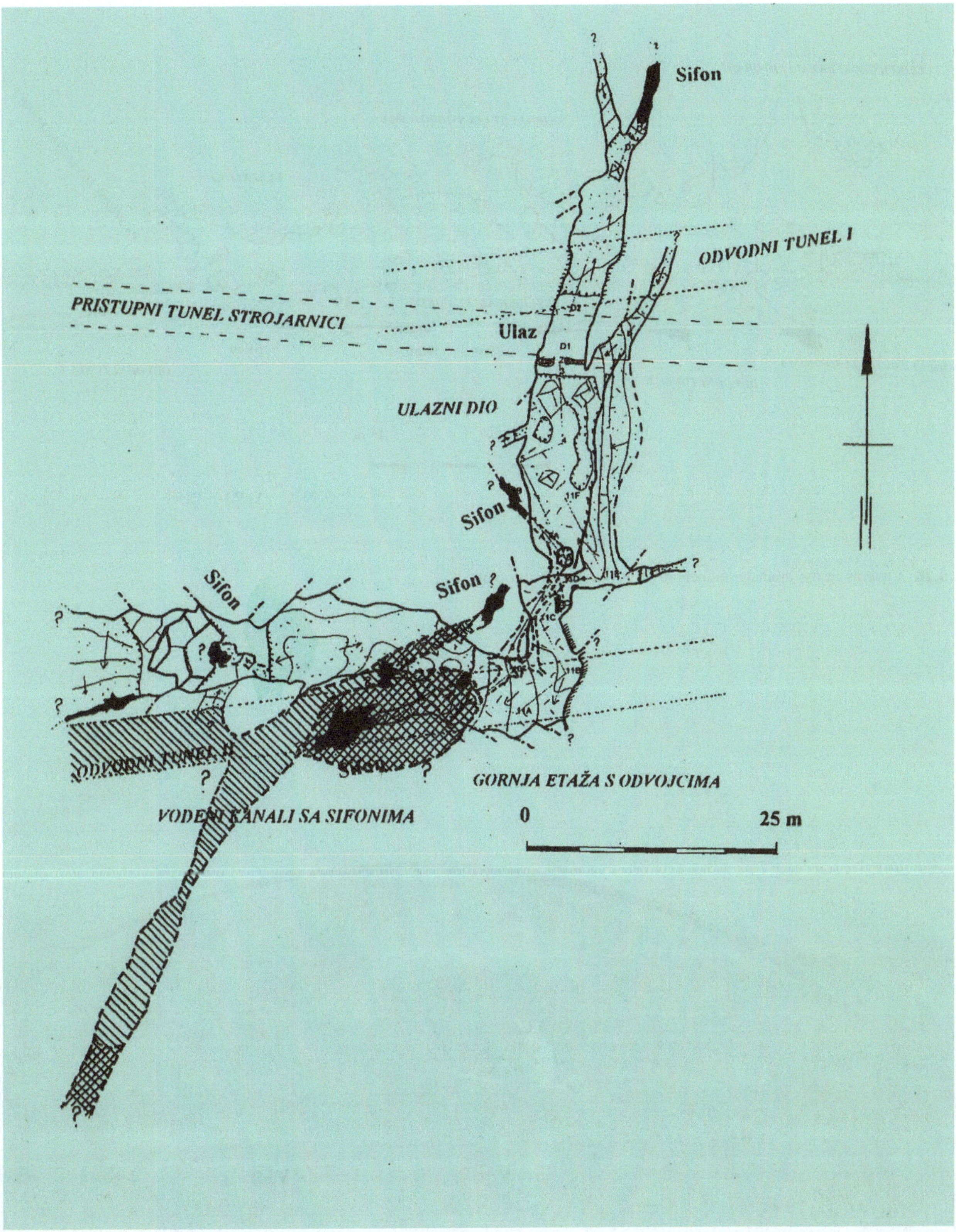

Fig. 5.25 Plan of caverns found in the drainage tunnels of HPP Dubrovnik in Plat

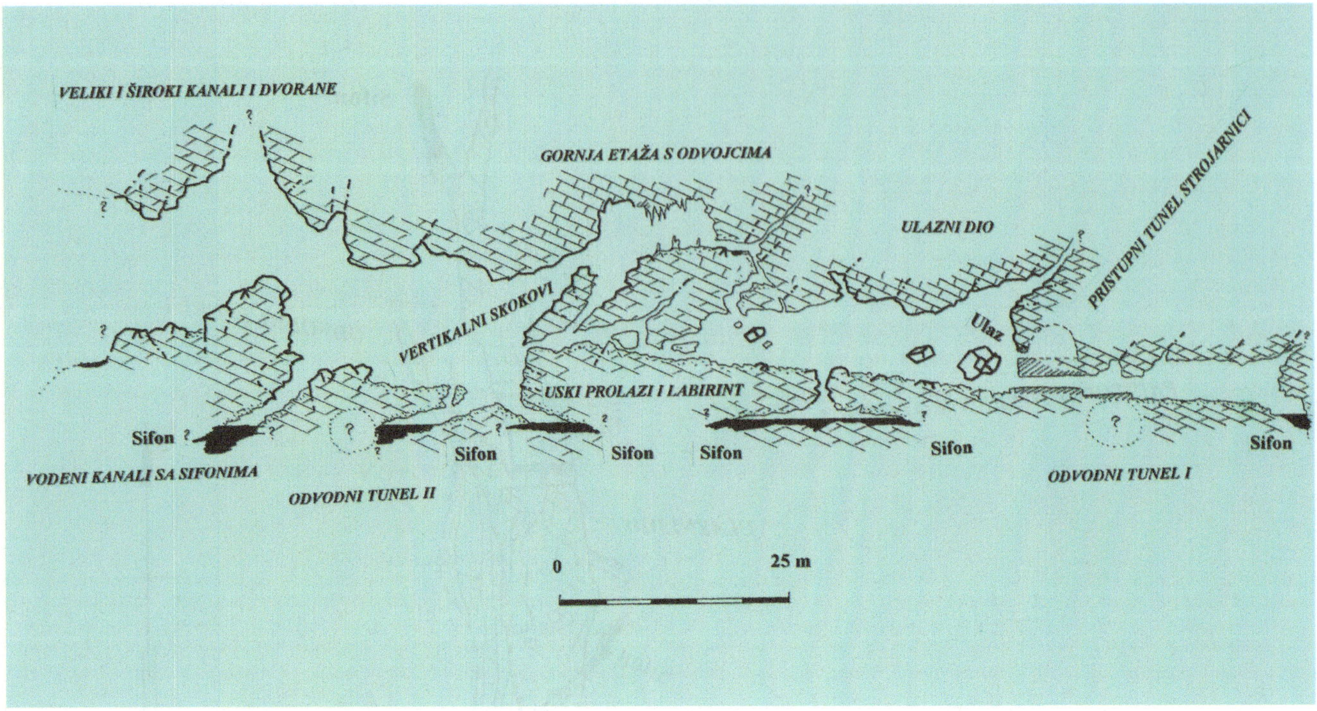

Fig. 5.26 Caverns in the drainage tunnels of HPP Dubrovnik (profile)

Fig. 5.27 Water-filled drainage tunnel during the inactive period of HHP Dubrovnik

6 Caverns Found and Explored During Construction of Road Tunnels in Istria and the Gorski Kotar Region

Abstract

Caverns discovered during the construction of road tunnels in the area of Istria (along the A-8 motorway) and Gorski Kotar (along the A-6 motorway) belong to the Adriatic Sea hydrological basin. However, some caverns in the interior of Gorski Kotar belong to the Black Sea Basin as well. Examples of caverns belonging to the Adriatic Sea Basin can be found in the tunnels "Učka", "Hrasten", "Tuhobić", "Vrata" and "Sljeme". Caverns belonging to the Black Sea hydrological basin were discovered in tunnels such as "Sopač", "Lučice", "Vršek", "Veliki Gložac" and "Čardak".

Caverns of different sizes and morphologies were discovered during the construction of almost all tunnels in the karst of Croatia. For example, during the construction of the 1,480 m long tunnel Pitve—Zavala in 1969, 16 smaller caverns were found (Fig. 6.1). Two of them were deeper (75 and 89 m) and vertical with interesting speleothems (similar to those in the nearby Kraljevska jama on Gulišina near Gromin Dolac on the island of Hvar). The tunnel was originally designed for water transportation to the south side of the island of Hvar and later transformed into a narrower road tunnel. Unfortunately, the caverns were later illegally buried.

6.1 Učka Tunnel

The Učka Tunnel is a 5,062 m long single-tube road tunnel that connects the city of Rijeka and Istria. It was built from 1976 to 1981. There are two small tunnels along the Učka Tunnel, Zrinščak 1 (185 m) and Zrinščak 2 (45 m). In these tunnels, the principles of the so-called New Austrian Tunnel Method were applied for the first time in Croatia. During the excavation, at a distance of 1,300 m from the Rijeka entrance, a very complexed system of large-scale underground spaces was discovered. The largest cavern consists of a series of small channels and large halls. One of the halls has a length of 175 m, a width of 70 m and a height of 65 m (Fig. 6.2). The other hall, located in the immediate vicinity of the tunnel, is 60 m long, 40 m wide and 55 m high (Hudec et al. 1980; Božičević 1985, 1993). The cavern system in the Učka Tunnel was discovered thanks to a small opening that appeared in the side of the tunnel (Fig. 6.3). If this had not happened, the construction workers would not have been aware that under the tunnels there were such large underground spaces and that the stability of the tunnel was in danger. This example shows that even the smallest speleological object must be explored in detail, because it can only be a small part of a large system. During speleological research, it was discovered that the water in this system flows 10–30 l / s in the dry period and over 1 m^3/s in the rainy period. During the remediation work, water was captured and is now used as an additional water supply for the city of Opatija (Figs. 6.4, 6.5 and 6.6).

Along the 5,062 m of tunnel, two caverns were found, investigated and connected into one system (Fig. 6.7). The most recent research connected the cave section from the surface of the terrain on Mount Učka with a cavern. The total length of that cave system is now 6,596 m, and a height difference is 430 m. This cave system is the longest and the deepest cave in Istria.

6.2 Tuhobić Tunnel

Thirty-eight caverns were found and investigated in the Tuhobić road Tunnel on the A-6 highway (Zagreb—Rijeka) during the construction of the first tunnel tube (1991–1993). 11 of them were larger. During the construction of the second tunnel tube (2005–2008), 7 caverns were found, two of which were of significant size. In total, 150 cavities were

M. Garasic, *The Dinaric Karst System of Croatia*, Cave and Karst Systems of the World,
https://doi.org/10.1007/978-3-030-80587-6_6

Fig. 6.1 In Pitve—Zavala tunnel, 16 smaller caverns were found in 1969

found during the excavation of Tuhobić Tunnel, but only the above-mentioned caverns were not completely filled with clay or calcite material.

The length of the Tuhobić 1 Tunnel is 2,141 m and Tuhobić 2 is 2,143 m, respectively, with a total of 45 caverns explored.

Tuhobić Tunnel, cavern at km 23 + 798; the paraclasis intersected by the tunnel, the entrance several metres horizontal, then a vertical difference of 41 m, continues deep into several parallel cracks, some of which extend below the tunnel (Figs. 6.8 and 6.9).

Cavern at km 24 + 340, with 3 fault plains, with a total height difference of 27 m (Figs. 6.10, 6.11 and 6.12).

Cavern at km 24 + 433, with a total height difference of 28 m (Figs. 6.13 and 6.14).

Cavern at km 24 + 544, with a height difference of 47 m, on several parallel faults (Figs. 6.15, 6.16 and 6.17).

In the half of Tuhobić Tunnel, at the turnpike, at km 24 + 707 with speleothems (Figs. 6.18, 6.19, 6.20 and 6.21).

Tunnel Tuhobic, cavern at the km 25 + 078, vertical pit, intersected by the tunnel, height difference 45 m, followed by crack extension of the pit in parallel faults (Figs. 6.22, 6.23 and 6.24).

Tuhobic Tunnel, cavern at km 25 + 257, cut in half, door set aside, height difference 33 m. Continued in several parallel fractures and fault plains (Figs. 6.25 and 6.26).

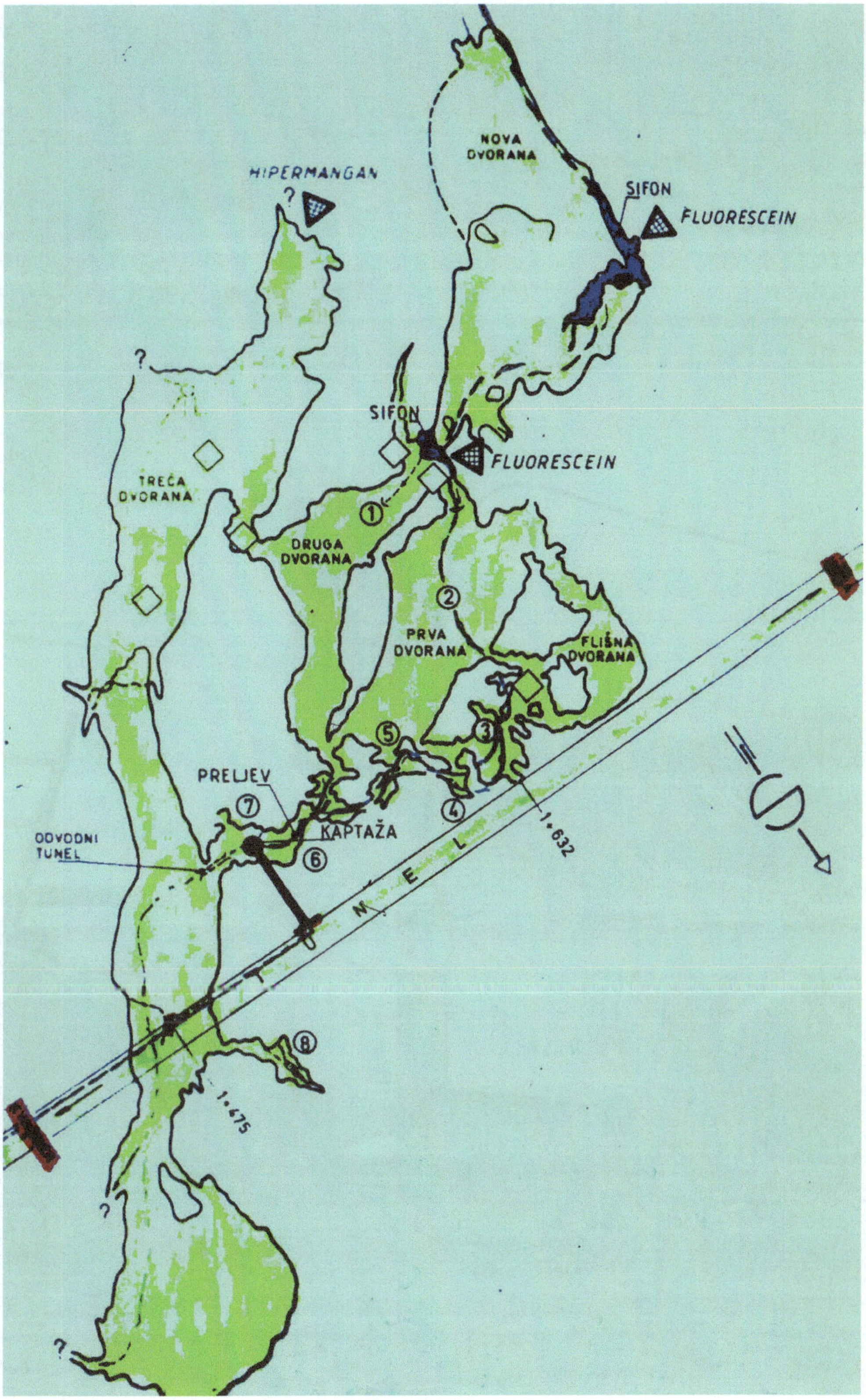

Fig. 6.2 Cavern in Učka Tunnel—plan—tracing of groundwater flow

Fig. 6.3 First entrance to the cavern in Učka Tunnel was sealed with concrete

Fig. 6.4 Cavern in Učka Tunnel—first tracing of groundwater flow

Fig. 6.5 Colour tracing in the upper parts of cavern in Učka Tunnel

Fig. 6.6 Colour trace in the lower parts of the cavern in Učka Tunnel

Fig. 6.7 Upper syphon—the author of this book after discovering the upper syphon (December 1977)

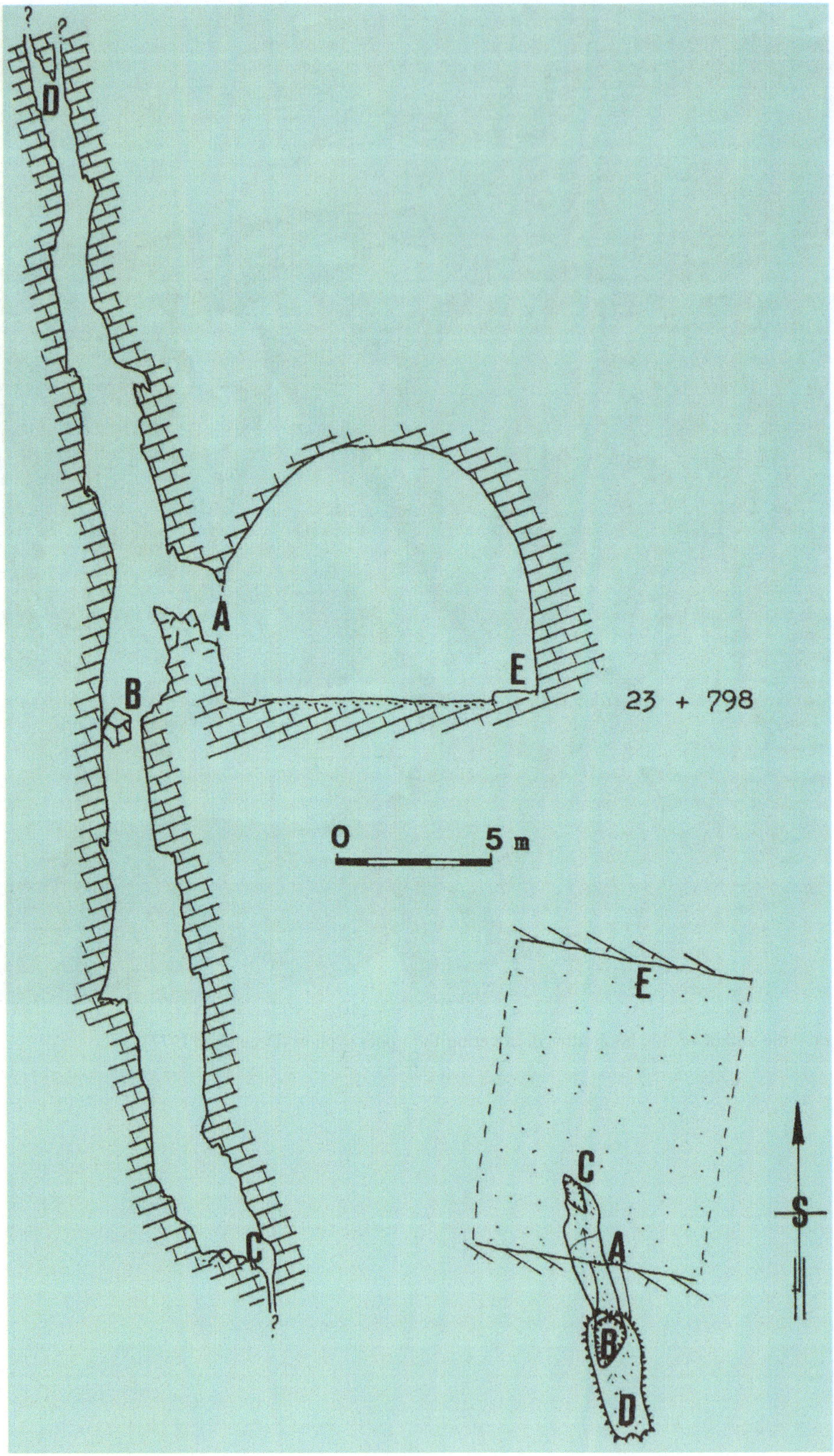

Fig. 6.8 Plan and profile of cavern at km 23 + 798 in Tuhobić Tunnel

Fig. 6.9 Vertical shaft in the cavern at km 23 + 798 in Tuhobić Tunnel

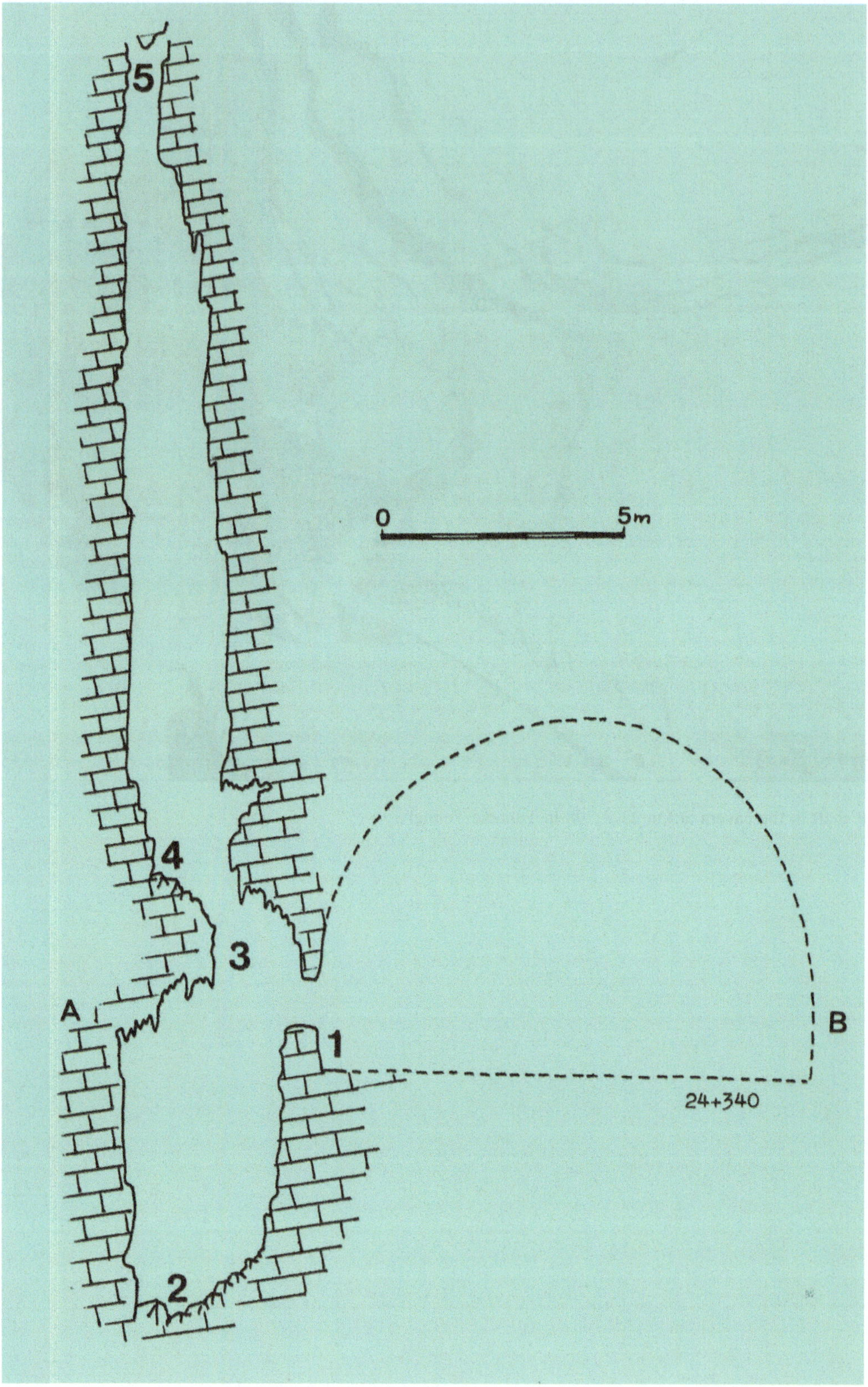

Fig. 6.10 Profile of cavern at km 24 + 340 in Tuhobić Tunnel

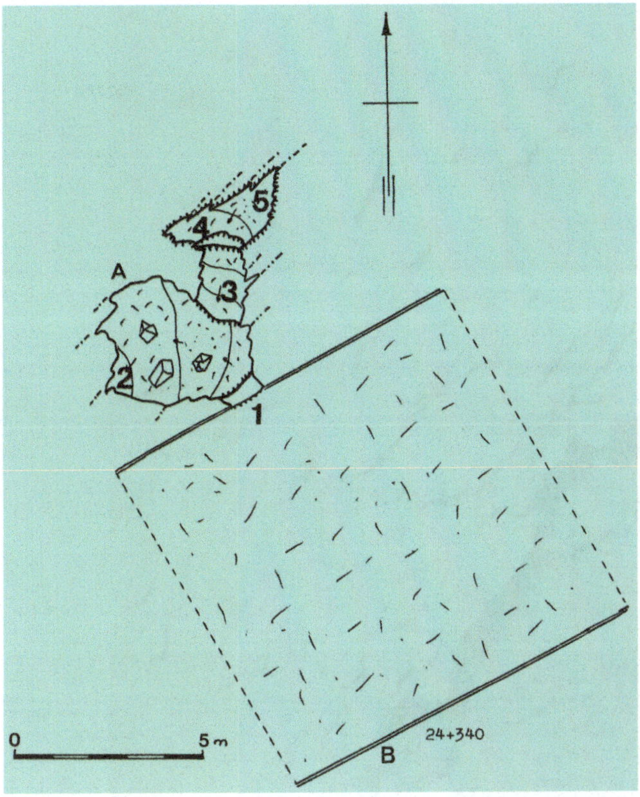

Fig. 6.11 Plan of cavern at km 24 + 340 in Tuhobić Tunnel

Figure 6.27 shows a fault zone with clay material and a cavern which is a rare occurrence.

Very strong corrosion effect was detected in caverns in the second tunnel tube of Tuhobić Tunnel (Figs. 6.28, 6.29, 6.30, 6.31, 6.32, 6.33 and 6.34).

6.3 Sleme Tunnel

The *Sleme* Tunnel on the A-6 Zagreb—Rijeka highway, near Lokve in Gorski Kotar, is 835 m long. 8 caverns were investigated in the left and two caverns in the right tunnel tube. The largest cavern is −128 m deep. The tunnel is located on the border of the Adriatic and Black Sea basins. Speleological objects located in the half of the tunnel, closer to Zagreb, belong to the Black Sea Basin (towards Lokve), and deep caverns closer to Rijeka belong to the Adriatic Basin. The tunnel is only 835 m long. Here will be mentioned: Cavern km 35 + 532 (Figs. 6.35, 6.36 and 6.37), height difference 128 m, Cavern km 35 + 270, height difference 26 m (Fig. 6.38), Cavern km 35 + 541, height difference 65 m (Figs. 6.39 and 6.40).

In Sleme Tunnel at km 35 + 585 (Figs. 6.41 and 6.42), there is a horizontal cavern 22 m long with a height difference of 11 m, in the vicinity of which there were two more caverns at the foot of the tunnel.

Caverns of fractural type tectonic genesis were found at km 35 + 599, 14 m deep and 27 m long (Fig. 6.43) and at km 35 + 755, 12 m deep and 5 m long (Figs. 6.44 and 6.45).

6.4 Sopač Tunnel

The Sopač Tunnel, on the A-6 motorway near Lokve in Gorski Kotar, is 752 m long. The Zagreb–Rijeka main road is only 2 m away. The road was not permanently closed at the time the tunnel was built. In total, 11 caverns were found in the Sopač Tunnel in two pipes. Here will be mentioned: Cavern at the km 38 + 513 station, with a height difference of 28 m, Cavern at km 38 + 513, Set of caverns in Sopač Tunnel at km 38 + 779 and 38 + 780, height difference 18 m with three vertical channels, Cavern at km 38 + 782, height difference 23 m (Figs. 6.46, 6.47, 6.48 and 6.49).

6.5 Vršek Tunnel

The Vršek Tunnel, on the A-6 motorway is 868 m long. Due to its distinct tectonics, the tunnel was difficult to build. Several times, rock-creep materials that sometimes filled the palaeo-caverns collapsed. In total, 17 caverns were explored in two tunnel tubes of Vršek Tunnel.

The cavern in the Vršek Tunnel, at km 47 + 749 (Figs. 6.50 and 6.51), is 23 m long with a height difference of 11 m. It is almost completely located on the tunnel route, cavern at km 47 + 407 (Figs. 6.52, 6.53, 6.54 and 6.55), there are several caverns with a total length of over 50 m with a height difference of 18 m. These are two large halls connected by a single channel. Most of the cavern is on the tunnel route and is filled with speleothem.

The cavern at km 47 + 412 (Figs. 6.56 and 6.57) in the Vršek Tunnel is 24 m deep and extends below the cavern at km 47 + 407, cavern km 47 + 365 (Figs. 6.58, 6.59, 6.60 and 6.61) is 25 m long and 15 m deep, Cavern at km 47 + 389 (Figs. 6.62 and 6.63) in the Vršek Tunnel, 37 m long and 7 m deep. The origin of cavern at km 47 + 743 was mainly the result of tectonics (Fig. 6.64).

Fig. 6.12 Cavern at km 24 + 340 in Tuhobić Tunnel

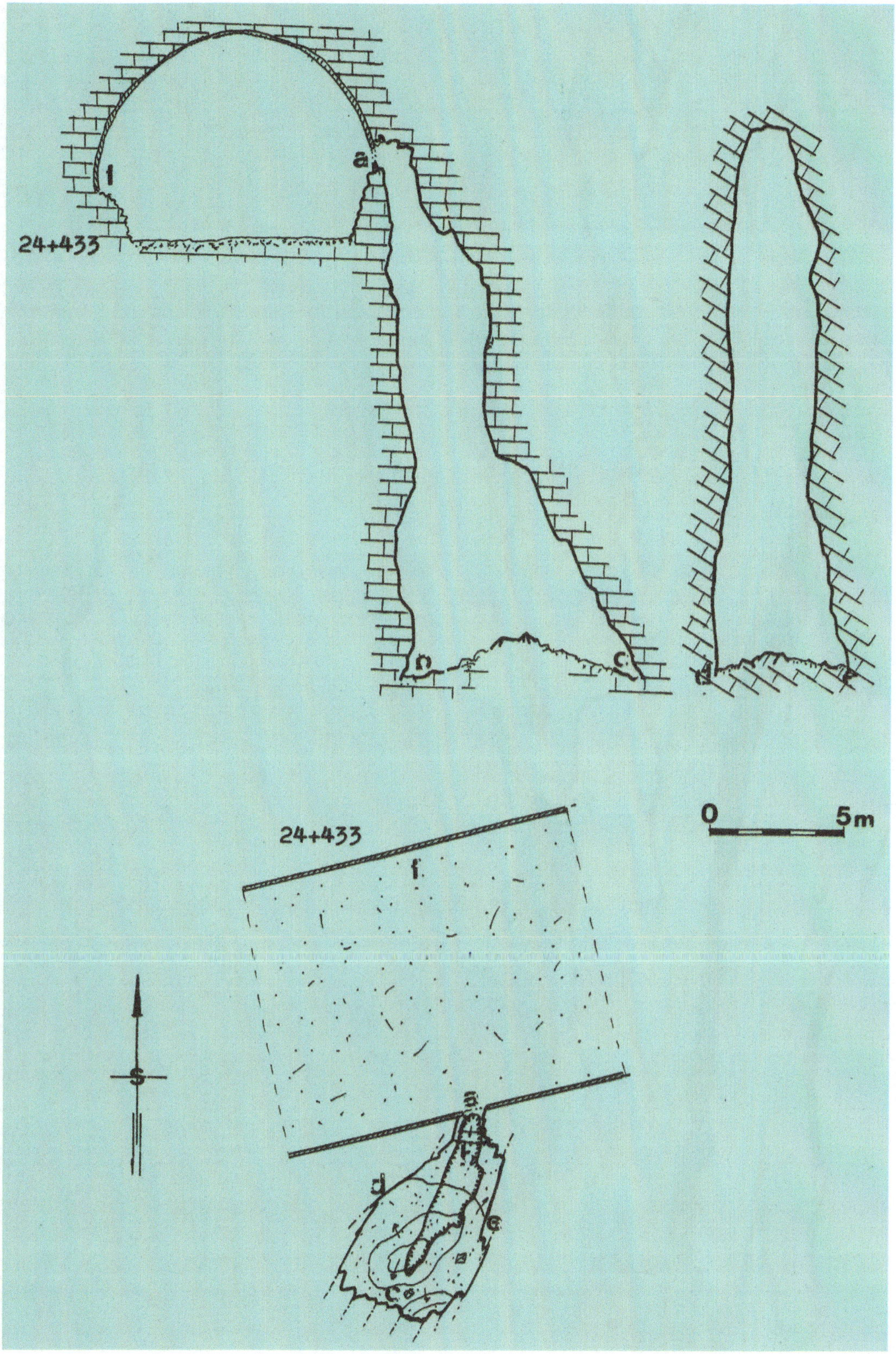

Fig. 6.13 Plan and profile of cavern at km 24 + 433 in Tuhobić Tunnel

Fig. 6.14 Vertical fissures in cavern km 24 + 433 in Tuhobić Tunnel

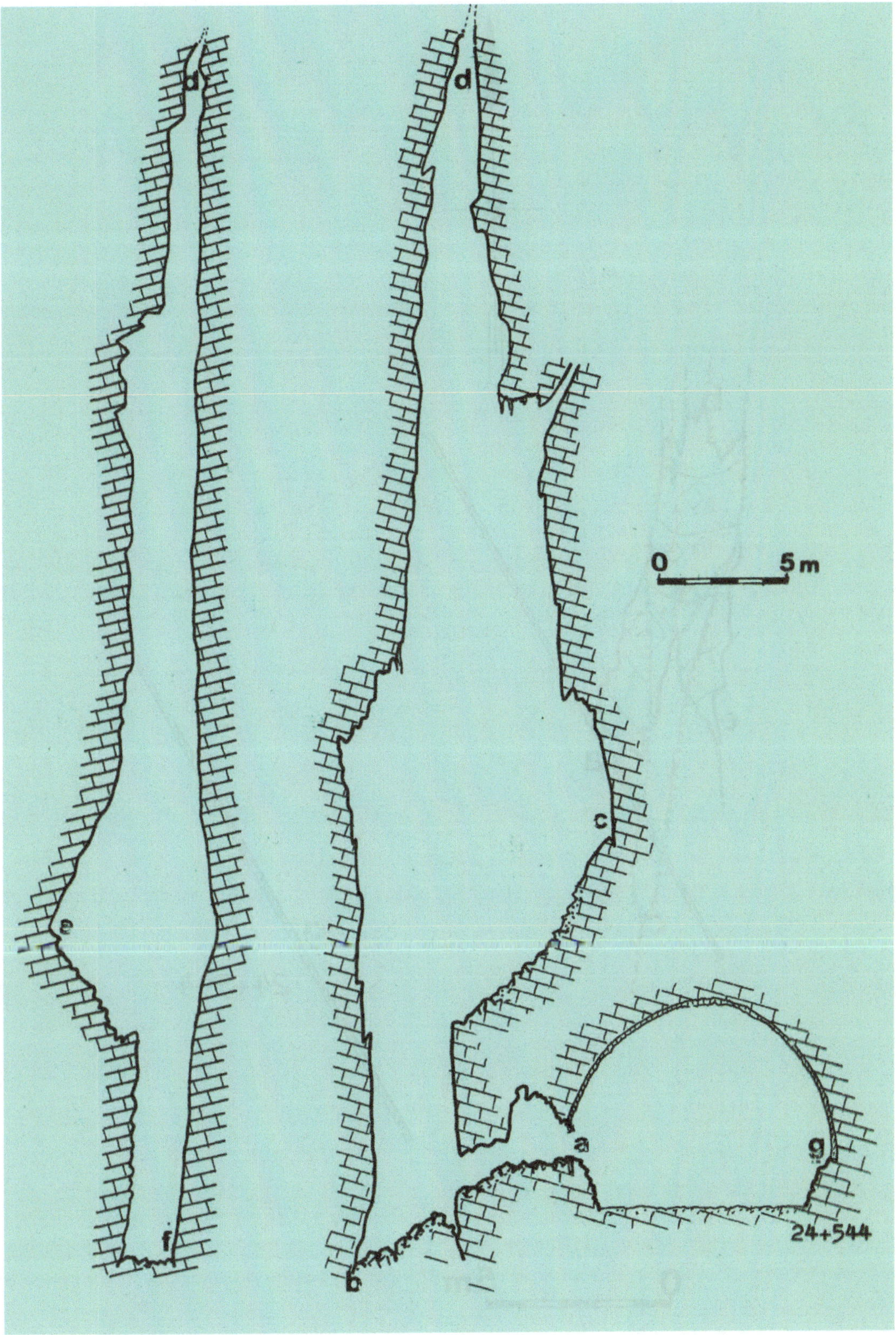

Fig. 6.15 Profiles of cavern at km 24 + 544 in Tuhobić Tunnel

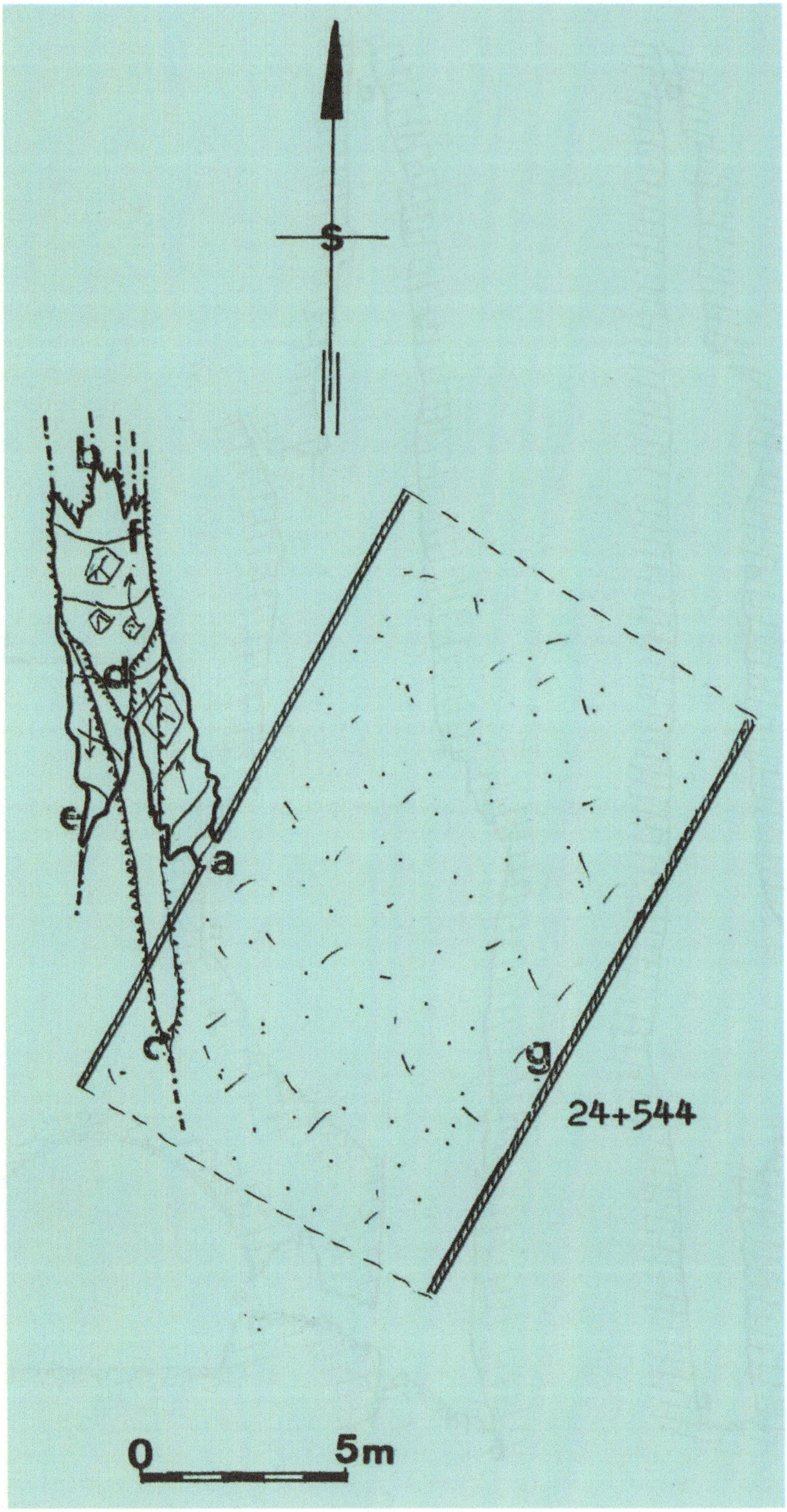

Fig. 6.16 Plan of cavern at km 24 + 544 in Tuhobić Tunnel

Fig. 6.17 Cavern at km 24 + 544 in Tuhobić Tunnel

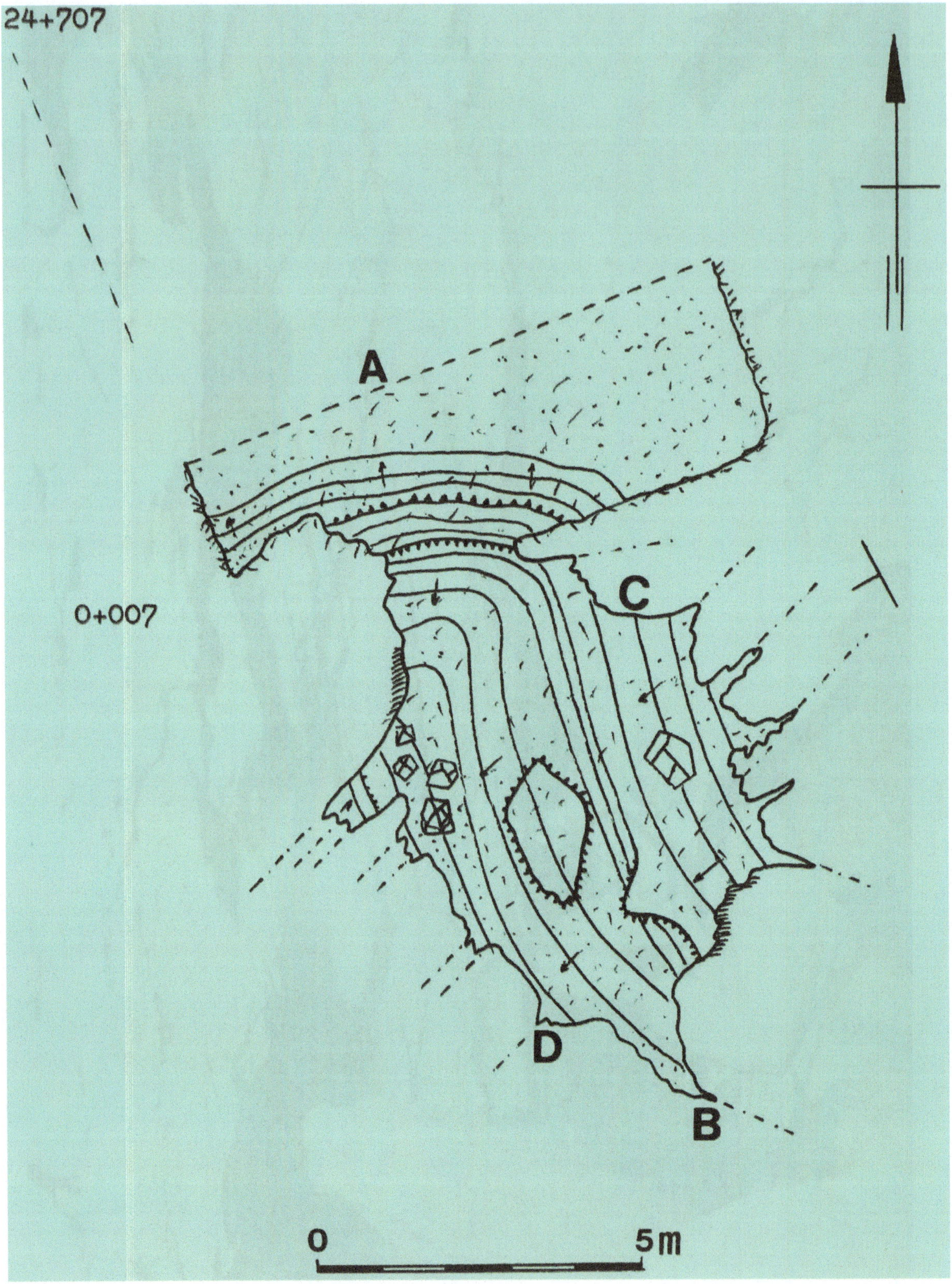

Fig. 6.18 Plan of cavern at km 24 + 707 in Tuhobić Tunnel

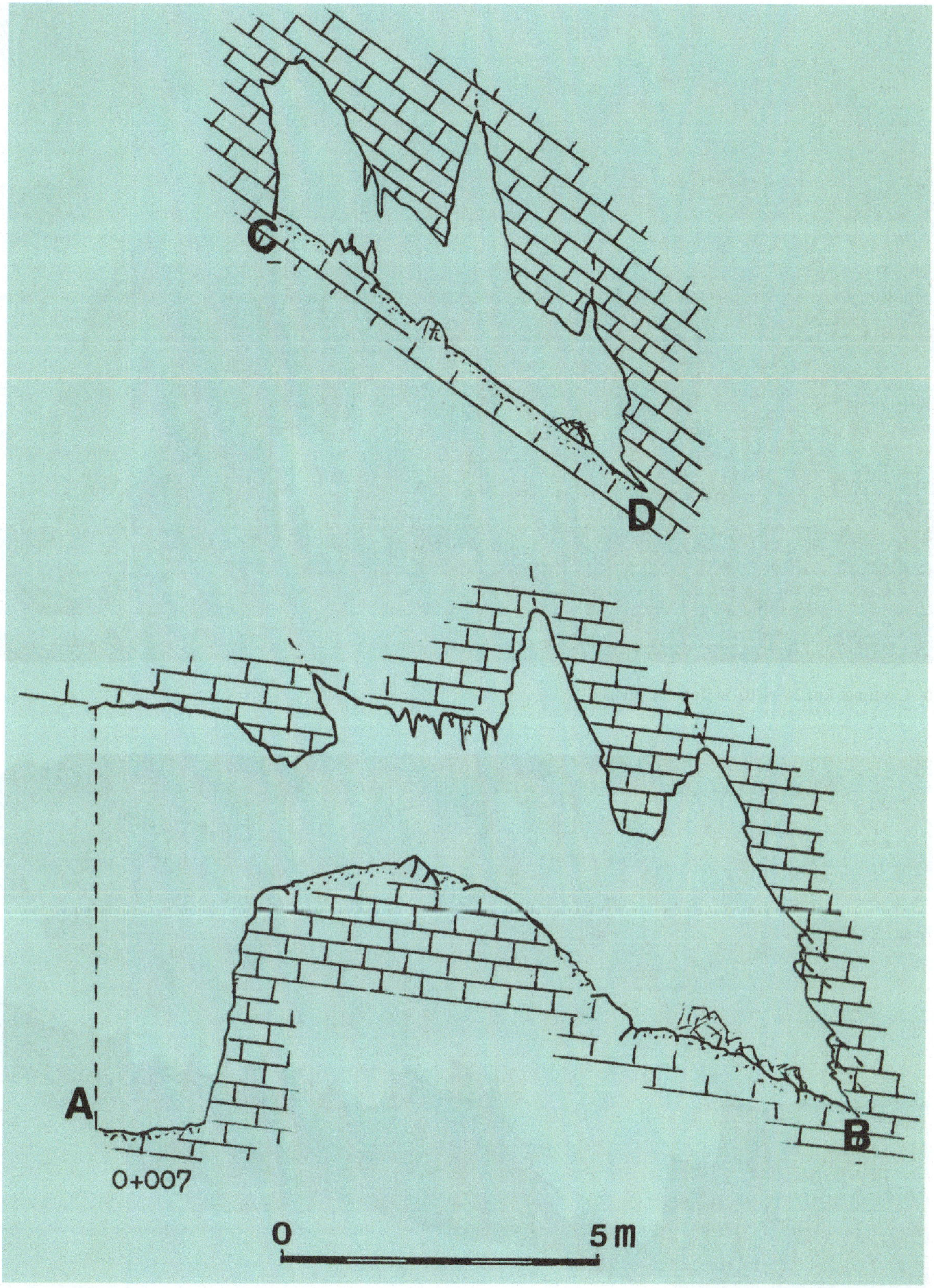

Fig. 6.19 Profile of cavern at km 24 + 707 in Tuhobić Tunnel

Fig. 6.20 Cavern at km 24 + 707 in Tuhobić Tunnel

Fig. 6.21 Speleothems in cavern at km 24 + 707 in Tuhobić Tunnel

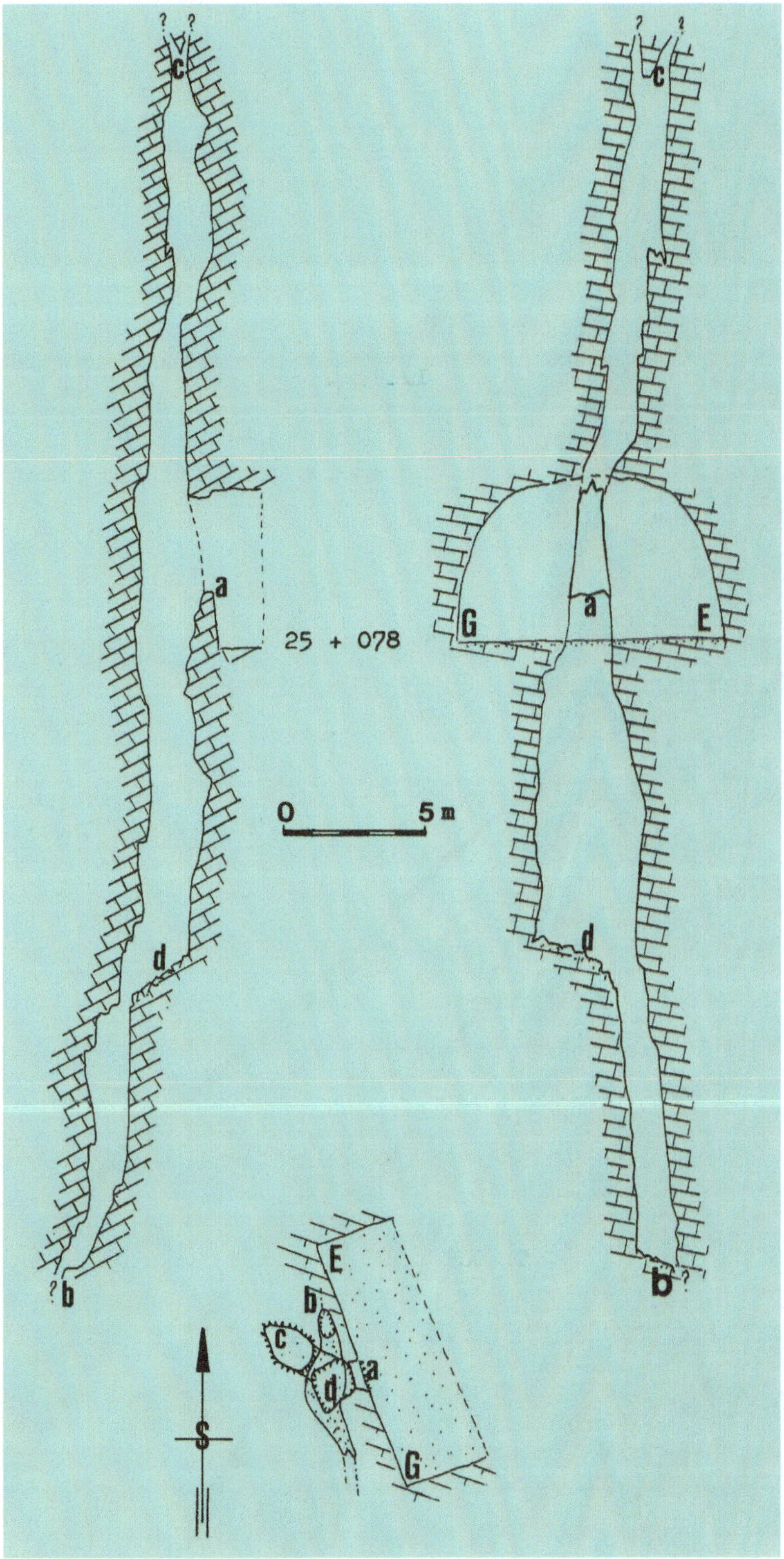

Fig. 6.22 Plan and profile of cavern at km 25 + 078 in Tuhobić Tunnel

Fig. 6.23 Upper parts of cavern at km 25 + 078 in Tuhobić Tunnel

Fig. 6.24 Lower parts of cavern at km 25 + 078 in Tuhobić Tunnel

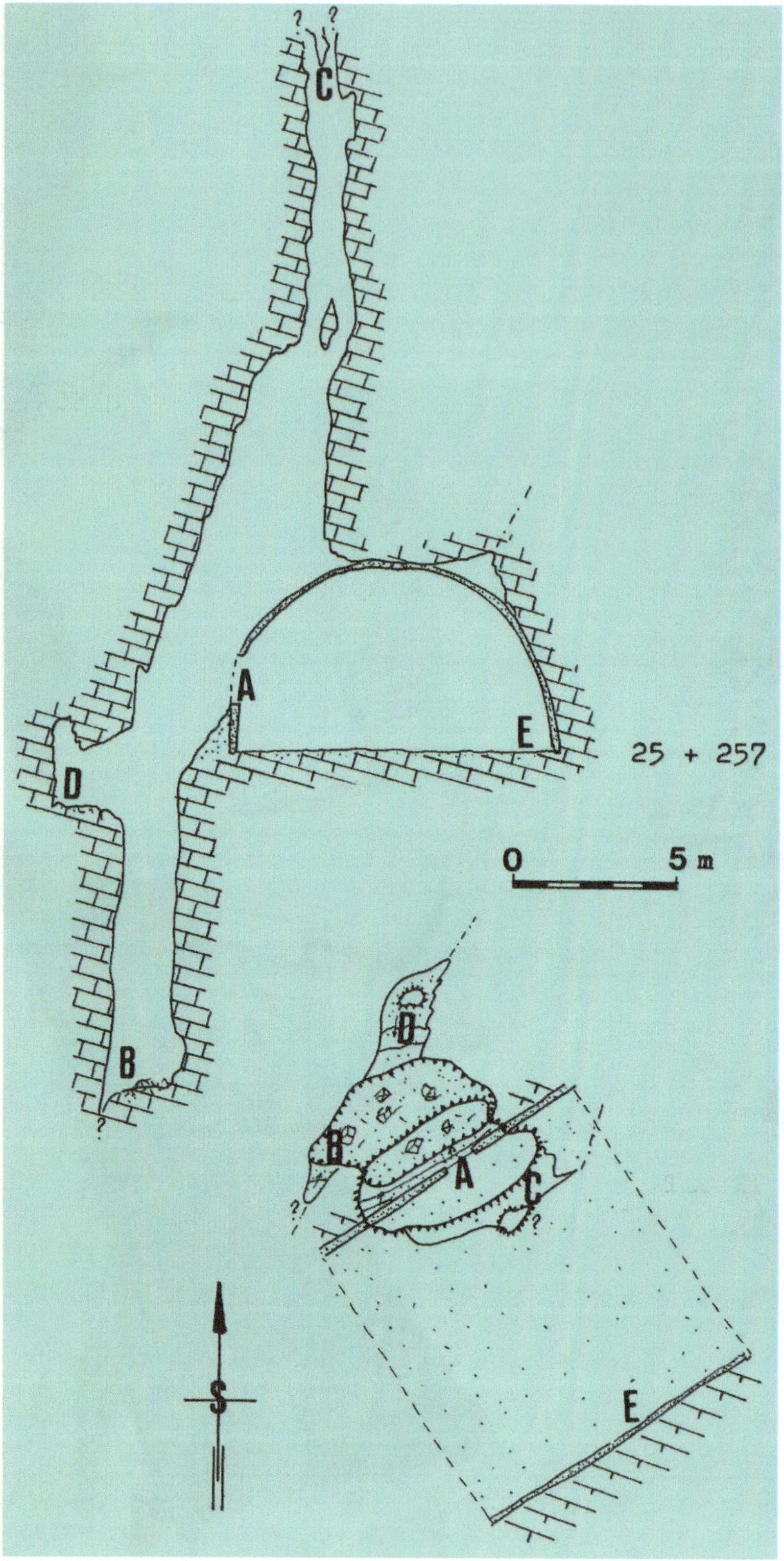

Fig. 6.25 Plan and profile of cavern at km 25 + 257 in Tuhobić Tunnel

Fig. 6.26 Cavern at km 25 + 257 in Tuhobić Tunnel

Fig. 6.27 Fault zone with clay material and cavern

Fig. 6.28 Water corrosion in caverns

Fig. 6.29 Fractures widened by the actions of water

Fig. 6.30 Cavern entrance covered with sprayed concrete

Fig. 6.31 Strong and intensive corrosion in Tuhobić Tunnel

Fig. 6.32 Razor sharp rocks

Fig. 6.33 Corrosion in Tuhobić Tunnel

Fig. 6.34 Different intensities of corrosion in Tuhobić Tunnel

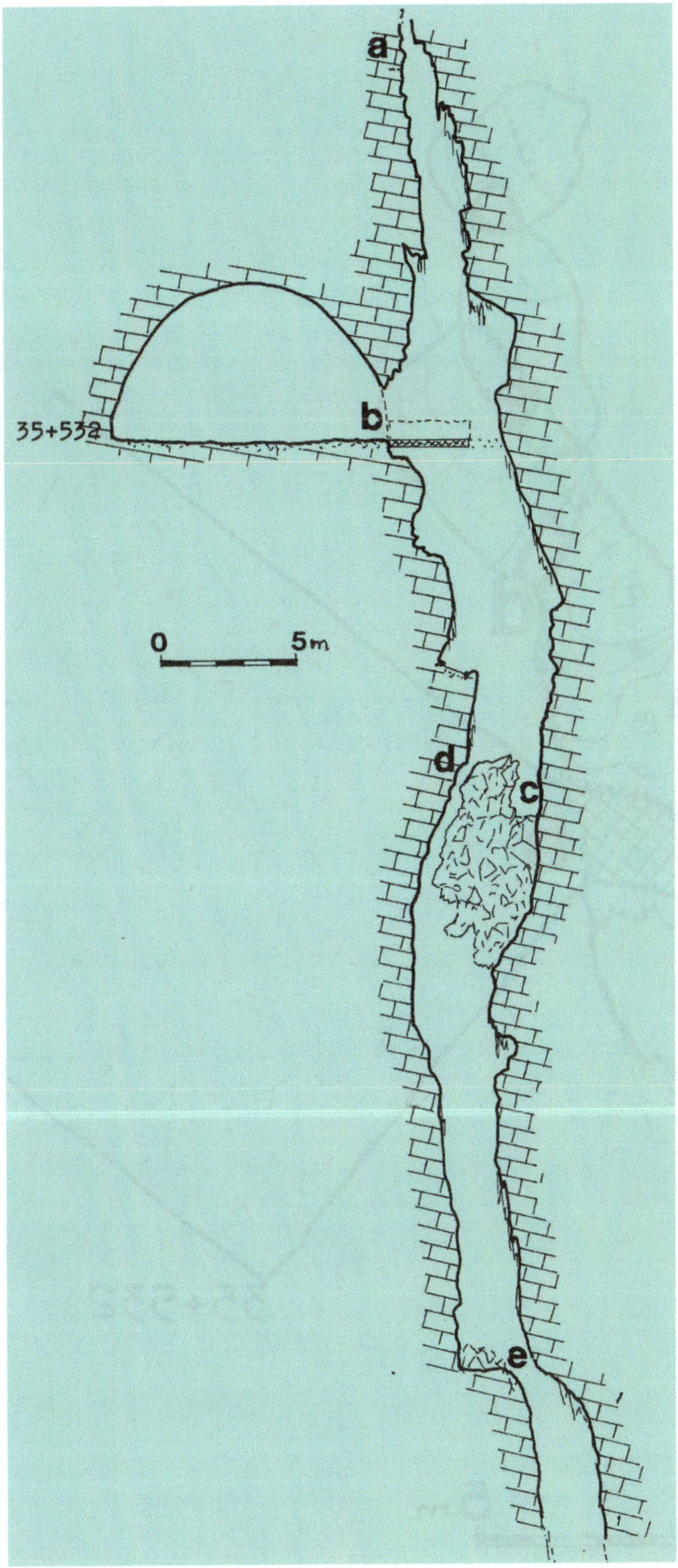

Fig. 6.35 Profile of cavern at km 35 + 532 in Sleme Tunnel

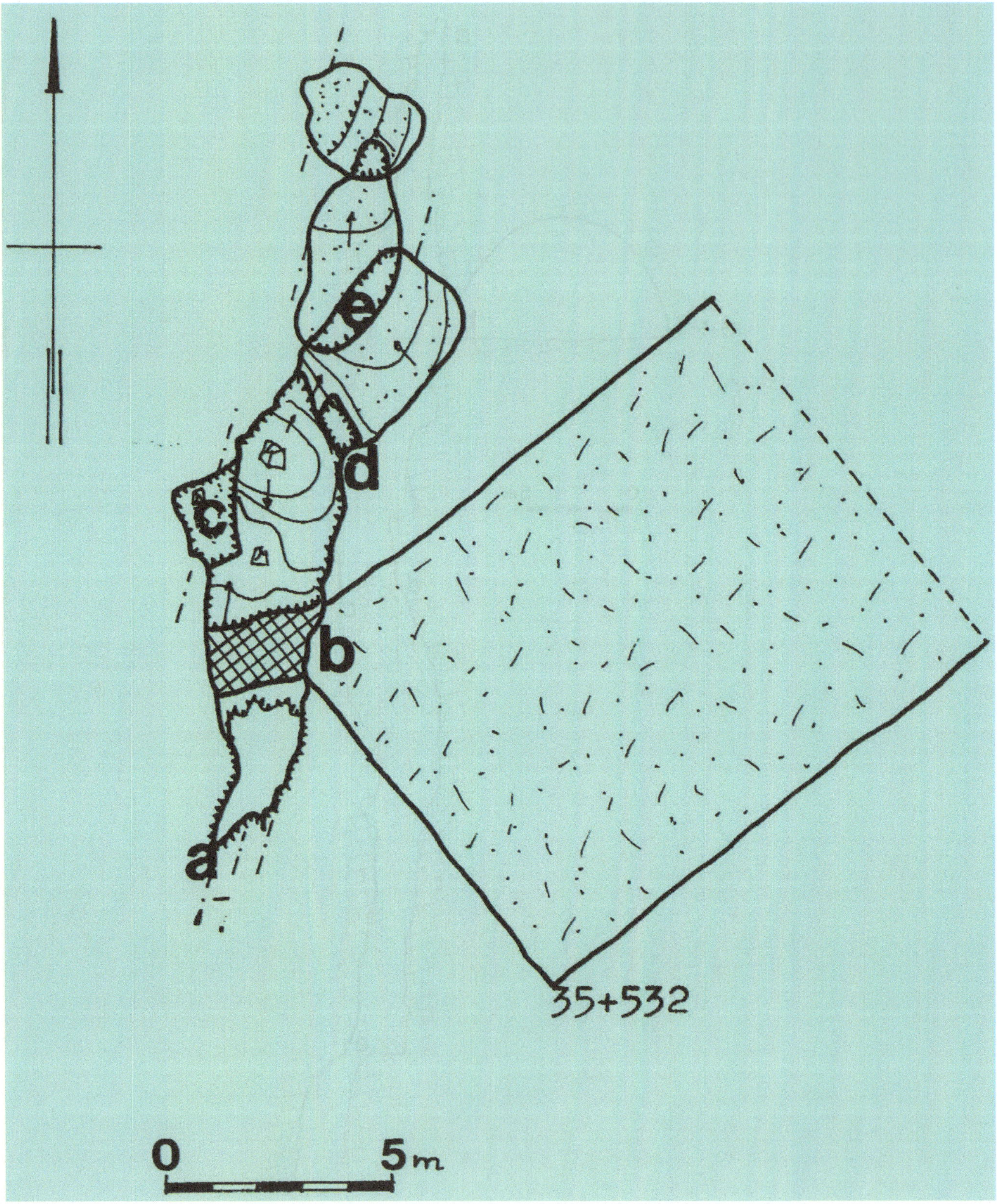

Fig. 6.36 Plan of cavern at km 35 + 532 in Sleme Tunnel

Fig. 6.37 Cavern at km 35 + 532 in Sleme Tunnel

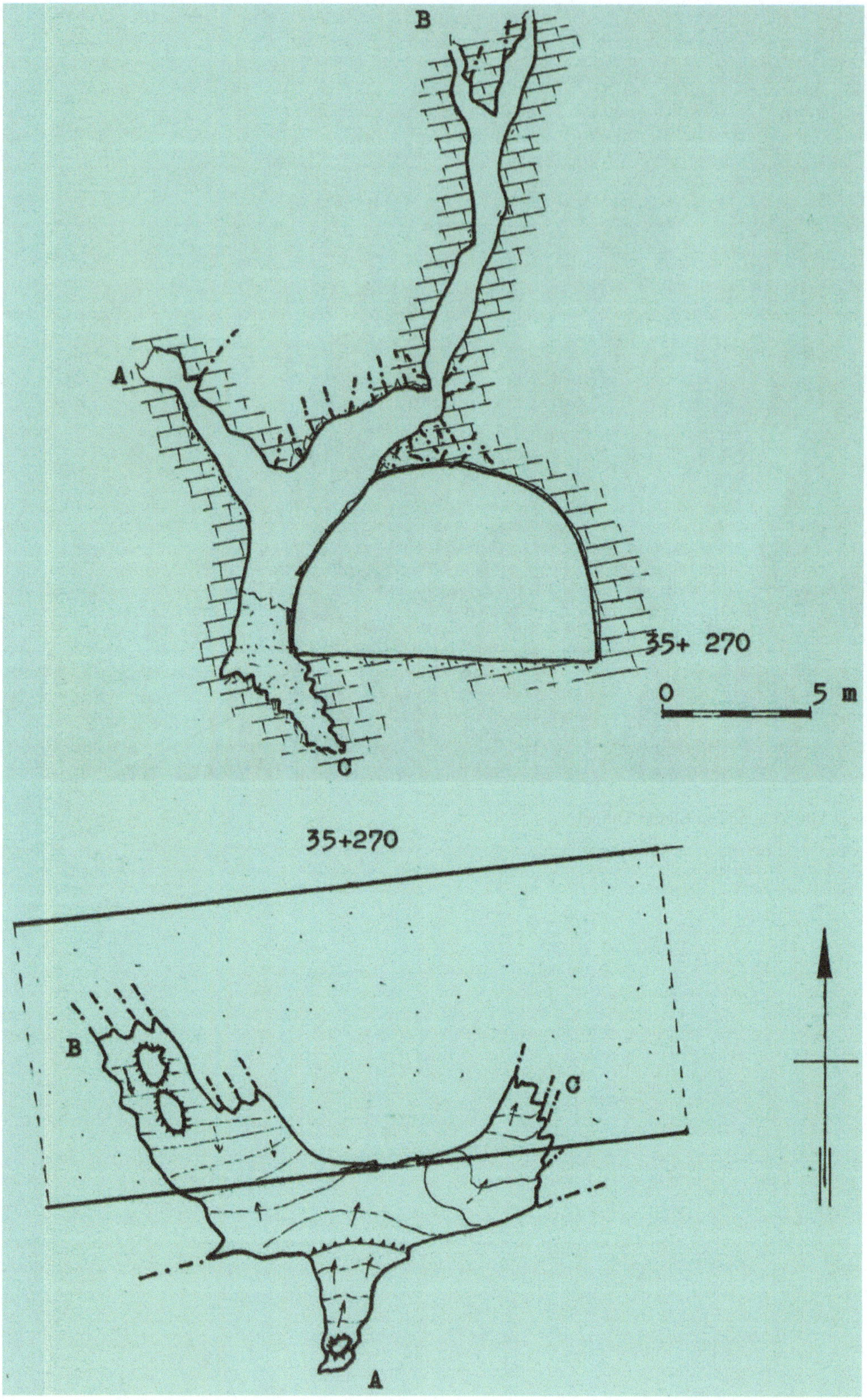

Fig. 6.38 Profile and plan of cavern at km 35 + 270 in Sleme Tunnel

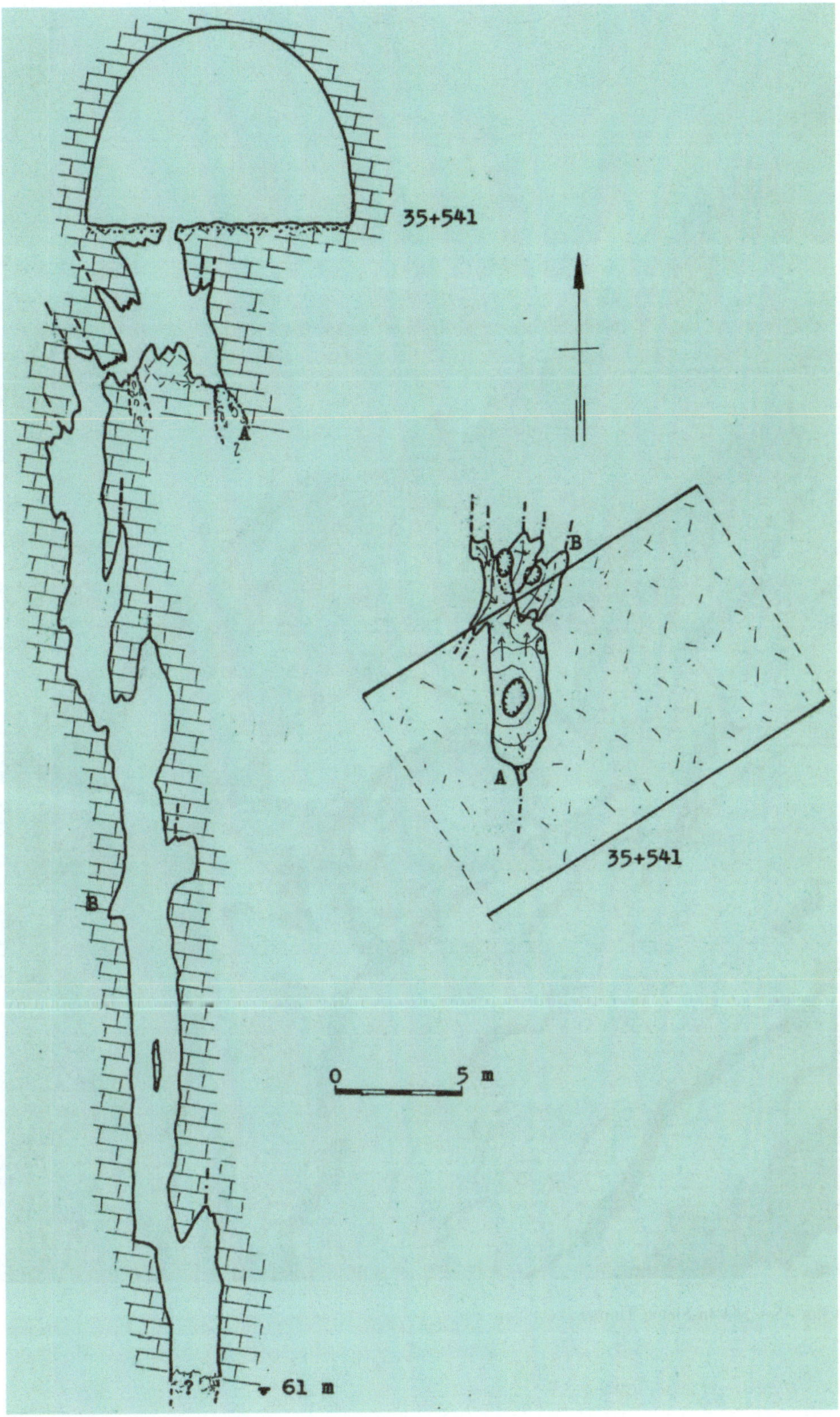

Fig. 6.39 Profile and plan of cavern at km 35 + 541 in Sleme Tunnel

Fig. 6.40 Cavern at km 35 + 541 in Sleme Tunnel

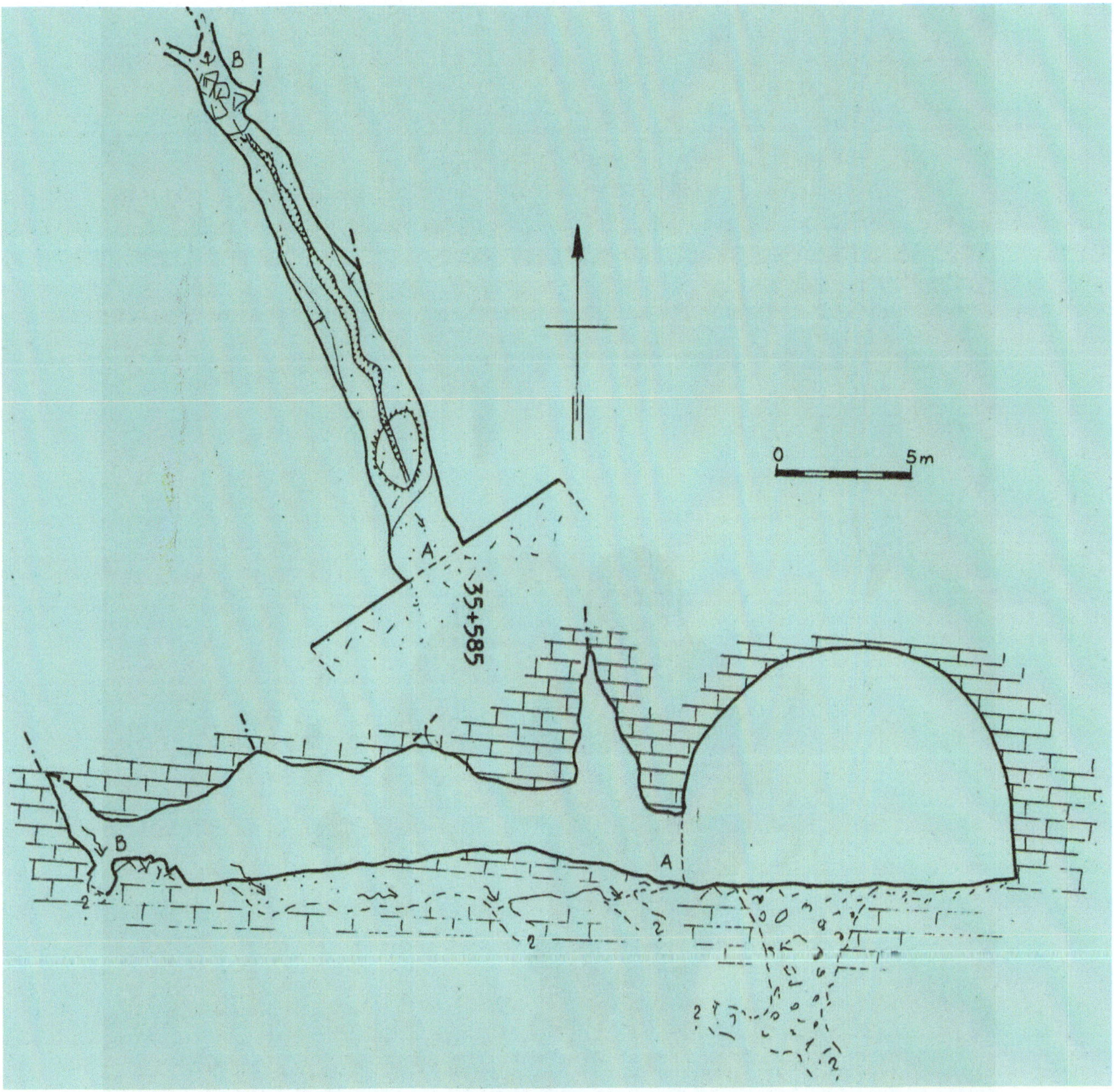

Fig. 6.41 Profile and plan of cavern at km 35 + 585 in Sleme Tunnel

Fig. 6.42 Cavern at km 35 + 585 in Sleme Tunnel

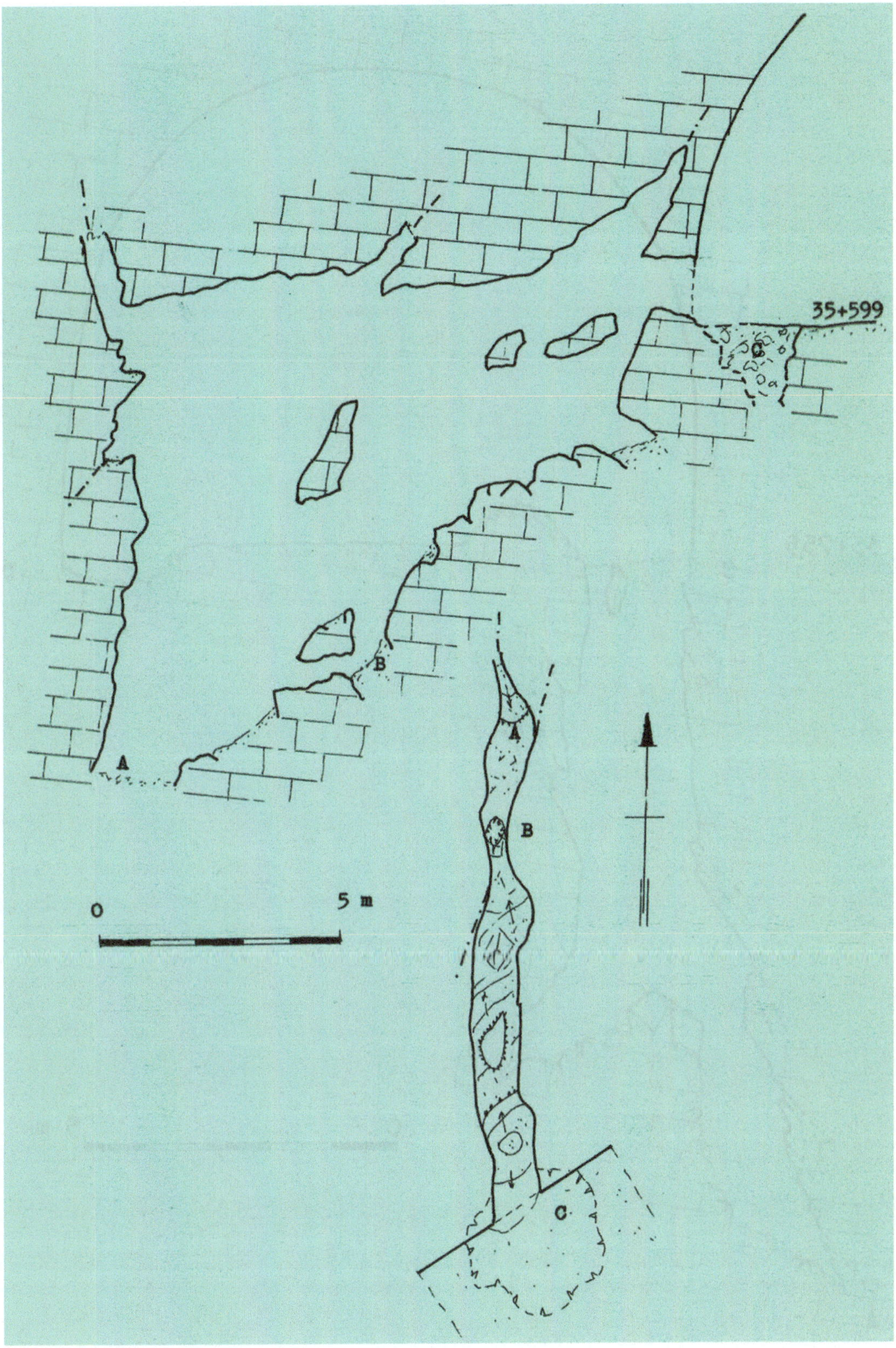

Fig. 6.43 Profile and plan of cavern at km 35 + 599 in Sleme Tunnel

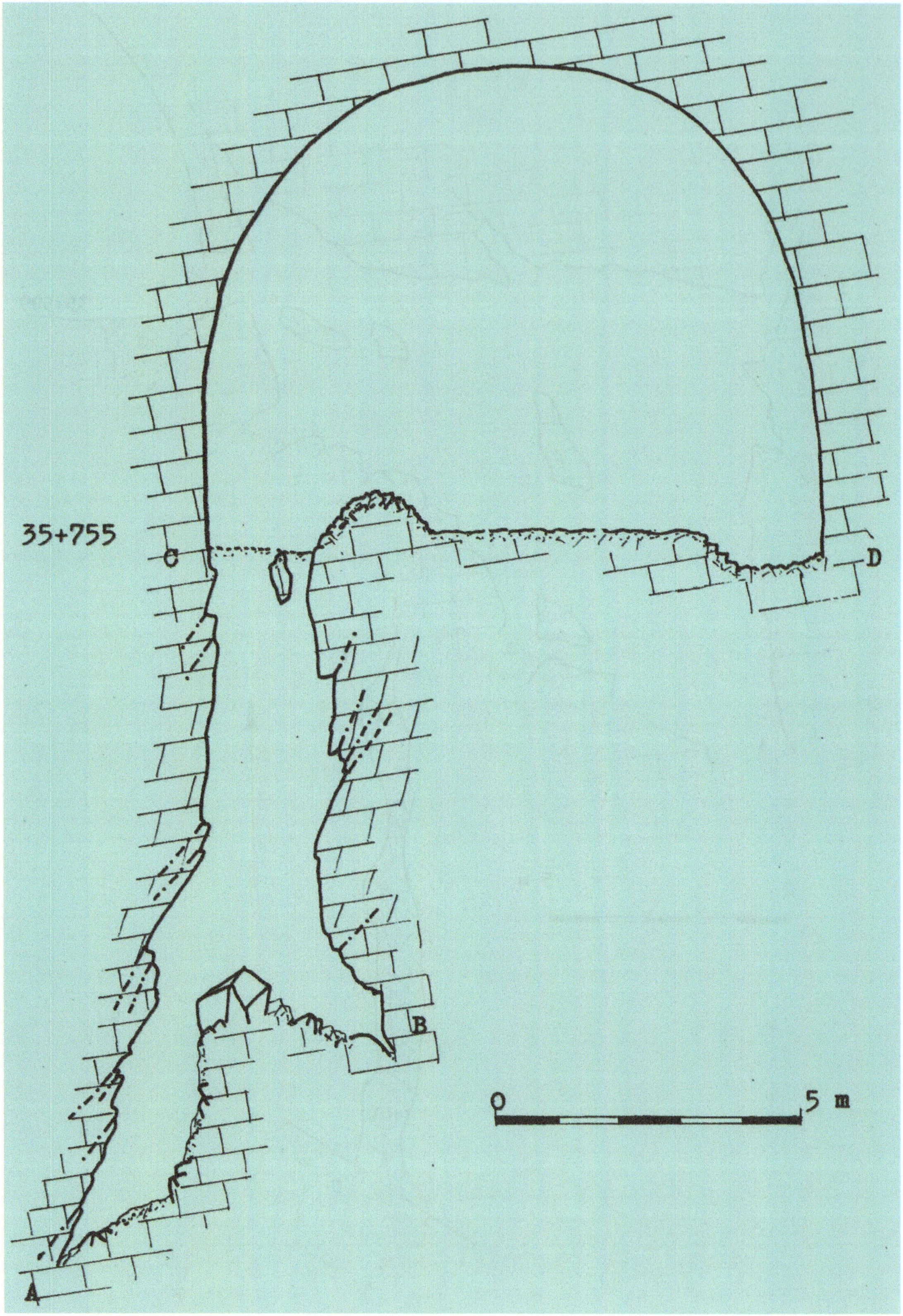

Fig. 6.44 Profile of cavern at km 35 + 755 in Sleme Tunnel

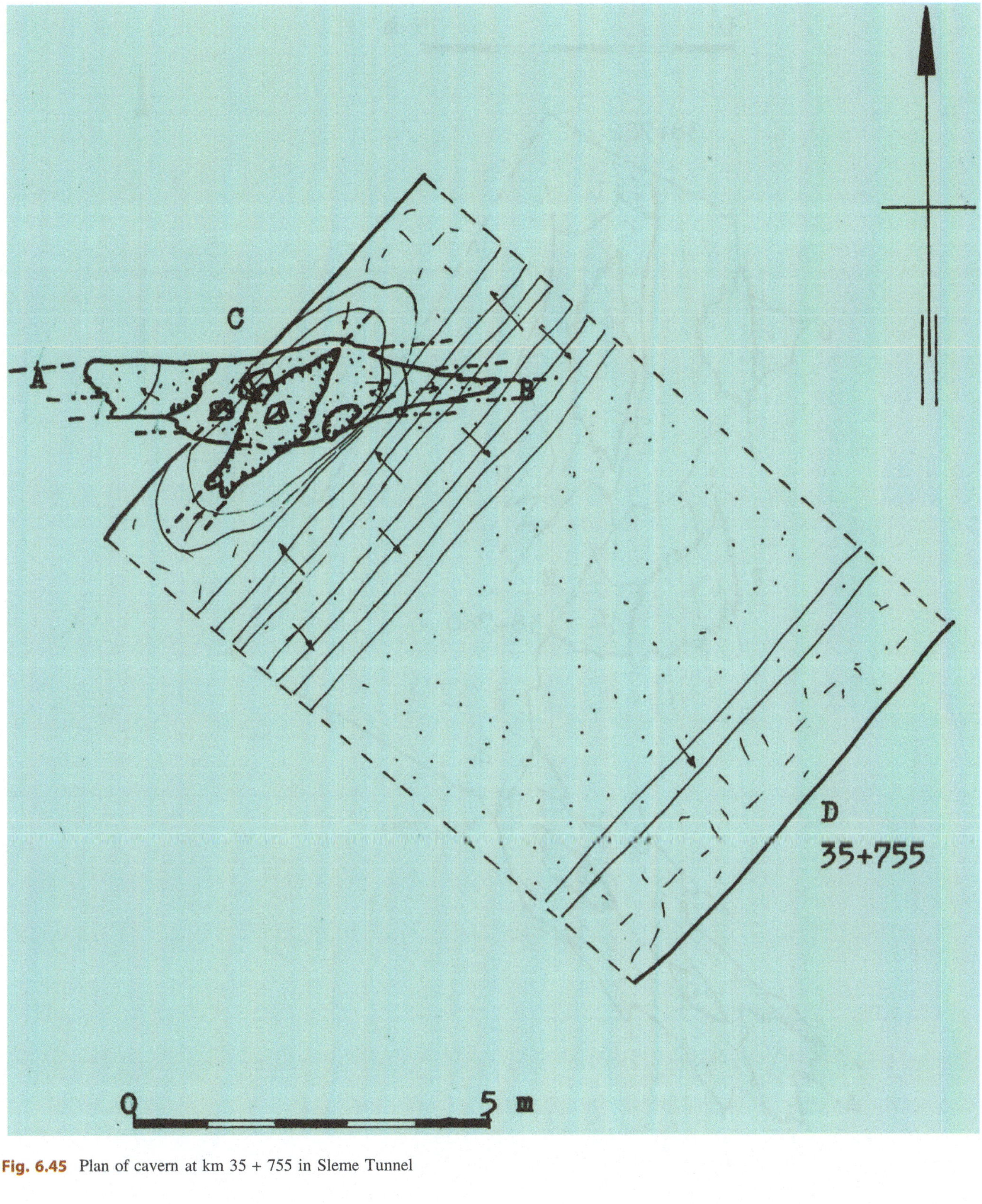

Fig. 6.45 Plan of cavern at km 35 + 755 in Sleme Tunnel

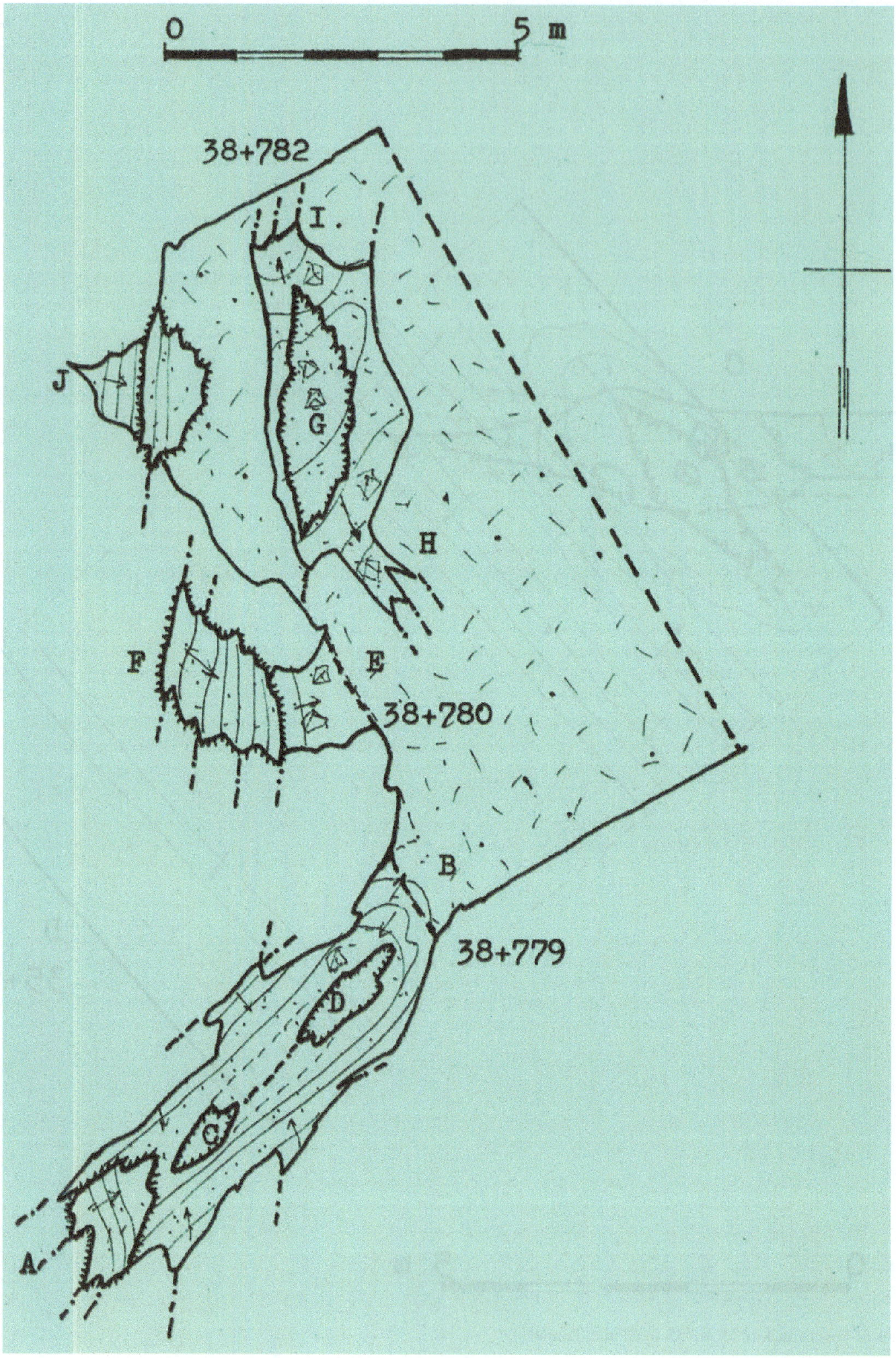

Fig. 6.46 Plan of caverns at km 38 + 779 and 38 + 780 in Sopač Tunnel

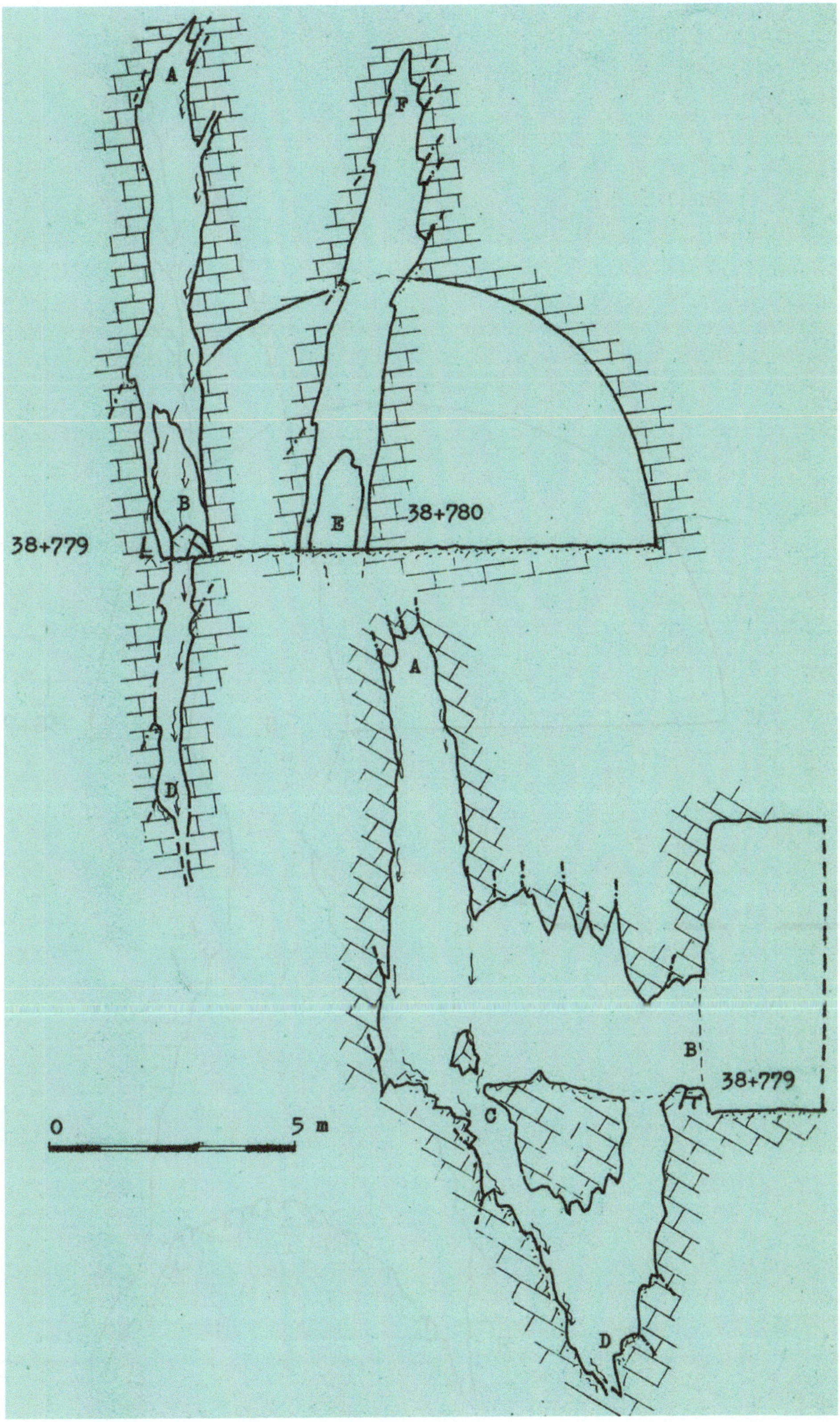

Fig. 6.47 Profiles of caverns at km 38 + 779 and 38 + 780 in Sopač Tunnel

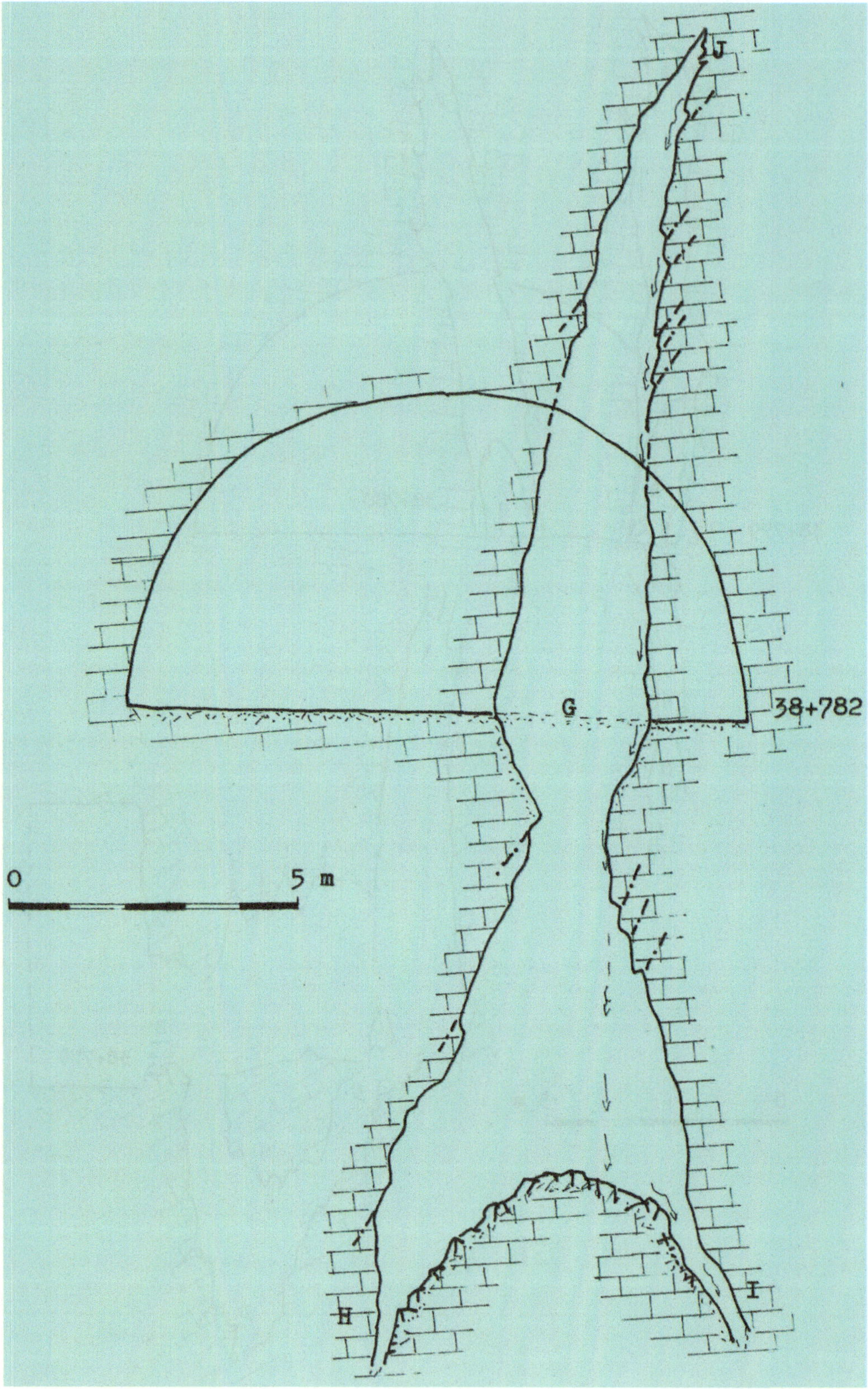

Fig. 6.48 Profile of cavern at km 38 + 782 in Sopač Tunnel

Fig. 6.49 Cavern at km 38 + 782 in Sopač Tunnel

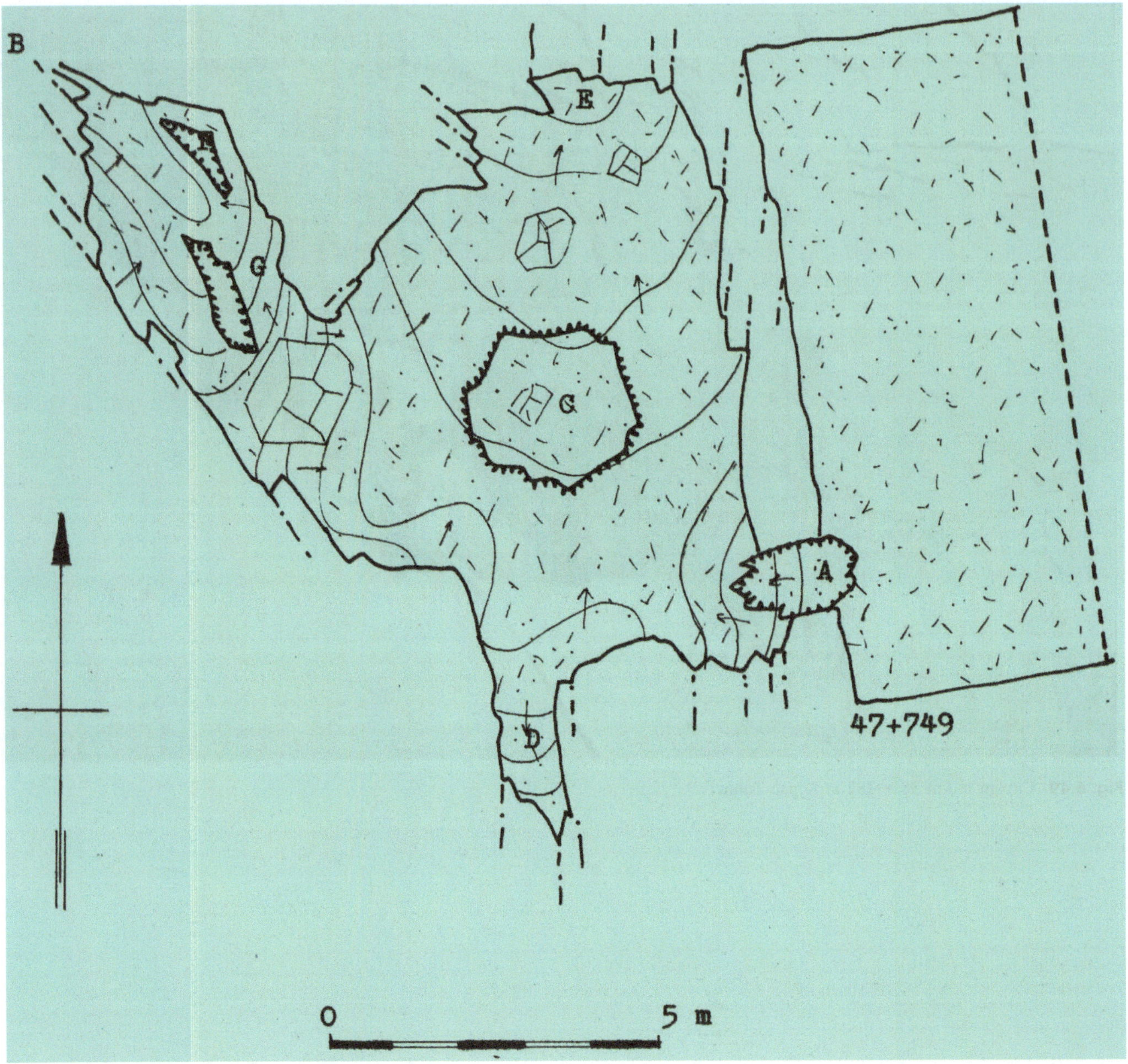

Fig. 6.50 Plan of cavern at km 47 + 747 in Vršek Tunnel

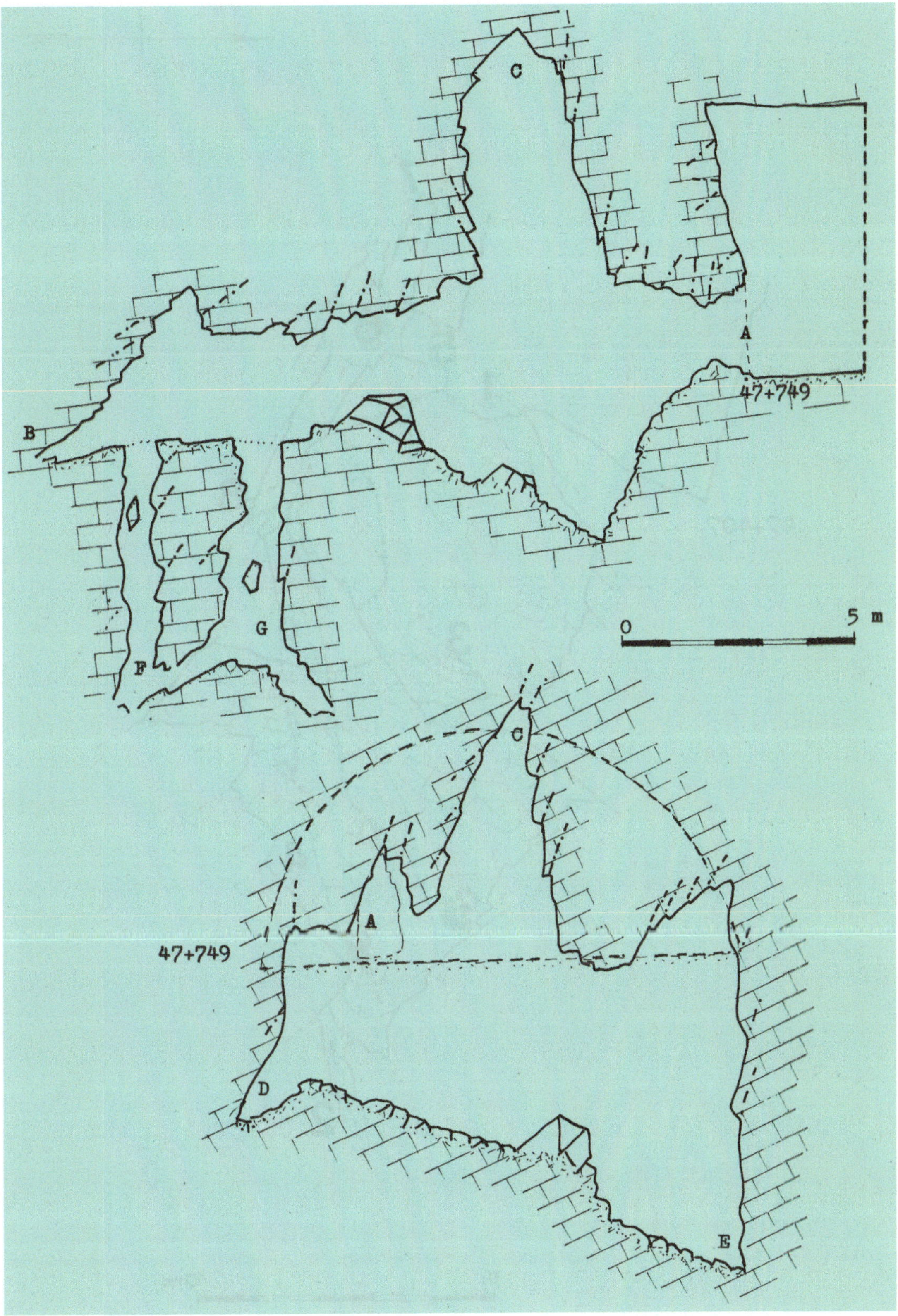

Fig. 6.51 Profile of cavern at km 47 + 749 in Vršek Tunnel

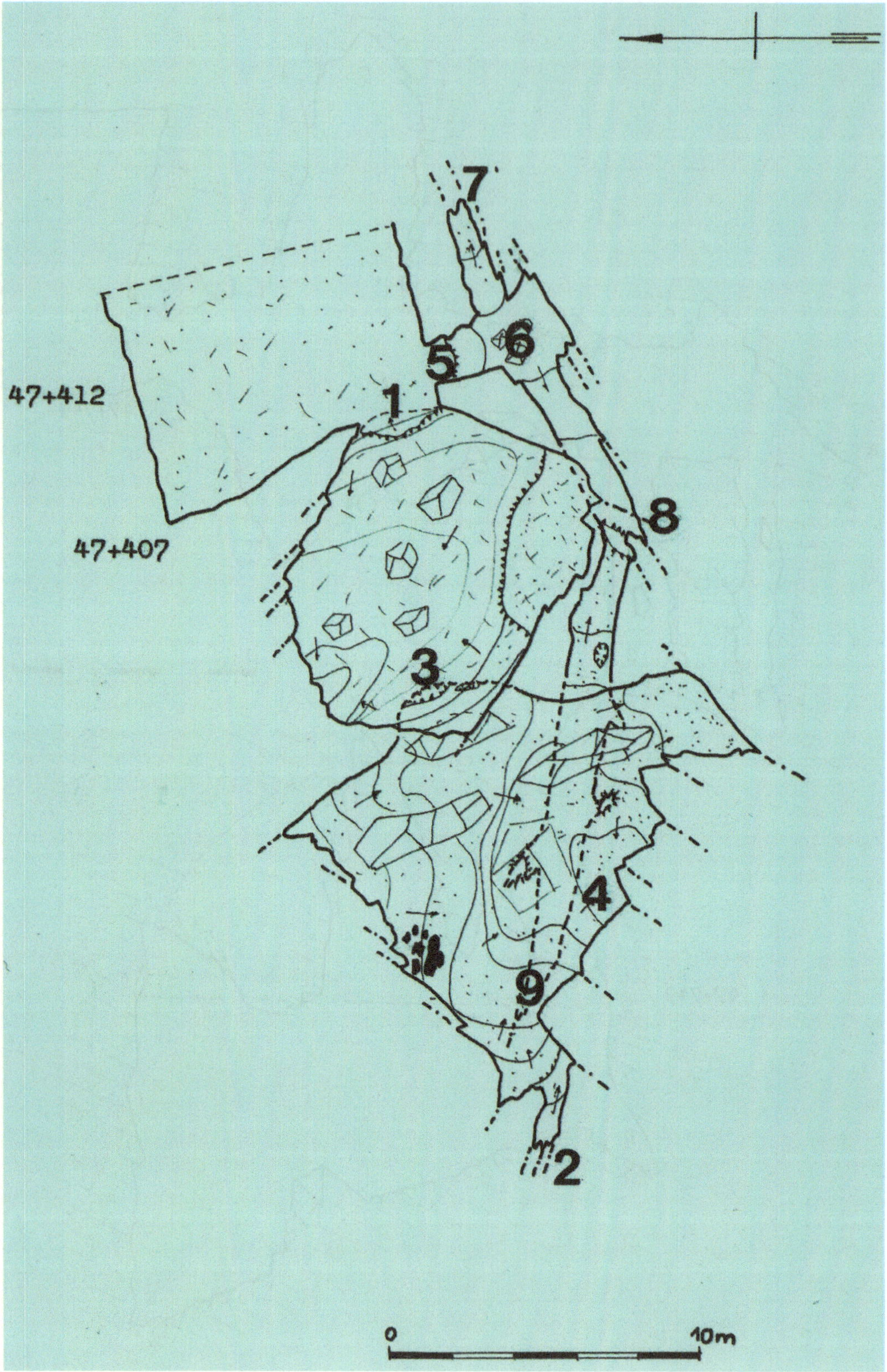

Fig. 6.52 Plan of cavern at km 47 + 407 in Vršek Tunnel

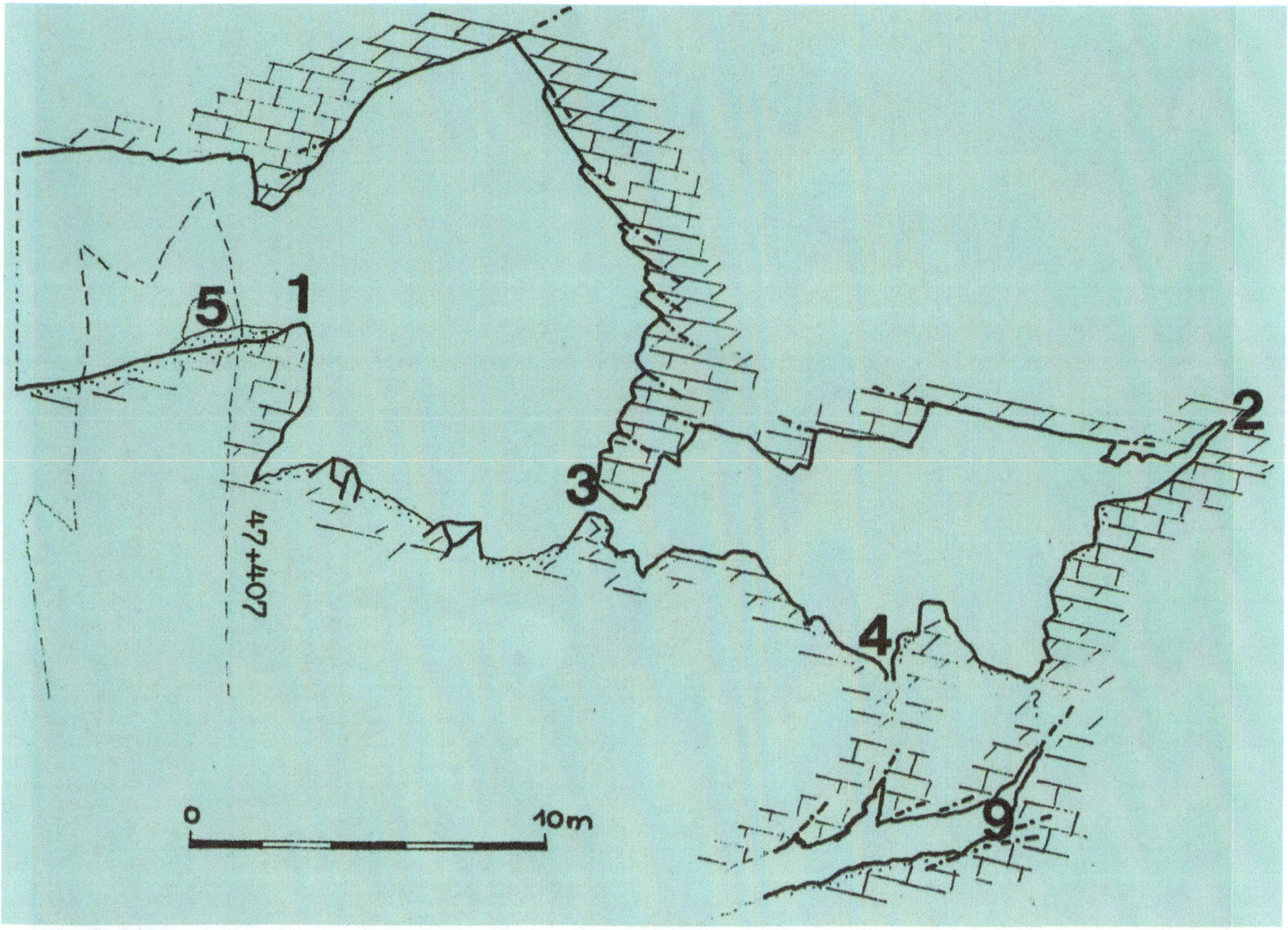

Fig. 6.53 Profile of cavern at km 47 + 407 in Vršek Tunnel

6.6 Vrata Tunnel

In the area of the "Vrata" Tunnel, which is only 257 m long, on the A-6 highway near Fužine in Goski Kotar, only one smaller cavern with a diameter of 1.4 × 1.4 m was found during the construction of the first tunnel tube (Garašić and Zanoškar 1992). However, a large cave hall (55 m x 60 m x 40 m) was found during the construction of the second tunnel tube at km 31 + 292 (Figs. 6.65, 6.66 and 6.67). Interestingly, it is one of the shortest tunnels on the A-6 motorway (Garašić 2006a).

In order to protect the underground flow and unstable dumping material, a 58 m long bridge (Figs. 6.68 and 6.69) was built in the cavern, which is considered to be the longest built on a highway within the cavern in the world (Garašić et al. 2010).

6.7 Veliki Gložac Tunnel

The Veliki Gložac Tunnel on the A-6 motorway in Gorski Kotar is 1130 m long. It belongs to high-profile tunnels, and 25 caverns have been found and explored in it.

Fig. 6.54 Speleothems in cavern at km 47 + 407 in Vršek Tunnel

Fig. 6.55 Entrance of cavern at km 47 + 407 in Vršek Tunnel

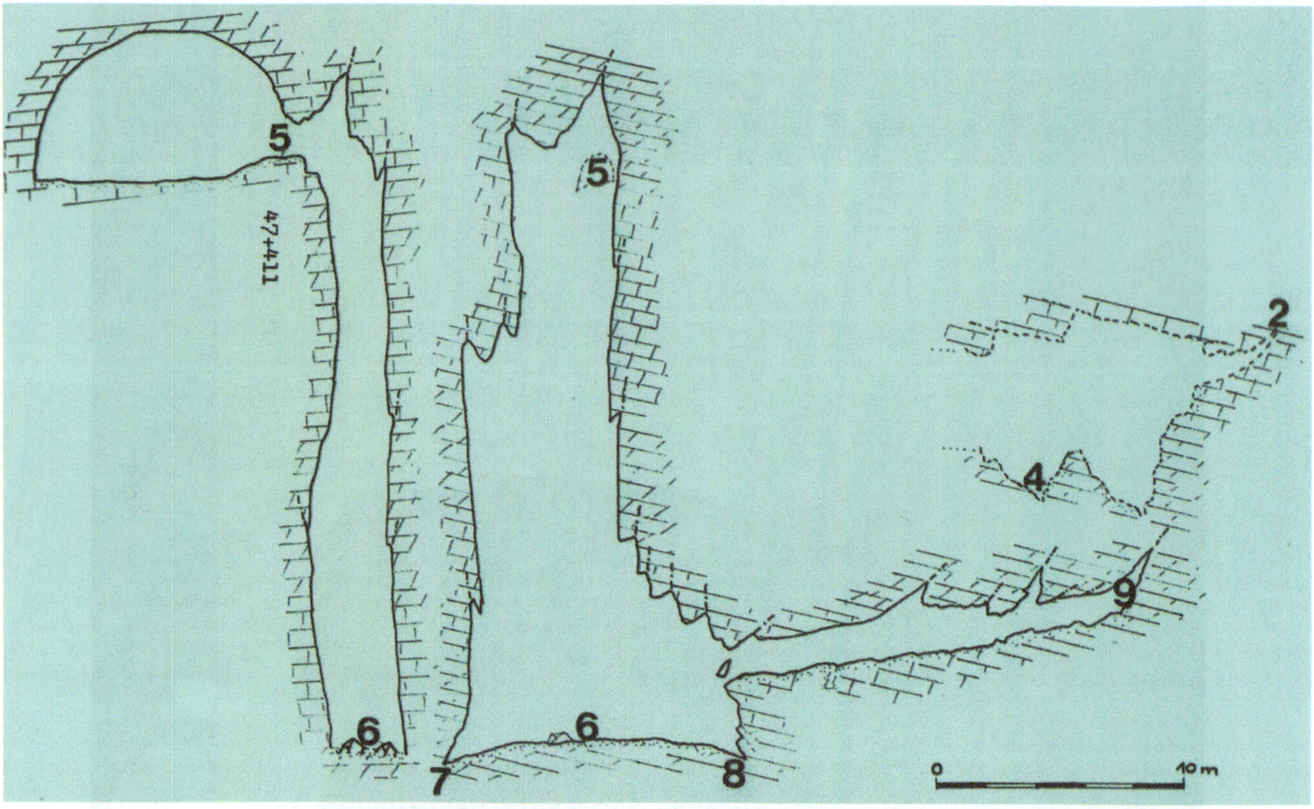

Fig. 6.56 Profile of cavern at km 47 + 412 in Vršek Tunnel

Fig. 6.57 Cavern at km 47 + 412 in Vršek Tunnel

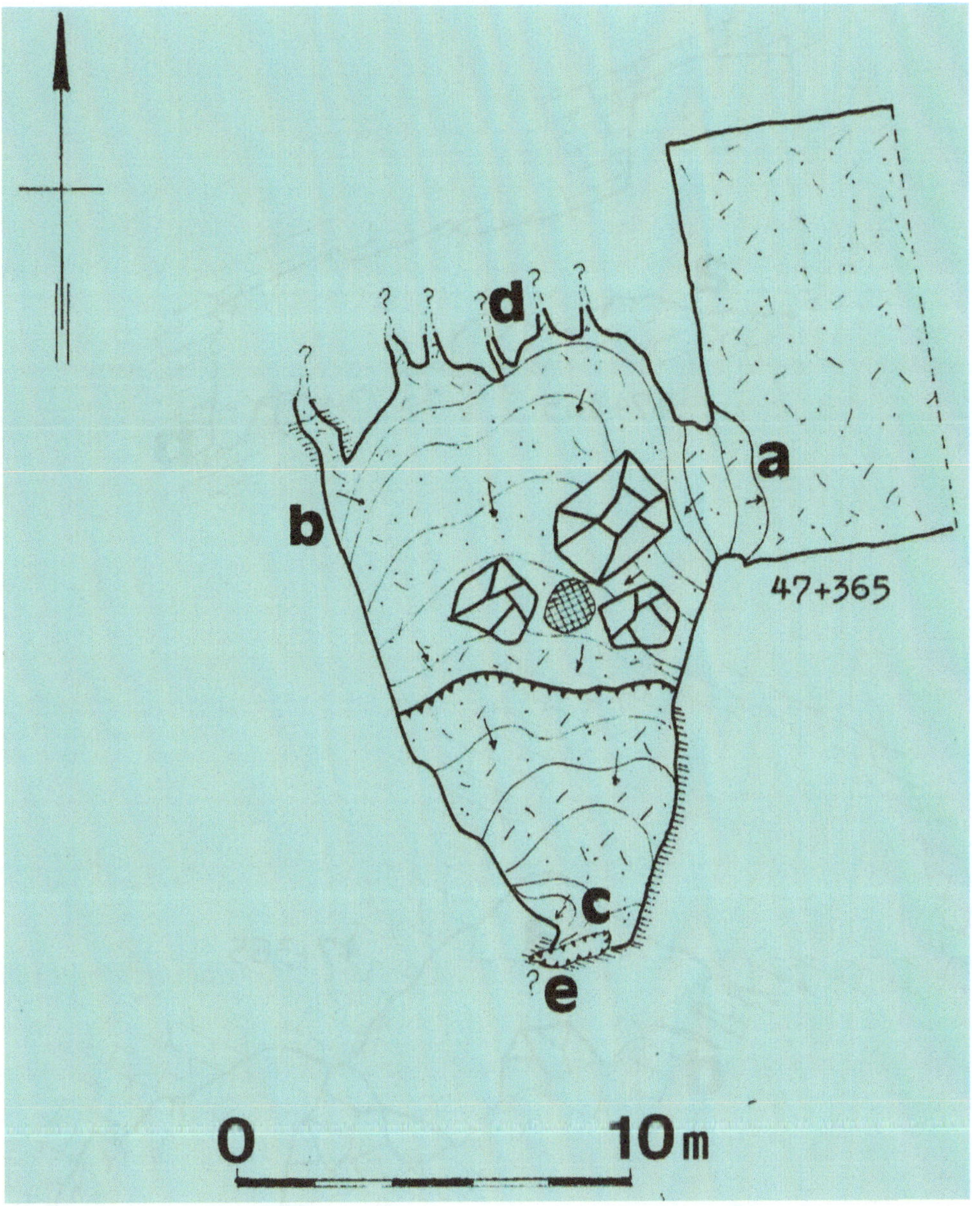

Fig. 6.58 Plan of cavern at km 47 + 365 in Vršek Tunnel

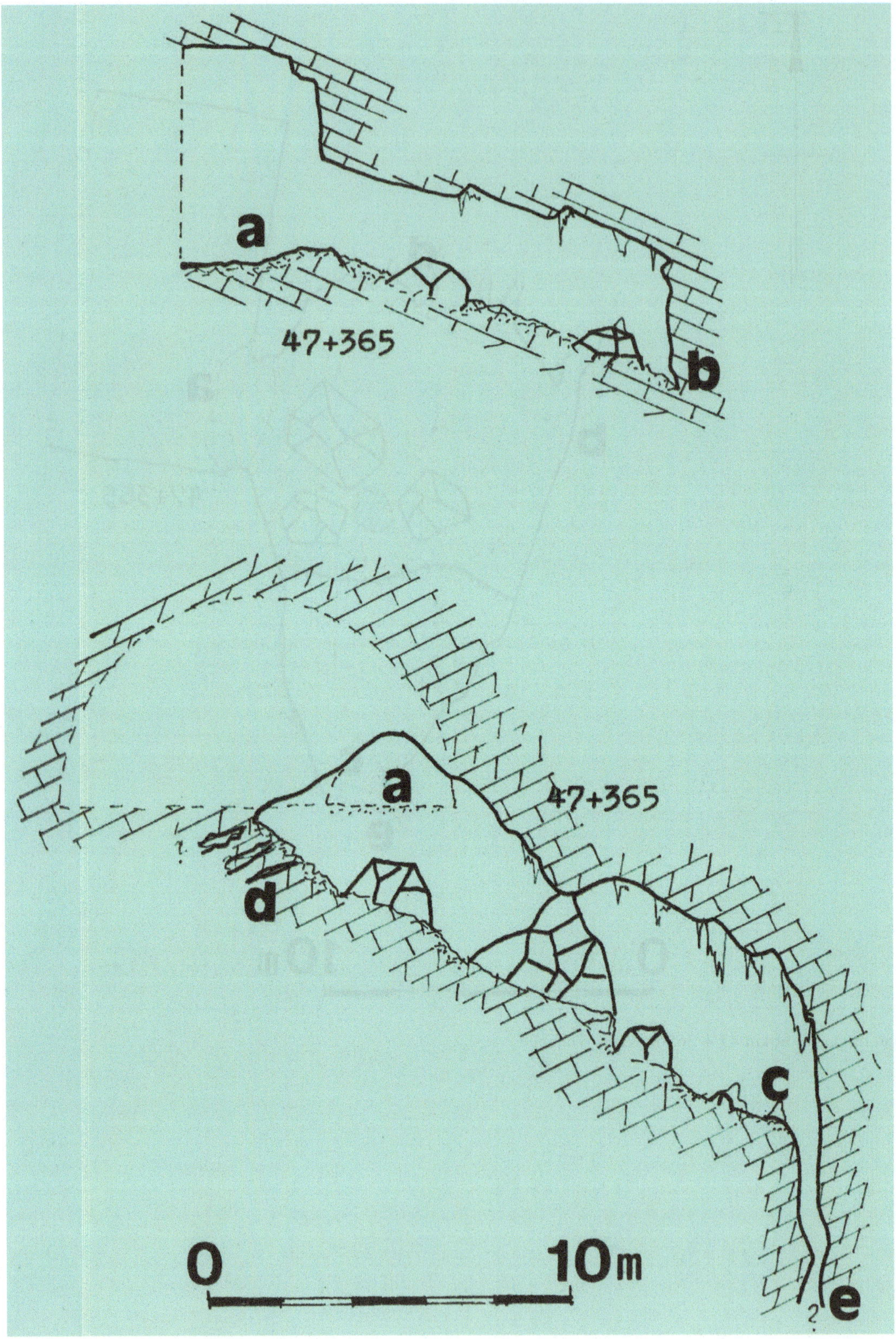

Fig. 6.59 Profile of cavern at km 47 + 365 in Vršek Tunnel

Fig. 6.60 Entrance of cavern at km 47 + 365 in Vršek Tunnel

Fig. 6.61 Speleothems in cavern at km 47 + 365 in Vršek Tunnel

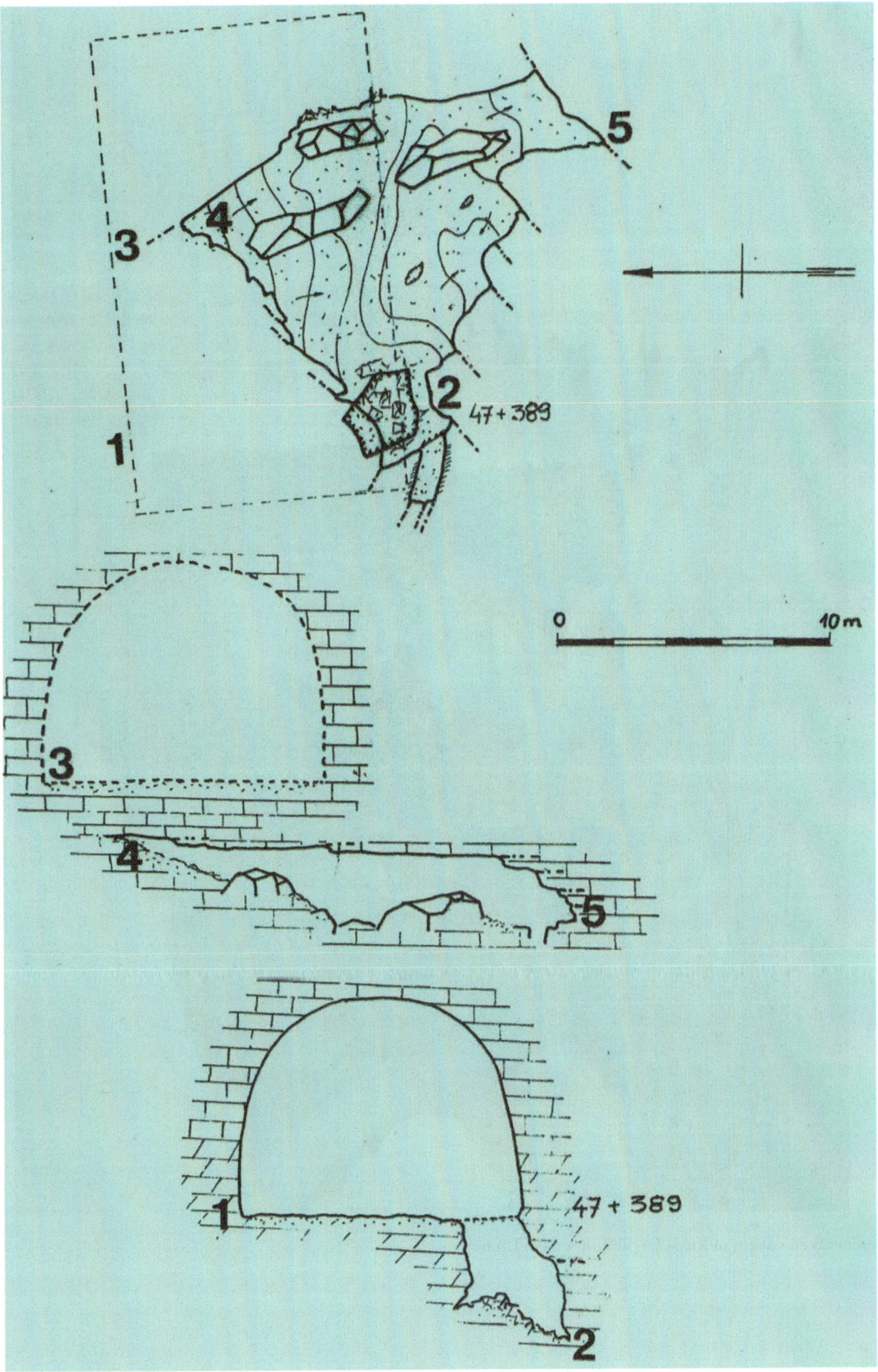

Fig. 6.62 Plan and profile of cavern at km 47 + 389 in Vršek Tunnel

Fig. 6.63 Speleothems in cavern at km 47 + 389 in Vršek Tunnel

Fig. 6.64 Cavern at km 47 + 743 in Vršek Tunnel was generated mostly by tectonic activity

The first bridge, smaller than the aforementioned bridge in the "Vrata" Tunnel, was designed and built in the "Veliki Gložac" Tunnel, which is 1,126 m long and is located on the A-6 motorway (Figs. 6.70 and 6.71). A very deep cavern (over 128 m) appeared in the southern tunnel tube about 250 m from the western portal of the tunnel throughout the profile. Its width at the entrance was between 4 and 7 m. It is bridged with a simple 9 m long bridge so the groundwater in the cavern is allowed to move smoothly (Fig. 6.72). More important caverns are at km 74 + 050, then cavern at km 73 + 638,, km 73 + 302, km 73 + 070, km 73 + 652, km 74 + 057, km 73 + 150, km 73 + 771 (Fig. 6.73), and cavern on km 73 + 167 (Figs. 6.74 and 6.75).

6.8 Čardak Tunnel

The Čardak Tunnel is 601 m long and is located on the A-6 motorway, with two tunnel tubes. Its overburden is maximum of 30 m thick. 22 caverns with predominantly clay infill were discovered. Figures 6.76 and 6.77 are cavern at km 59 + 166, and Figs. 6.78 and 6.79 are cavern on km 59 + 313.

Fig. 6.65 Cavern at km 31 + 292 in Vrata Tunnel

Fig. 6.66 Large cavern in Vrata Tunnel

Fig. 6.67 Dimensions 55 m × 60 m × 40 m chamber in Vrata Tunnel

Fig. 6.68 A 58-m-long bridge was built across the cavern at km 31 + 292 in Vrata Tunnel

Fig. 6.69 Doors in the tunnel lining leading to the cavern

Fig. 6.70 Veliki Gložac Tunnel

Fig. 6.71 Cavern in Veliki Gložac Tunnel

Fig. 6.72 Cavern with the bridge in Veliki Gložac Tunnel

Fig. 6.73 Cavern at km 73 + 711 in Veliki Gložac Tunnel

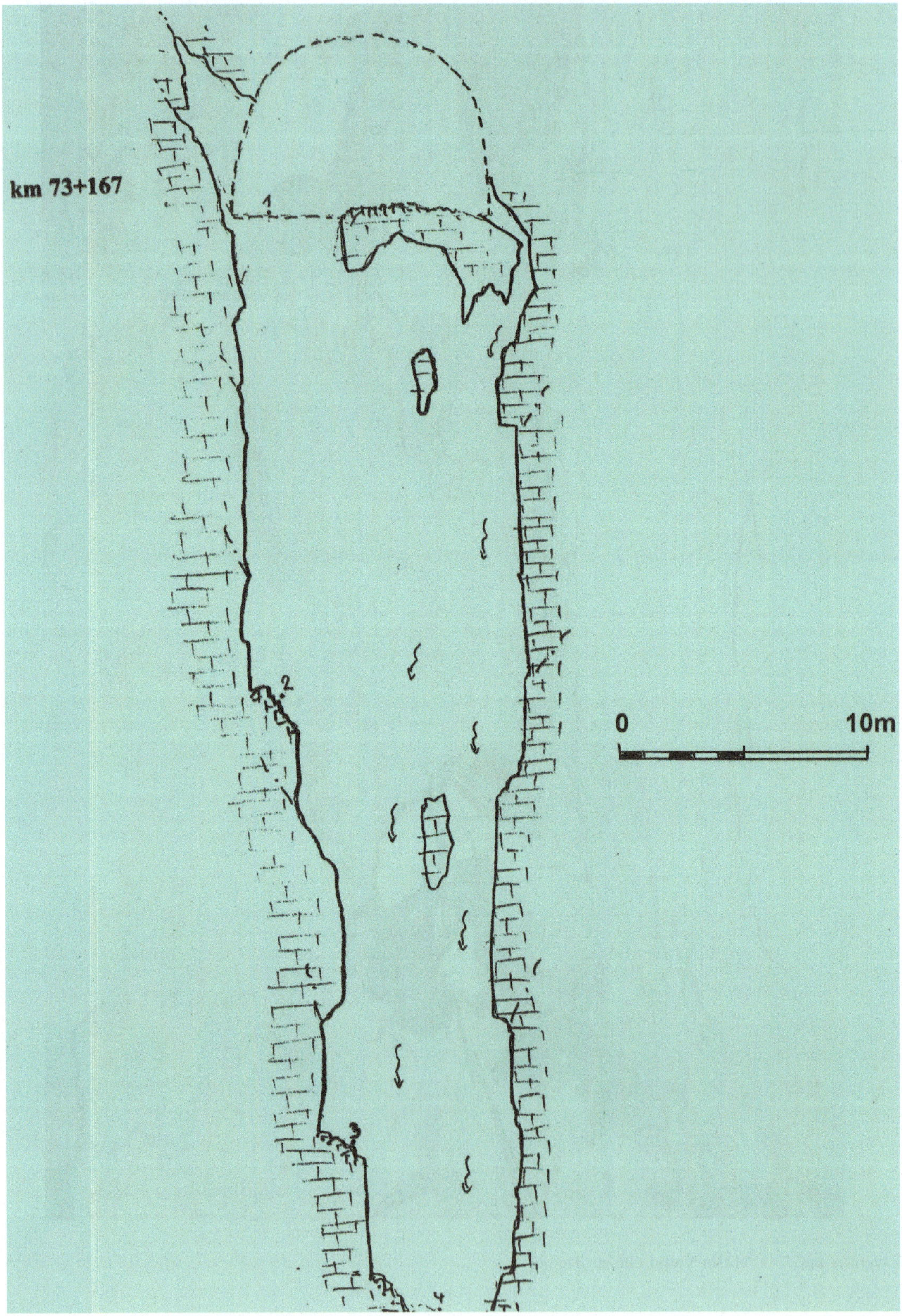

Fig. 6.74 Profile of cavern at km 73 + 167 in Veliki Gložac Tunnel

Fig. 6.75 Cavern at km 73 + 167 in Veliki Gložac Tunnel

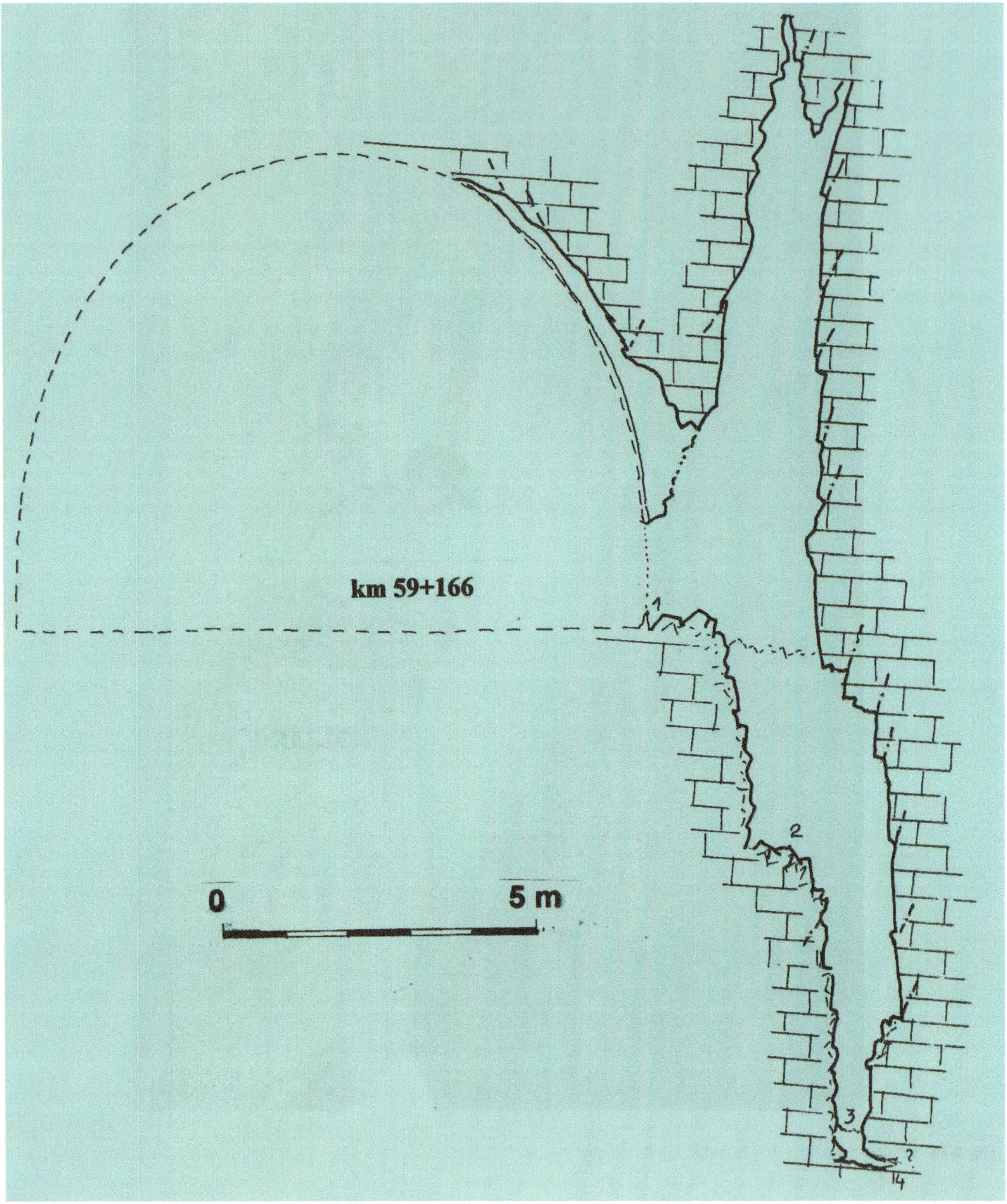

Fig. 6.76 Profile of cavern at km 59 + 166 in Čardak Tunnel

Fig. 6.77 Clay in cavern at km 59 + 166 in Čardak Tunnel

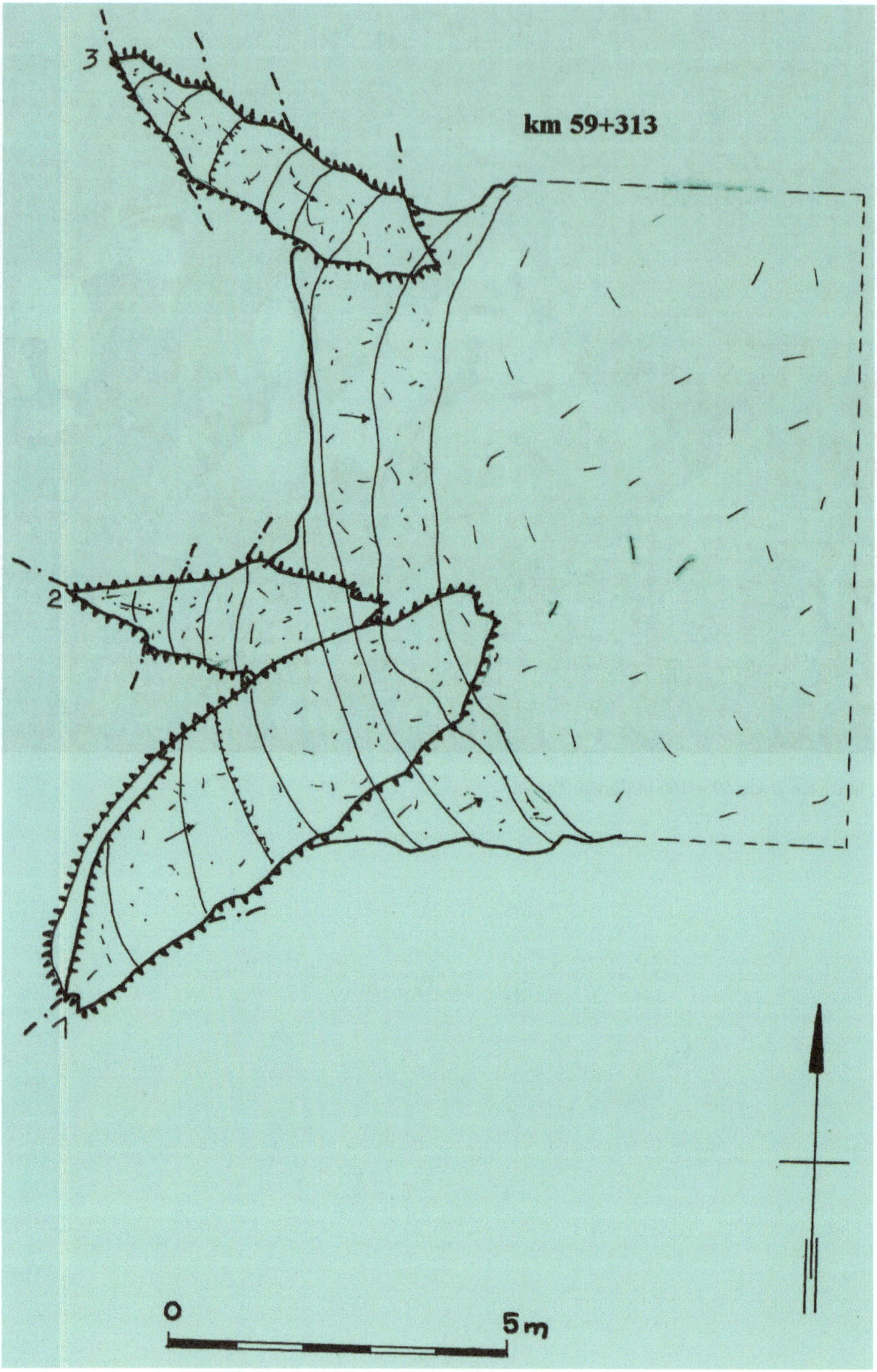

Fig. 6.78 Plan of cavern at km 59 + 313 in Čardak Tunnel

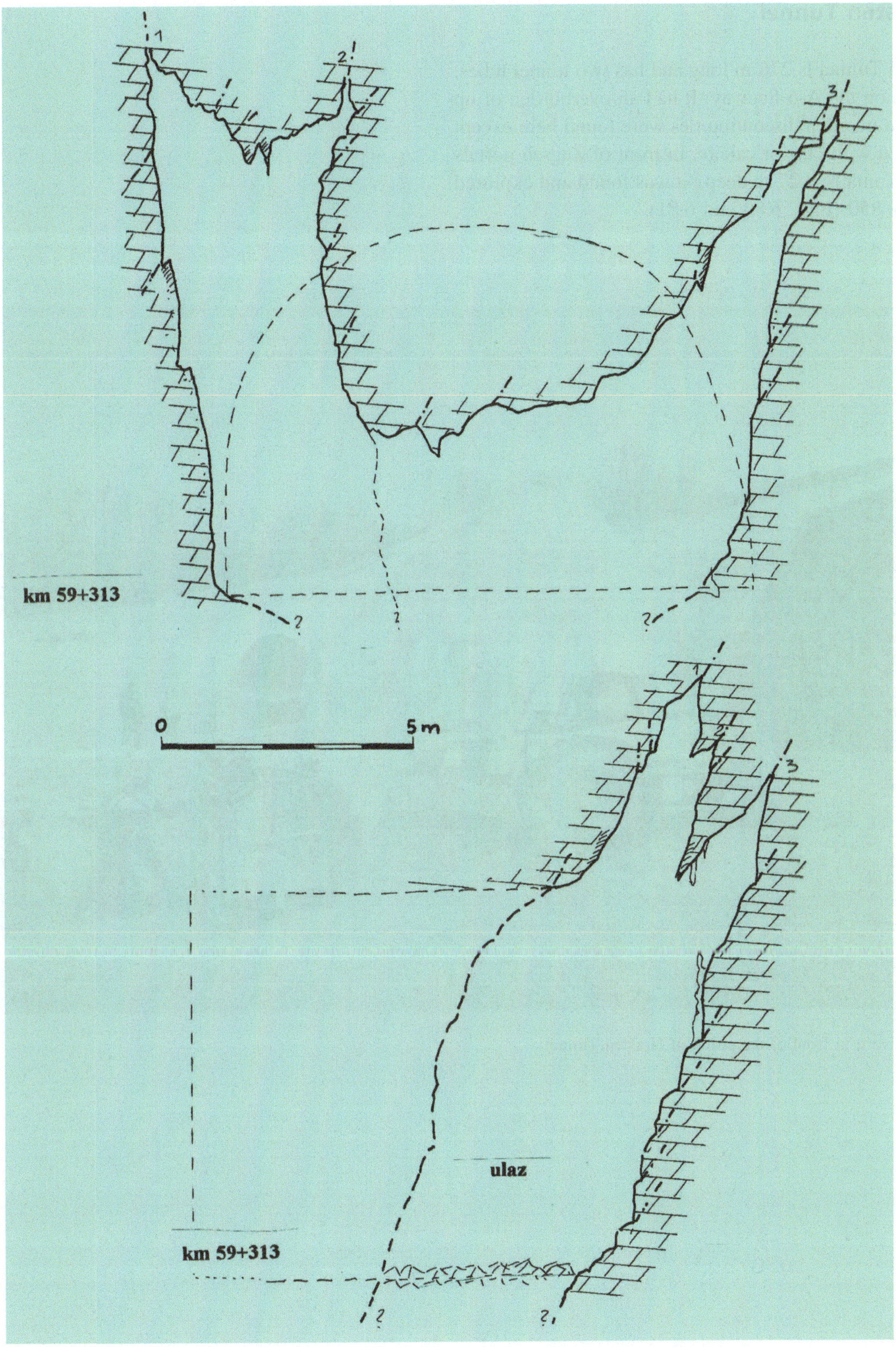

Fig. 6.79 Profile of cavern at km 59 + 313 in Čardak Tunnel

6.9 Hrasten Tunnel

The Hrasten Tunnel is 276 m long and has two tunnel tubes. It is located on the A-6 freeway. It had an overburden of up to 10 m, and no open discontinuities were found here except caverns filled with clay or calcite. In front of Zagreb portals of Hrasten Tunnels, a 27 m deep pit was found and explored on km 24 + 850 (Figs. 6.80 and 6.81).

Fig. 6.80 Cavern in front of the portal of Hrasten Tunnel

Fig. 6.81 Cavern at km 24 + 850 near Hrasten Tunnel

6.10 Lučice Tunnel

The tunnel "Lučice" is 576 m long and has two tunnel tubes. It is located on A-6 in Gorski Kotar near Delnice. The proximity of the railway and the underpass made this construction work particularly difficult. Several caverns were found, the most interesting being in front of the Lučice Tunnel at km 41 + 262 (Fig. 6.82), and 37 m deep and 18 m long, with a very narrow entrance.

Fig. 6.82 Cavern at km 41 + 262 in front of the Lučice Tunnel

7 Caverns Found and Explored During the Construction of Road Tunnels in Lika and Dalmatia Regions and the City of Rijeka

Abstract

Caverns discovered during the construction of road tunnels in the area of Lika, Dalmatia and around the city of Rijeka belong mainly to the Adriatic hydrological basin. However, in the interior of Lika, some caverns also belong to the Black Sea basin. Numerous examples of caverns were found in the tunnels "Sveti Rok", "Mala Kapela", "Brinje", "Plasina", "Grič", "Bristovac", etc., on the A-1 highway and the tunnel "Sveti Ilija" in Dalmatia. In "Pećine" Tunnel in the city of Rijeka, a connection with the Adriatic Sea was found in a cavern. In these areas, there are numerous deep and complicated caves that sometimes have watercourses in their interior. Hundreds of caverns in tunnels in the Dinaric Karst in Croatia were explored and documented.

7.1 Sveti Rok Tunnel

During the construction of the Sveti Rok Tunnel (two tunnel tubes—west and east 5679 and 5670 m), a hundred caverns were found, 48 of which were thoroughly investigated and documented. The others were smaller in size or did not present a problem during construction. All the caverns were repaired so that the flow of groundwater was not disturbed, and in some of them the door was left for further exploration. As in all other tunnels in Croatia, this was done at the explicit request of the speleologists–geologists, and we thank them for that (Ereš et al. 1999).

Caverns in Sveti Rok Tunnel were found in both tunnel tubes. Caverns mentioned here are at km 196 + 741 (Figs. 7.1 and 7.2), km 196 + 890 (Fig. 7.3), km 201 + 126.5 (Figs. 7.4 and 7.5), km 201 + 058 (Figs. 7.6, 7.7, 7.8 and 7.9), km 199 + 820, km 198 + 645 (Figs. 7.10, 7.11 and 7.12), caverns at km 199 + 898 (Figs. 7.13 and 7.14), km 199 + 922, km 198 + 934, km 198 + 950, km 199 + 345, km 199 + 334, km 200 + 525 (Figs. 7.15, 7.16, 7.17, 7.18, 7.19 and 7.20) and cavern km 200 + 515 (Fig. 7.21).

An interesting phenomenon where three caverns are fused into one due to parallel karstification at parallel faults was detected in Sveti Rok Tunnel. Similar occurrences are rarely or in no way visible on the surface of the terrain. Speleogenesis occurs at the same time, but at slightly altered speeds and intensities (Garašić 1995). Multiple caverns 2–6 can be connected, but not necessary if they are not nearby. The thicker the overburden, the greater the possibility of caverns merging (Lugomer et al. 1999). When a cavern is formed on a fault plain (or between layers) whose parallel forms are found nearby in the tunnel, one can always expect more speleological objects within a few metres or several tens of metres.

In cases of large overburdens, for example, in Sveti Ilija Tunnel, when the overburden was between 1000 and 1300 m, the occurrence of parallel speleological objects may be several hundred metres. In the Sveti Ilija Tunnel, the caverns were 222 m apart.

After the 5062–m-long Učka Tunnel, which was opened for traffic in 1981, Sveti Rok Tunnel was the longest road

M. Garasic, *The Dinaric Karst System of Croatia*, Cave and Karst Systems of the World,
https://doi.org/10.1007/978-3-030-80587-6_7

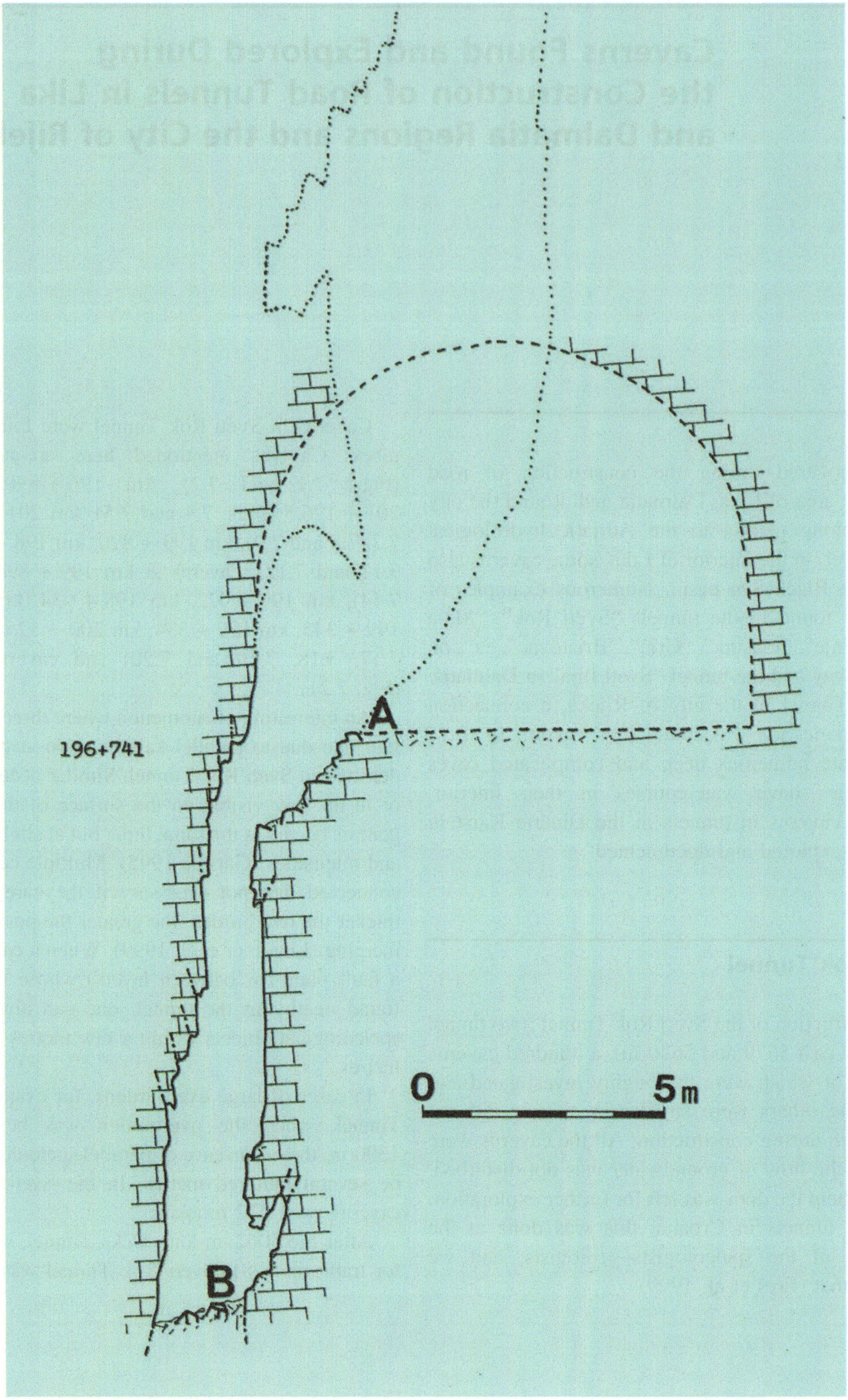

Fig. 7.1 Profile of cavern at km 196 + 741 in Sveti Rok Tunnel

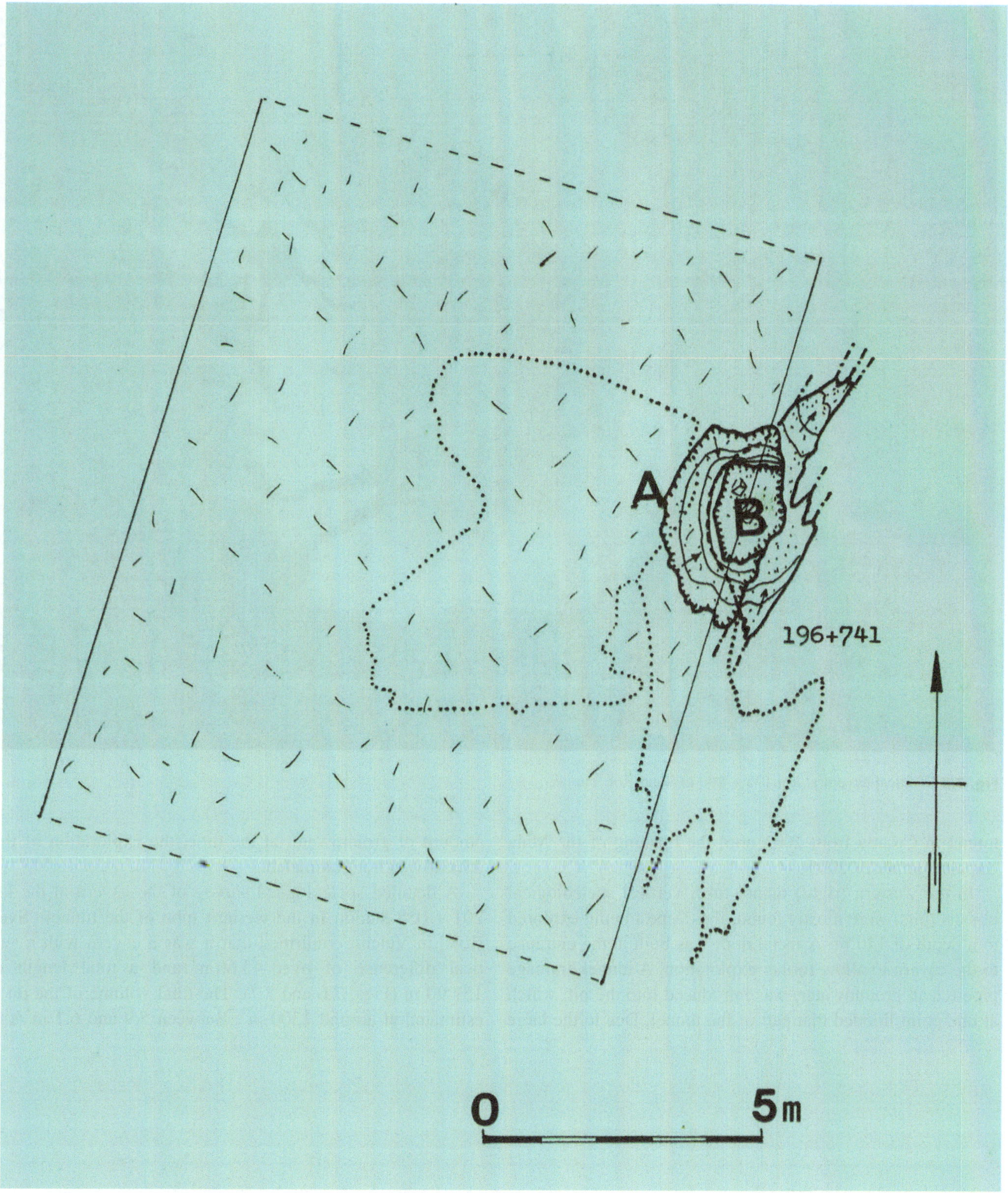

Fig. 7.2 Plan of cavern at km 196 + 741 in Sveti Rok Tunnel

Fig. 7.3 Vertical cavern at km 196 + 890 in Sveti Rok Tunnel

tunnel in Croatia from 2003 until the opening of the Mala Kapela Tunnel in 2005.

In the western (right) tunnel tube, vertical speleological objects (pits) were mostly found, the deepest being explored to a depth of 140 m. A metal door was built at the entrance to the cavern to allow further exploration. A tunnel drainage pipe (clean groundwater) was introduced into the pit, which at one point flooded that part of the tunnel. Due to the large amount of groundwater at the time, the exploration of this cavern was not completed.

A detailed speleological survey of the cavern at the km 201 + 058 station in the western tube of the tunnel "Sveti Rok" in Velebit confirmed that it was a cavern with a vertical difference of over 43.60 m and a total length of 157.90 m (Figs. 7.6 and 7.7). The total volume of the pit is estimated at around 1300 m^3. Between 5.9 and 6.1 m^3/s of

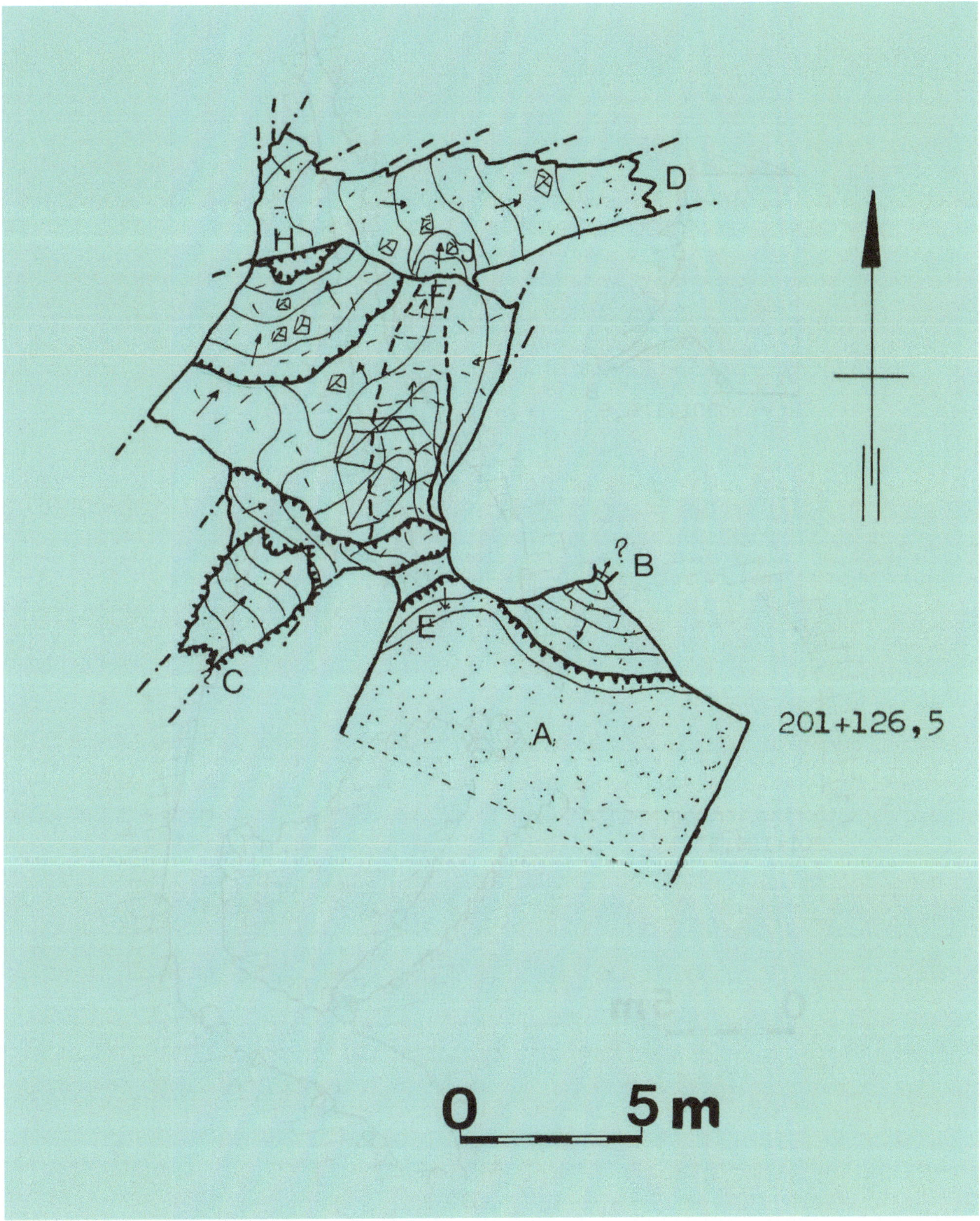

Fig. 7.4 Plan of cavern at km 201 + 126 in Sveti Rok Tunnel

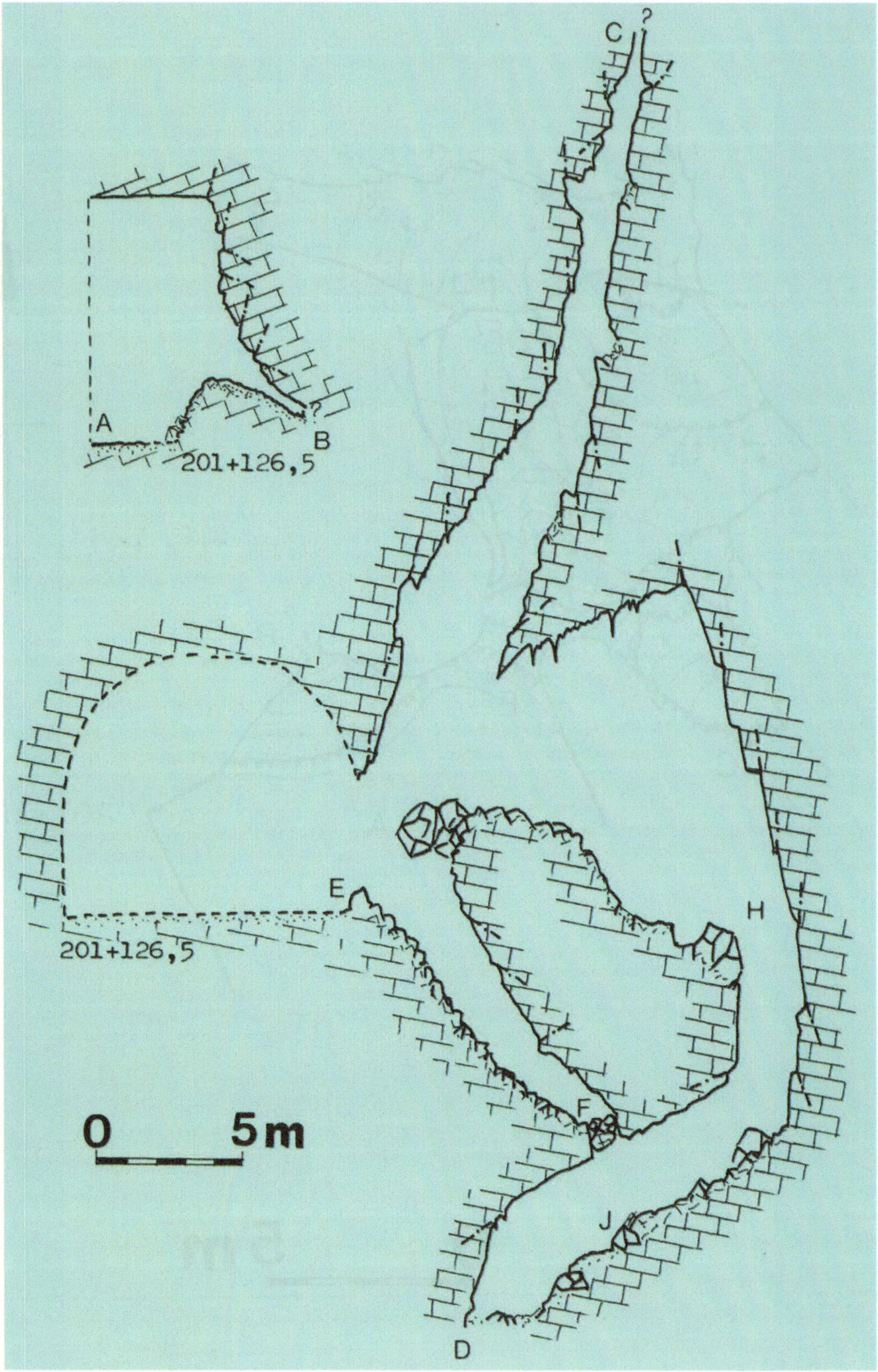

Fig. 7.5 Profile of cavern at km 201 + 126 in Sveti Rok Tunnel

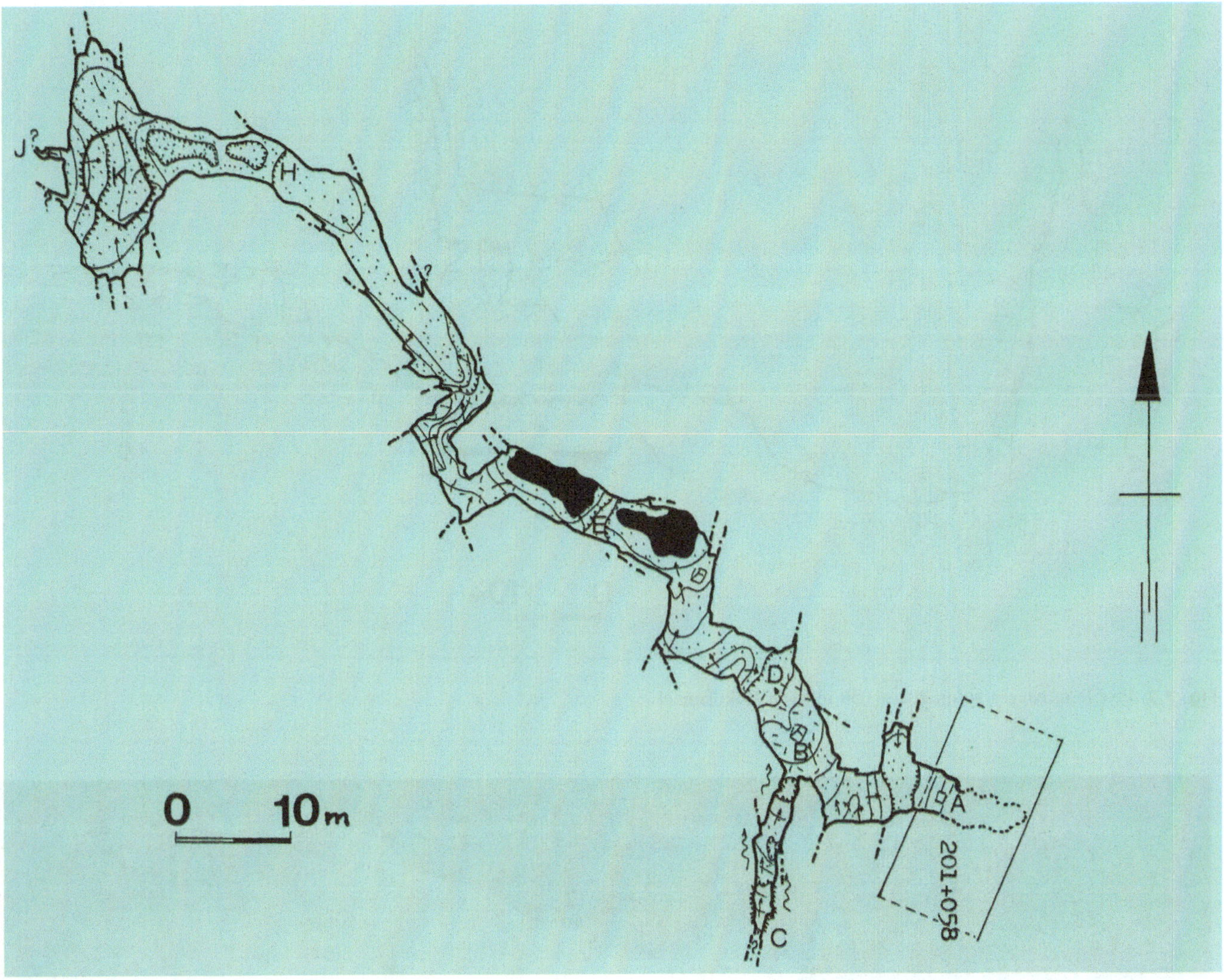

Fig. 7.6 Plan of cavern at km 201 + 058 in Sveti Rok Tunnel

air flowed from the cavern to the tunnel at the time of the survey on 26 June 1998, with an average velocity of 5–6 m/s, as measured by a digital anemometer. Airflow existed inside the cavern even before it was connected to the tunnel, but in the lower parts of the cavern. Anemolites, that is, speleothems that are bent under the influence of the wind during their genesis have been detected there. Bats have been spotted in the middle parts of the cavern, which proves beyond doubt that the cavern is connected to the surface through so far unknown channels. Overburden in this area is about 500 m thick.

Occasional active water flow (during heavy rainfall) was observed. In the upper parts of the cavern along with waterfalls, the so-called cave corals are deposited. In the middle parts of the cavern, two lakes were found with a total length of about twenty metres, up to 1.2 m deep, where botroids were deposited. Botroids or clustered speleothems crystallize only in still cave water, which means that there is no turbulent run-off in this part of the cavern. The end of the cavern is characterized by an occasional outlet syphon completely covered with clay materials. The water in these parts rises to a maximum of 8.0 m. The cavern is corroded especially in the upper parts. Speleothems occur especially in the upper and middle portions of the cavern, with water dripping through them. In the lower part of the cavern (Final Hall), there is a thicker layer (min 2.0–6.0 m) of clay material, below which the rock material is probably located and the so-called false bottom, which is sedimented here by a long process of karstification. High-plastic brown clay is present in the lowest parts of the cavern. The cavern is generally developed in a northwest–southeast direction, which corresponds to a fault orientation that played a significant role in its genesis. In addition to this fault, a significant north–south fault and a fracture system that is almost

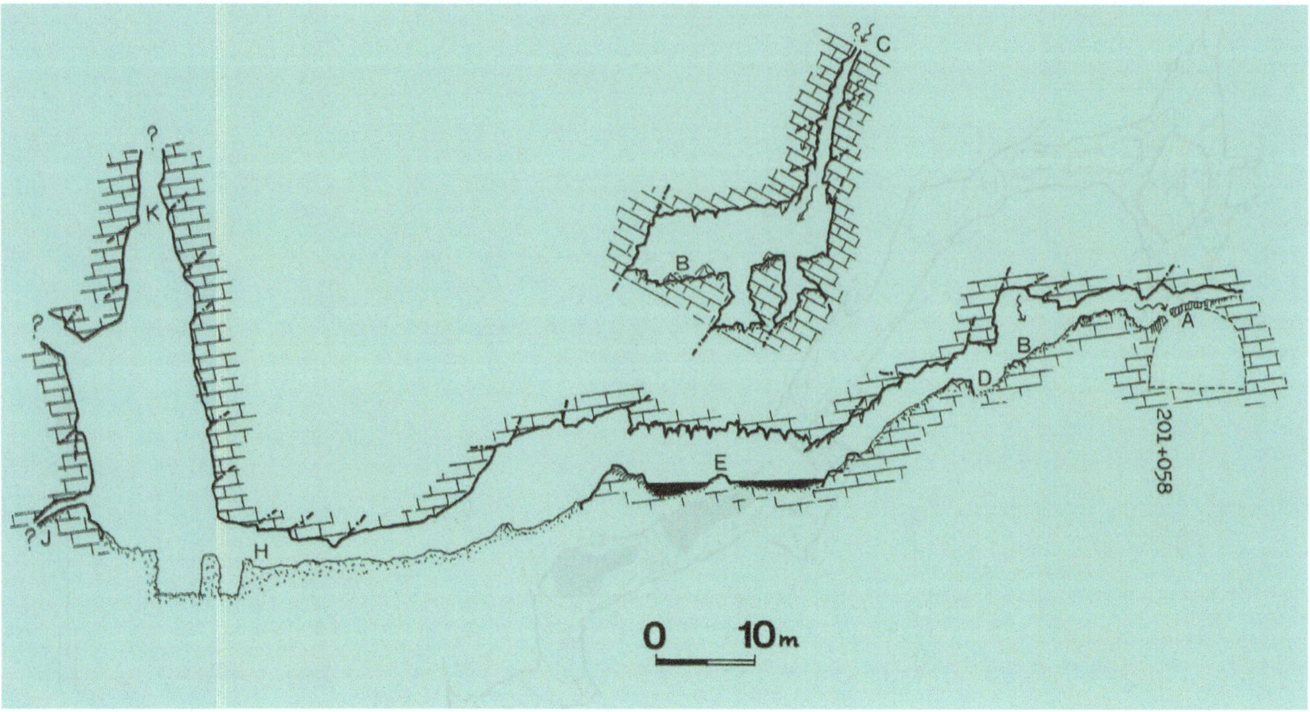

Fig. 7.7 Profile of cavern at km 201 + 058 in Sveti Rok Tunnel

Fig. 7.8 Speleothems in cavern at km 201 + 058 in Sveti Rok Tunnel

Fig. 7.9 Cavern at km 201 + 058 in Sveti Rok Tunnel

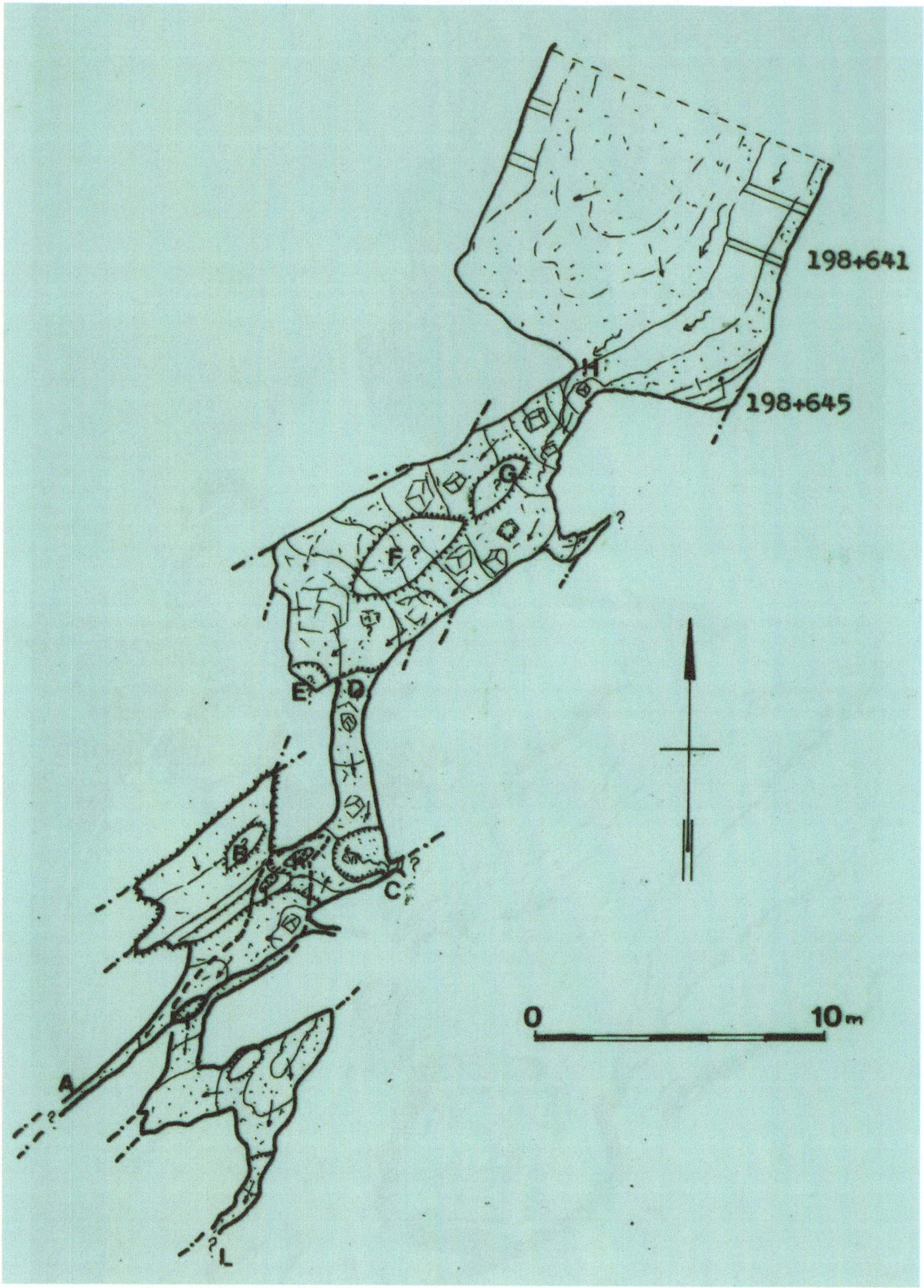

Fig. 7.10 Plan of cavern at km 198 + 645 in Sveti Rok Tunnel

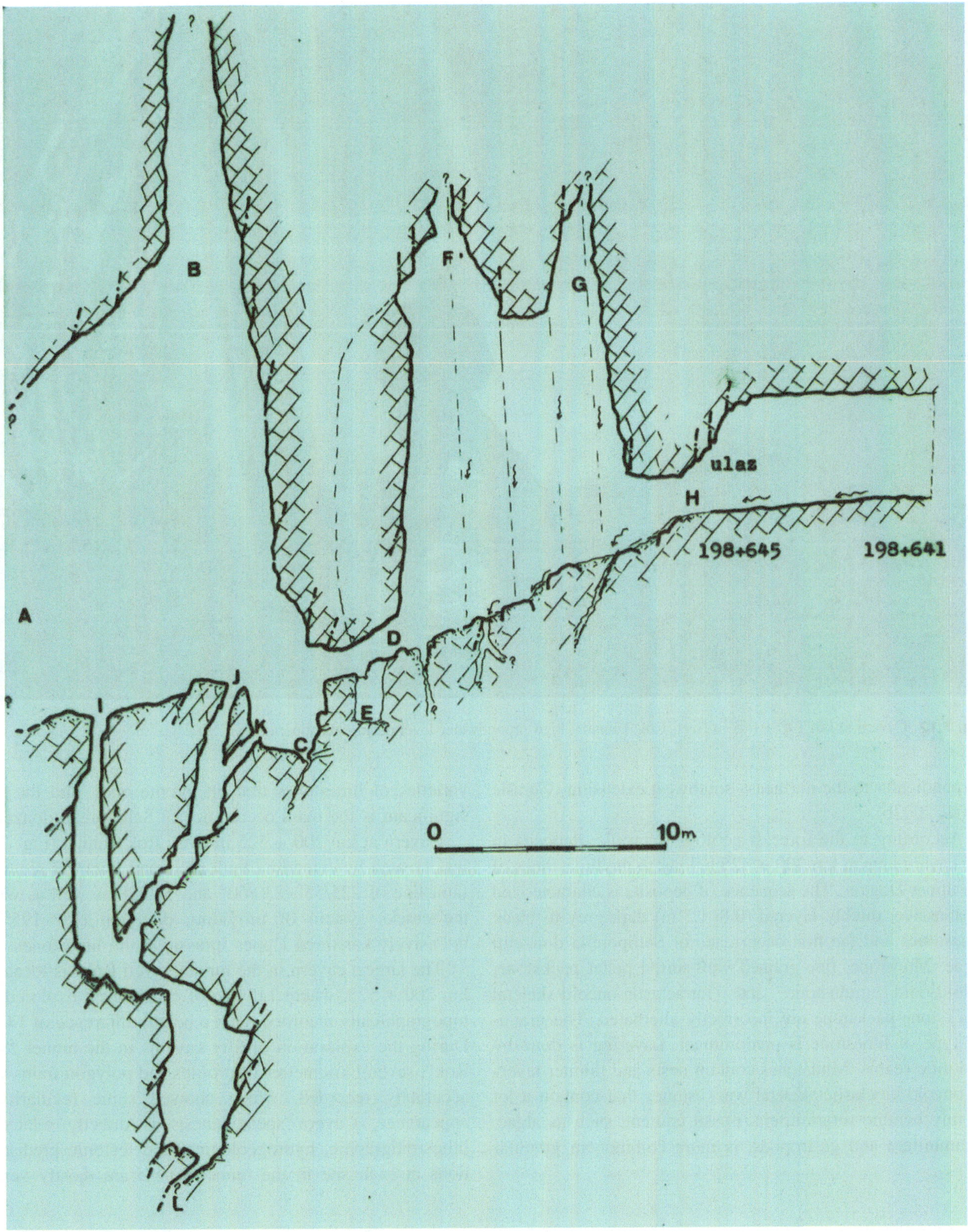

Fig. 7.11 Profile of cavern at km 198 + 645 in Sveti Rok Tunnel

Fig. 7.12 Cavern at km 198 + 645 in Sveti Rok Tunnel—high groundwater level during rainy periods

perpendicular to the northeast–southwest extension (Garašić 2004, 2007b).

According to the forecast geological profile, deposits in the cavern area at km 200 + 525 belong to the limestones of the upper Dogger. The sequence of deposits is characterized by massive, thickly layered (0.8–1.5 m) dark grey to black limestones and the first occurrence of Selliporella donzellii algae. Mudstone, fine-grained well-sorted pellet packstone, peloid-ooid grainstone and intraclastic-oncoid-skeletal weckstone-packstone are incorrectly alternated. The granular type of limestone is predominant. Layering is done by stylolitic seams. Mudstones contain nests and thinner layers of oncoid-bioclastic-skeletal wackestones that contain a lot of tiny benthic foraminifera. Fossil content, such as algae, foraminifera and gastropods, is more common in granular varieties of limestones than in micrite ones, and the most significant is the mass occurrence of Selliporella donzellii.

Cavern at km 200 + 525 in Sveti Rok Tunnel (Fig. 7.15) was created by extending fault plain and fractures along the fault line of 225/55°–230/60° and 140/88 as well as within the cracks system of providing direction (15°–195°) in intensively karstified Upper Jurassic Malm limestones.

The largest cavern in the tunnel "Sveti Rok" is located at km 200 + 525, where 1137 m of canals were explored and topographically recorded, with a height difference of 147 m. During the exploration of fifty caverns in the tunnel "Sveti Rok", several kilometres of profiles and polygon trains were accurately recorded, which showed some regularity in appearance. Cavern speleogenesis is directly related to lithostratigraphic, hydrogeological and tectonic predispositions at each site in the tunnel. These are mostly vertical

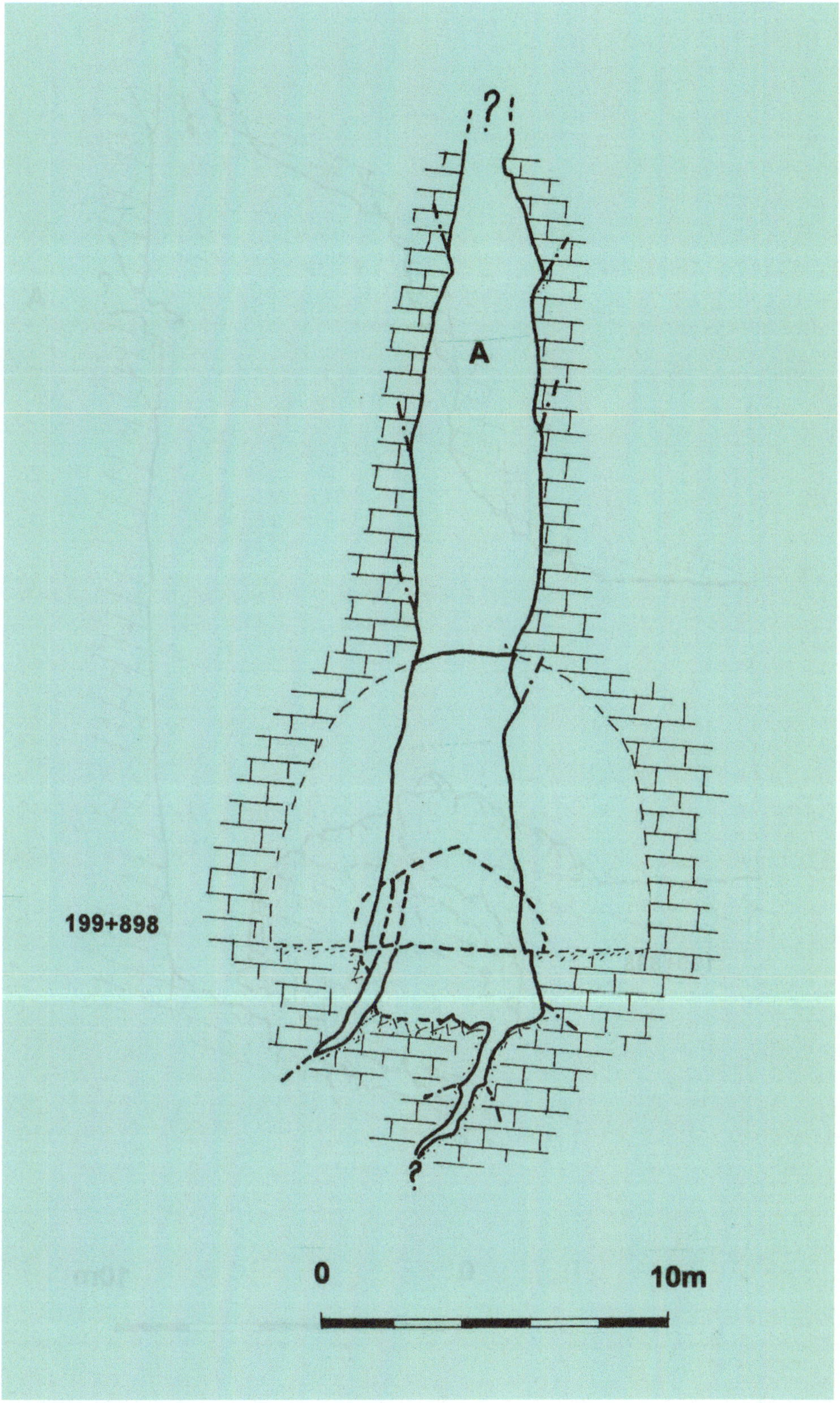

Fig. 7.13 Profile of cavern at km 199 + 898 in Sveti Rok Tunnel

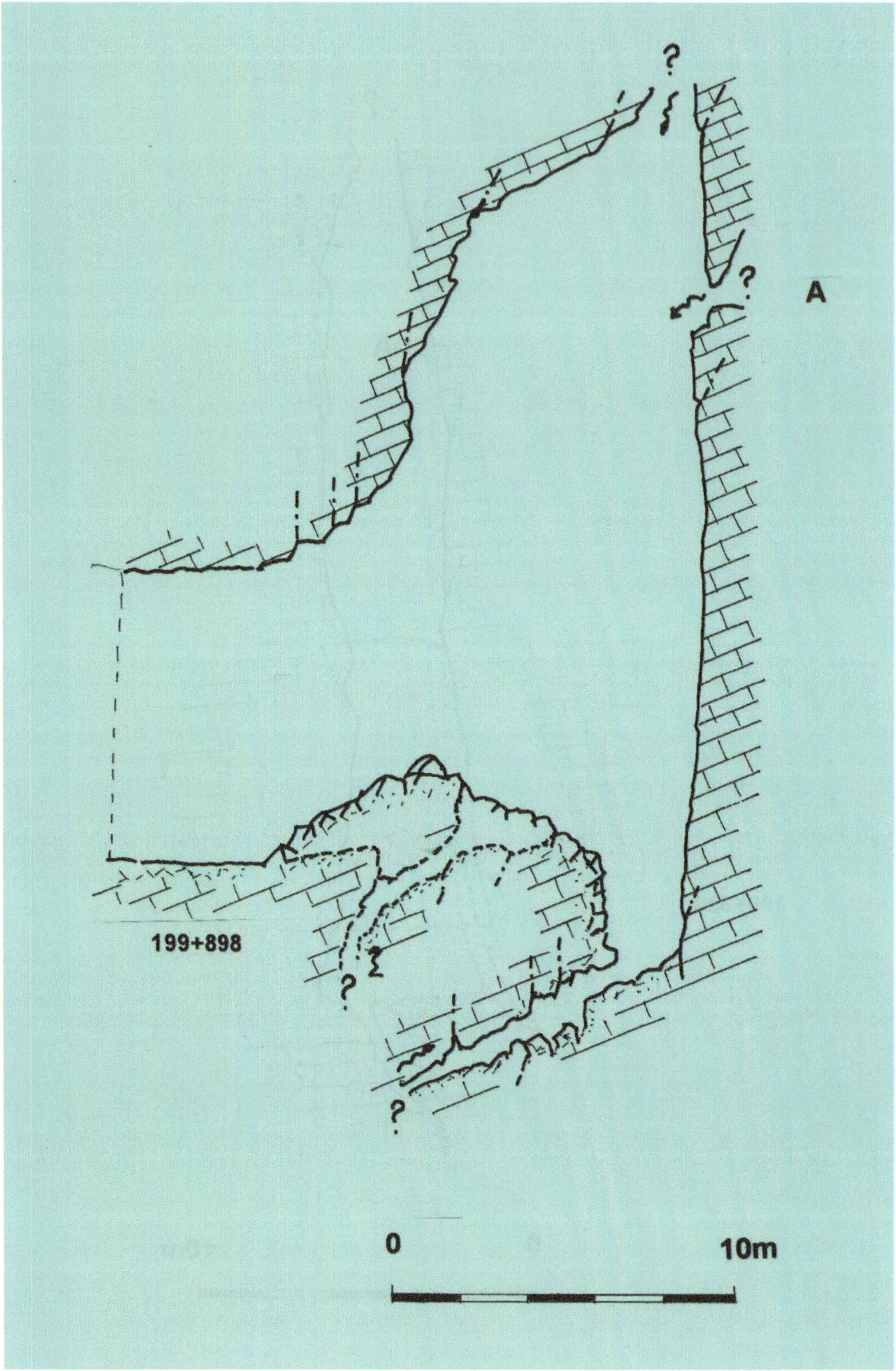

Fig. 7.14 Profile 2 of cavern at km 199 + 898 in Sveti Rok Tunnel

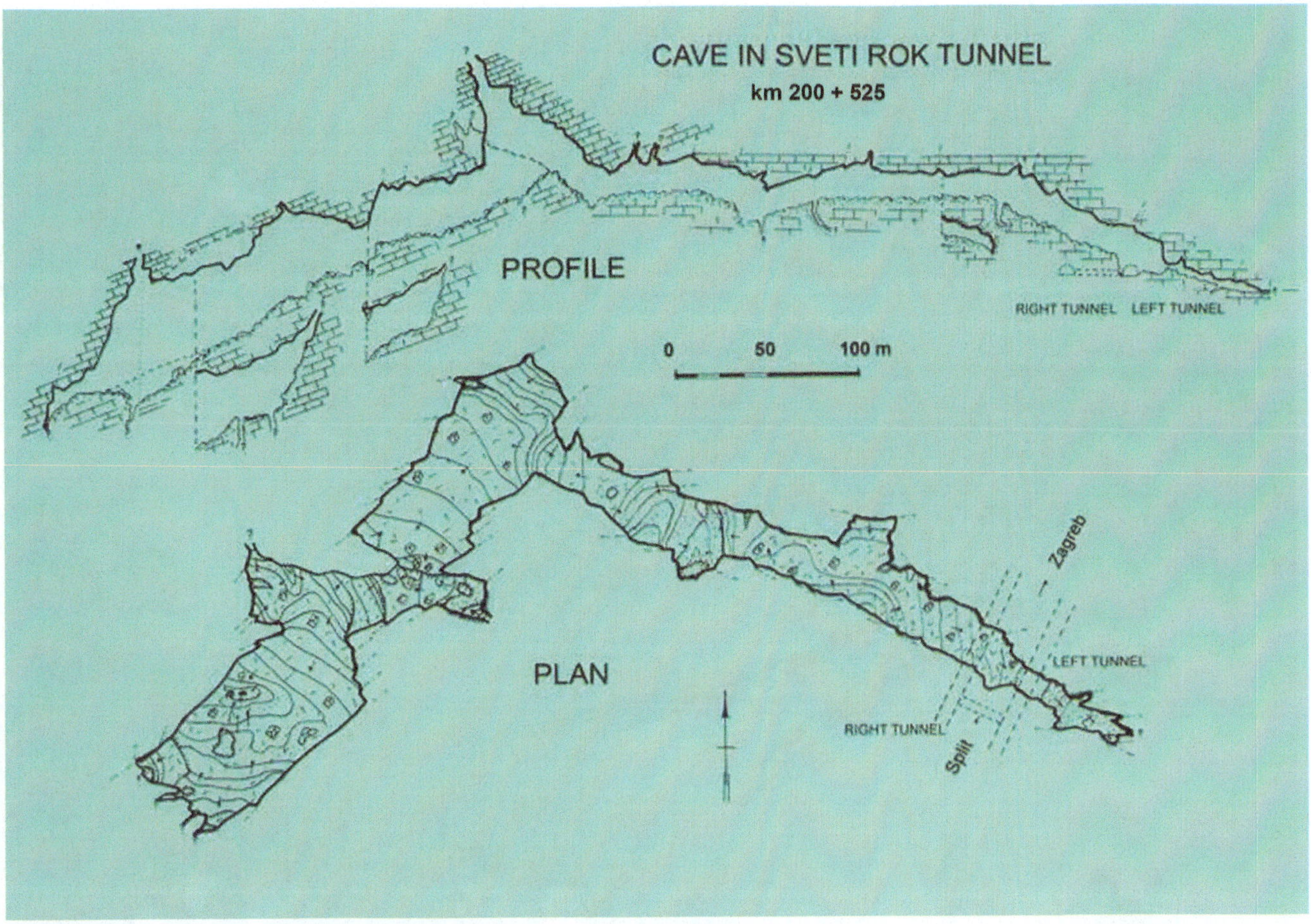

Fig. 7.15 Profile and plan of cavern at km 200 + 525 in Sveti Rok Tunnel

speleological objects (pits), of knee or columnar morphological type, with hydrogeological function of intermittent or permanent springs. The only real exception is the cavern at km 200 + 525, where the hall measuring 148 × 53 × 62 m was explored (Fig. 7.20). Research continues.

7.2 Mala Kapela Tunnel

Tunnel Mala Kapela is the longest road tunnel in Croatia, constructed in two tunnel tubes. The tunnel tubes are not equally long—the northbound tube is 5821 m long, while the southbound one is somewhat shorter, as it is 5780 m long. The tunnel tubes are excavated 25 m (82 ft) m apart and are linked by six vehicle passages and 14 pedestrian passages. In tunnels, many caves were explored. Here it will be mentioned cavern on km 14 + 229 (Figs. 7.22 and 7.23). In Mala Kapela Tunnel, seven caves were explored: at km 13 + 955 (Fig. 7.24), caverns at km 13 + 911 (Figs. 7.25, 7.26 and 7.27) and cavern at km 14 + 229 (Figs. 7.28, 7.29, 7.30 and 7.31).

The rocks in which the cavern is formed belong to the Lower Cretaceous Neocomian limestones, dolomitic limestones and dolomites K_1^{1-3}, whose layer thickness varies

Fig. 7.16 Cavern at km 200 + 525 in Sveti Rok Tunnel (the first contact with cavern)

Fig. 7.17 Main channel in cavern at km 200 + 525 in Sveti Rok Tunnel

Fig. 7.18 Speleothems in cavern at km 200 + 525 in Sveti Rok Tunnel

Fig. 7.19 Cavern at km 200 + 525 in Sveti Rok Tunnel

Fig. 7.20 Final chamber in cavern at km 200 + 525 in Sveti Rok Tunnel

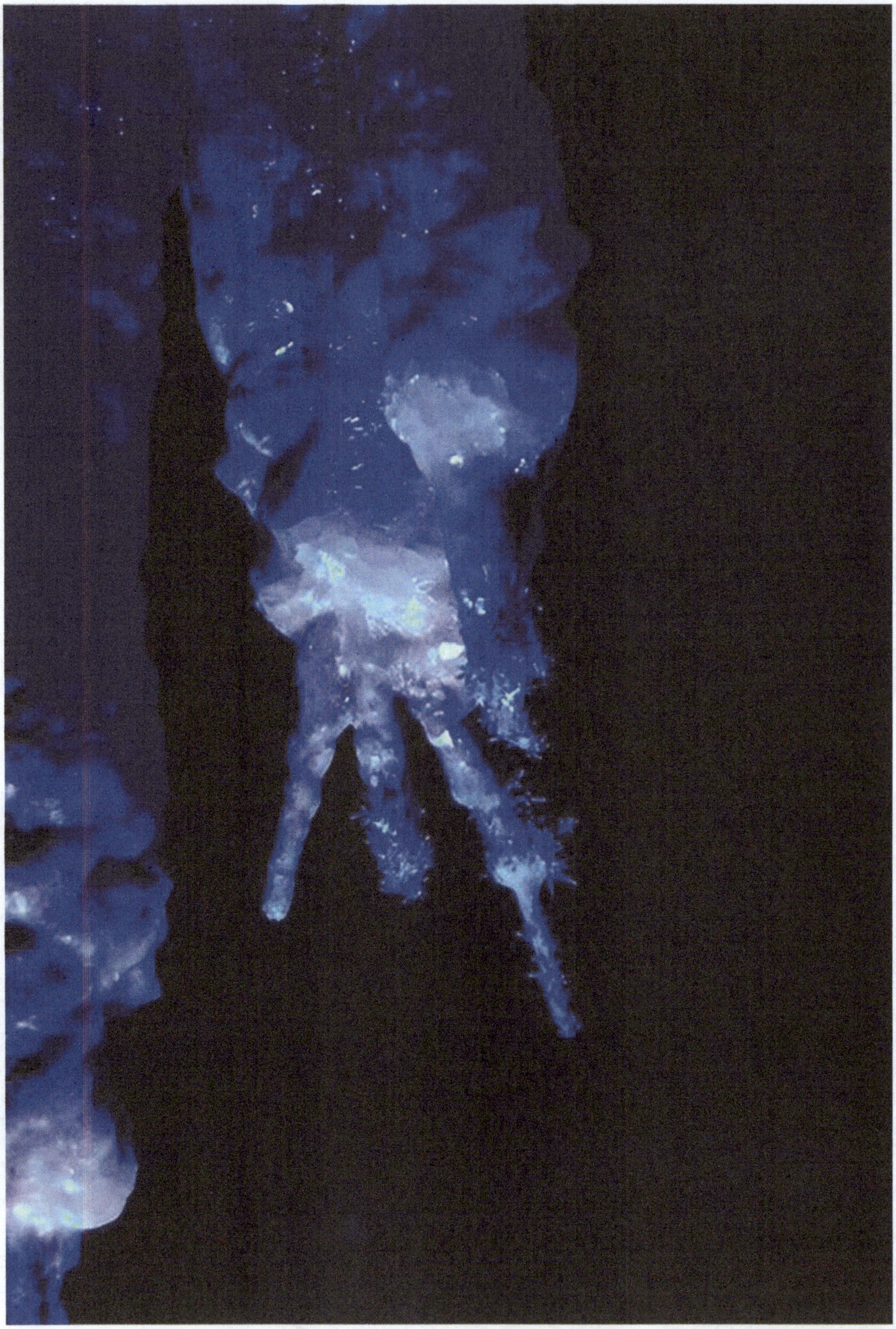

Fig. 7.21 Blue fingers—speleothems from cavern at km 200 + 525 in Sveti Rok Tunnel

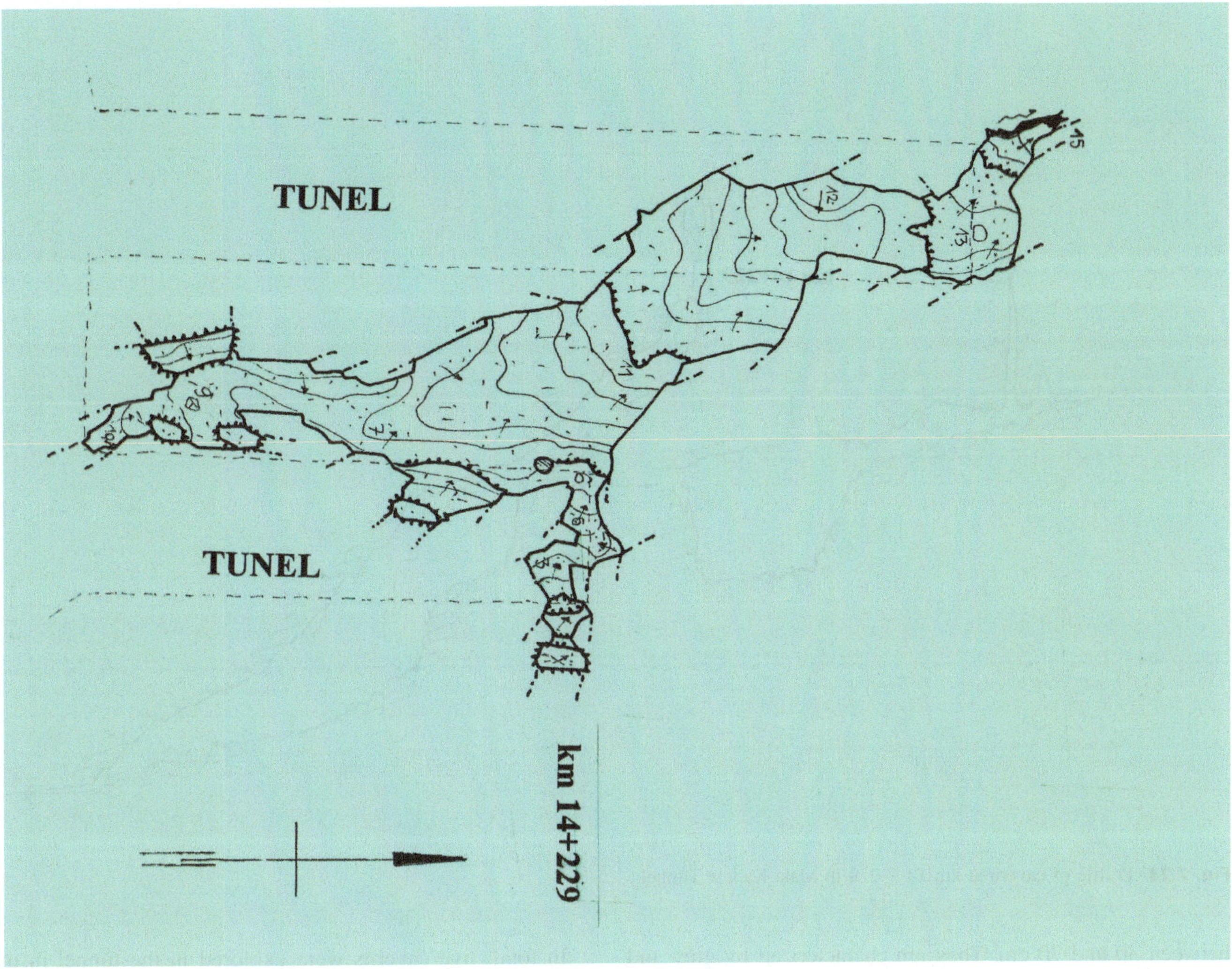

Fig. 7.22 Plan of cavern at km 14 + 229 in Mala Kapela Tunnel

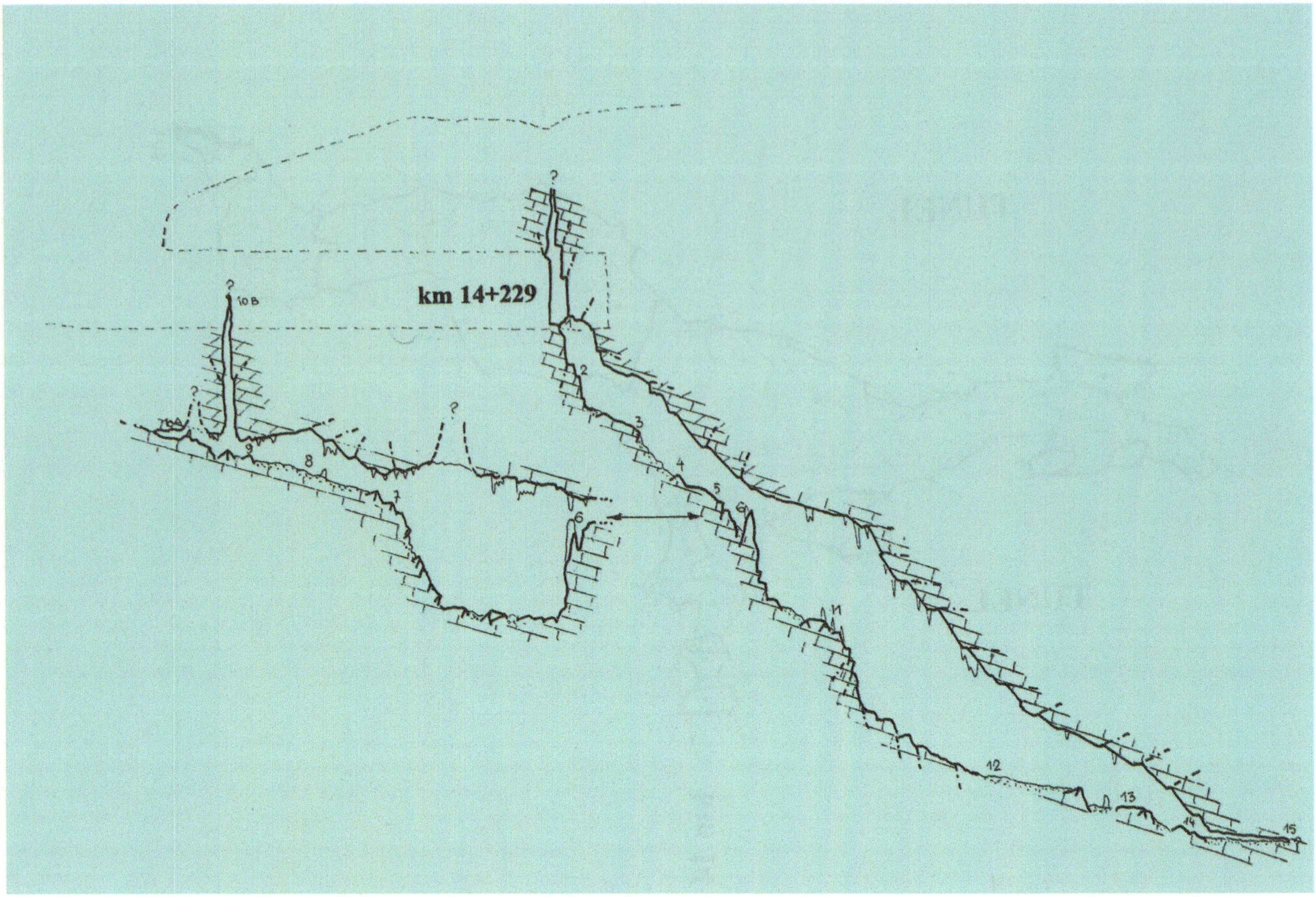

Fig. 7.23 Profile of cavern at km 14 + 229 in Mala Kapela Tunnel

between 30 and 70 cm. They are characterized by grey and light grey colour and good layering.

The Neocomian deposits are located north and northwest of Jezerane, where they continuously lie on the dolomites of upper Malm as well as in the area northwest of Razvala. The Neocomian deposits are limestones, while at the bottom the alteration of limestones and dolomites is only a few metres thick. Limestones are of structural-type mudstone, wackestone containing foraminifera and paxton-grainstone with favreins. The granular structure of the limestones over the muddy mudstone limestones prevails. In granular limestones, the particles are densely packed and in contact with each other. In addition to favrein and foraminifera, we find ostracodes and green and blue-green algae. Grainy limestones often lack a silt base, and the particles are bonded with sparite cement (grainstone). The deposits are well layered. The thickness of the layers varies from 0.30 to 0.70 m, and the colour is grey–light grey.

In total, five caverns were explored in the tunnel in the first few hundred metres of the tunnel on the south side. The interesting thing is that the total length of these explored caverns was greater than the total length of the tunnel excavated up to that point. Occurrence of large caverns was expected in the tunnel "Mala Kapela", mostly with groundwater.

Some interesting and important measurements were done there. The microgravimetry method was used for the first time in tunnels from 2006 to 2008 when exploring caverns in the tunnel "Mala Kapela". It helped finding new caverns and proving the existence of already know ones (Markovinović and Garašić 2010).

For the purpose of defining the points at which gravimetric measurements were performed in the tunnel itself, the points were stabilized using a steel bolt. A total of 71 gravimetric points were set. The points included the left and right sides of the tunnel, and the raster of points was 5 m.

Fig. 7.24 Upper part of cavern at km 13 + 955 in Mala Kapela Tunnel

Fig. 7.25 Cavern at km 13 + 911 in Mala Kapela Tunnel

Fig. 7.26 Speleothems in cavern at km 13 + 911 in Mala Kapela Tunnel

Fig. 7.27 Lower part of cavern at km 13 + 911 in Mala Kapela Tunnel

Fig. 7.28 Speleothems in cavern at km 14 + 229 in Mala Kapela Tunnel

Fig. 7.29 Lower part of cavern at km 14 + 229 in Mala Kapela Tunnel

Fig. 7.30 Upper part of cavern at km 14 + 229 in Mala Kapela Tunnel

The total observation area covered the first 400 m from the entrance to the tunnel on the south side.

For the first time, independent speleological and gravimetric surveys have been conducted on the territory of the Republic of Croatia. Previous speleological research has resulted in finding and defining a number of speleological objects in the Mala Kapela Tunnel.

Precise relative gravimetric measurements were made at the initial 400 m of the southern entrance to the Mala Kapela Tunnel. The measurements and subsequent calculations of Bouguer force acceleration anomalies more difficultly detected the locations of caverns and subsurface structures. The results of the conducted research indicate excellent integration of these two research methods. Unification and joint interpretation make it possible to simplify and accelerate the research process, as well as its economic cost-effectiveness and the advantage of gravimetric research before using other geotechnical methods. This is an example to investors of how investing in science and new technologies brings concrete results for economic purposes. It is important to note that the gravimetric method is one of the most effective means of mapping subsurface geology and is used in regional geological exploration, oil and gas exploration, minerals and archaeological research (Markovinović and Garašić 2010).

7.3 Bristovac Tunnel

The Bristovac Tunnel is 688 m long and has two tunnel tubes. Interestingly, there were three caverns that connected both tunnels. Very nice speleothems are in cavern at km 207 + 425 (Figs. 7.32, 7.33, 7.34 and 7.35).

7.4 Plasina Tunnel

The Plasina Tunnel, on the A-1 highway above Gacko Polje in Lika, is 2300 m long. It has two tunnel tubes. Over 40 caverns were found, some of which were filled with clay. In some of the caverns, the presence of groundwater was discovered. This is very important because of proximity to Gacka River and the possible dangers of flooding. All the

Fig. 7.31 Upper part of cavern at km 14 + 229 in Mala Kapela Tunnel

Fig. 7.32 Cavern at km 207 + 425 in Bristovac Tunnel

Fig. 7.33 Horizontal part of cavern at km 207 + 425 in Bristovac Tunnel

caves and water are protected for future research. Caverns will be mentioned here at km 22 + 035, km 20 + 630, km 19 + 939 (Fig. 7.36), km 20 + 470 (Fig. 7.37) and km 20 + 578 (Fig. 7.38).

7.5 Grič Tunnel

The Grič Tunnel, on the A-1 motorway near Ličko Lešće in Lika, is a two tube, 1244-m-long tunnel in which 34 caverns were found. The deepest cavern (−79 m) is at km 0 + 634 (Figs. 7.39 and 7.40).

7.6 Brinje Tunnel

The "Brinje" Tunnel on the A-1 motorway near Brinje in Lika is 1561 m long, with two tunnel tubes. The most caverns (over 70) discovered in a single tunnel were found in tunnel "Brinje". Caverns will be mentioned here at km 13 + 850 (Figs. 7.41 and 7.42), km 14 + 065, km 14 + 027 (Figs. 7.43 and 7.44), km 14 + 184, km 14 + 300 (Fig. 7.45) and cavern at km 14 + 964 (Fig. 7.46).

The tunnel was excavated in full profile, and the thickness of the overburden was 43 m. When the tunnel entered the zone of dusty clay of medium to high plasticity, the contractor continued the excavation in full profile and the tunnel collapsed all the way to the surface of the terrain (Figs. 7.47, 7.48 and 7.49). The contractor tried to continue the work through the collapsed material, but failed. After that, about 3,200 m^3 of destroyed material was excavated, creating a chimney all the way to the surface. The work continued after stabilization of the excavation face with spray concrete and self-drilling anchors of 9 and 12 m length.

7.7 Sveti Ilija Tunnel

Sveti Ilija Tunnel is located in Dalmatia, near A-1 highway on road from Zagvozd to Baška voda. The main tunnel tube is 4249 m long, while the service tube is 4255 m long.

A detailed speleological investigation of the newly opened caverns, vertical speleological objects (pits) was conducted. Pits were situated in the so-called service tube of the Sveti Ilija Tunnel at km 1 + 415 (Fig. 7.50), km 1 + 193 and km 1 + 637 (Figs. 7.51 and 7.52) on the route of the

Fig. 7.34 Vertical part of cavern at km 207 + 425 in Bristovac Tunnel

Fig. 7.35 Speleothems in cavern at km 207 + 425 in Bristovac Tunnel

Fig. 7.36 Cavern at km 19 + 939 in Plasina Tunnel

Fig. 7.37 Cavern at km 20 + 470 in Plasina Tunnel

Fig. 7.38 Cavern at km 20 + 578 in Plasina Tunnel

Fig. 7.39 Cavern at km 0 + 634 in Grič Tunnel

Fig. 7.40 Descending to cavern at km 0 + 634 in Grič Tunnel

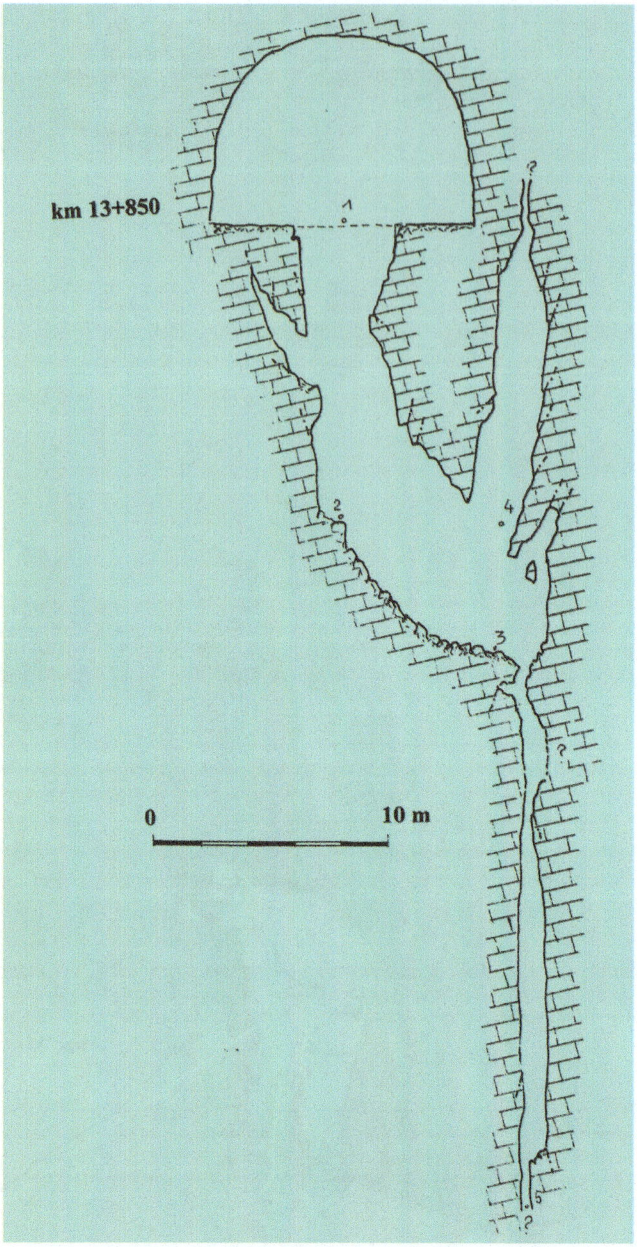

Fig. 7.41 Profile of cavern at km 13 + 850 in Brinje Tunnel

Zagvozd—Baška Voda road in Dalmatia (Garašić 2009a, b, c, Garašić and Garašić 2013).

The rocks in which the caverns were formed belong to the Lower Cretaceous Baremian-Aptian limestones $K_1^{4,5}$, and these formations vary between 15 and 75 cm in layer thickness. The limestones are light to dark grey in colour, and brown intercalations of carbonate calcitic binder can locally be observed between the layers. In the immediate vicinity of the cave, and in the site itself, the inclination of carbonate formations varies between 45° and 55°, and they strike towards the northeast (from 42° to 48°). The fissures noted in the site and in its immediate vicinity are up to 10 cm wide, and they are planar and smooth, very rarely filled with clayed material, and are very firm. They are more often filled with carbonate calcitic binder in crystalline or amorphous form.

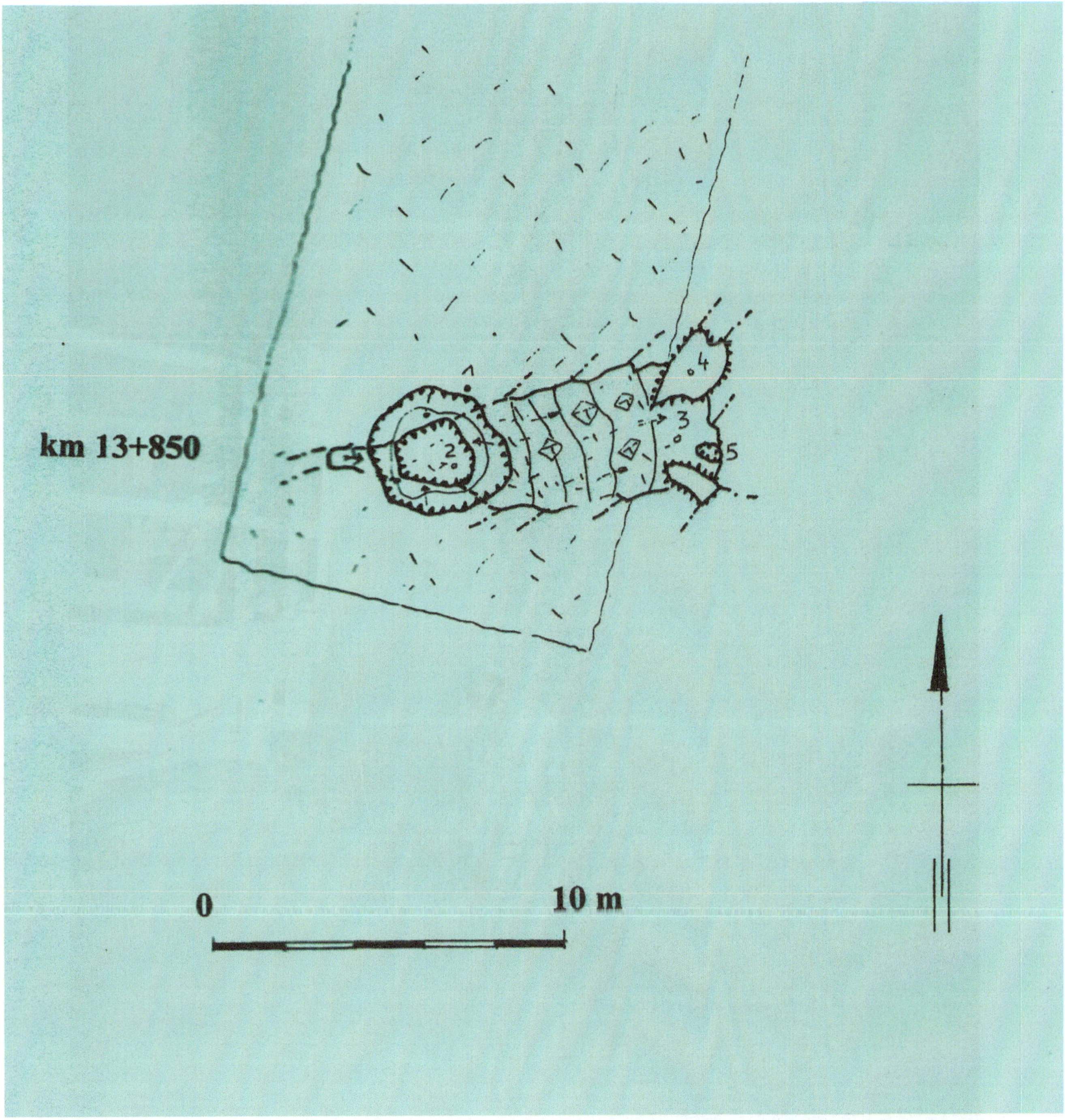

Fig. 7.42 Plan of cavern at km 13 + 850 in Brinje Tunnel

Fig. 7.43 Descending to cavern at km 14 + 027 in Brinje Tunnel

Fig. 7.44 Vertical cavern at km 14 + 027 in Brinje Tunnel

The tunnel runs through central lower parts of all three caverns. Thus, if viewed from the tunnel axis, the caverns spread vertically towards the ground surface, but also vertically downwards. Dolomitic intercalations locally appear as "shelves" in vertical cave segments.

These caverns show signs of karstification by gravity in their upper parts, while there are traces of regressive karstification, i.e. aggressive karstification towards the ground surface, in the bottom parts of the caverns.

Caverns in Sveti Ilija Tunnel at km 1 + 415, km 1 + 193 and km 1 + 637 are large-sized vertical caves of an knee-shaped morphological type, characterized by the occurrence of the so-called false bottoms, with height differences of about 297 m, 268 m and 203.5 m, respectively. This is the zone of the highest tunnel overburden (from 1300 to 1380 m). This means that the deepest parts of these caverns (accessed by bats from the ground surface rather than from the tunnel) sometimes extend to 200–250 m below the tunnel line. Therefore, from the standpoint of geology, these sites can be classified among the deepest speleological sites discovered so far in Dinaric karst in Croatia. The depths from the ground surface range from 1350 to possible 1650 m (Garašić and Garašić 2013).

New studies in last few years (2016–2019) showed that some of these caves were filled with rock material as a consequence of blasting. However, cavers climbed to the new parts of the caverns. In the cavern at km 1 + 637 (Figs. 7.53 and 7.54), they climbed more than 315 m. Therefore, today that cavern has a vertical difference of approximately 459 m.

Fig. 7.45 Cavern at km 14 + 300 in Brinje Tunnel

Fig. 7.46 Cavern at km 14 + 964 in Brinje Tunnel

Fig. 7.47 Brinje Tunnel collapsed all the way to the surface of the terrain

Fig. 7.48 Large hole on terrain surface

Fig. 7.49 Inside the Brinje Tunnel in time of collapsing

7.8 Pećine Tunnel

The Pećine Tunnel is located in the city of Rijeka, about 100 m from the sea. It is the widest tunnel in Croatia. The tunnel passes under the city district of the same name and is part of the city highway D404 in Rijeka. The total length of the tunnel is 1258.50 m. Of this, 60% of the total length (751.50 m) was derived as a three track tunnel and the remaining 40% (507 m) as a four track tunnel. At the junction, the tunnel profile becomes hexagonal with a span of 28.35 m, a height of 13.60 m and an excavation area of up to 310 m^2. The tunnel runs below the urban area, with an overburden of about twenty metres, mostly through extremely karstified rock mass, which makes this tunnel particularly challenging in terms of design and construction. The main rock mass through which the tunnel route passes is composed of Cretaceous deposits (transitional carbonate breccias, dolomites and limestones in alternation and rudist limestones) and Paleogene deposits (Foraminifera limestones and limestone breccias). The rock mass is classified in category IV according to Bieniawski's RMR classification. However, due to the small overburden (20 m) and numerous surface structures, construction conditions belong to the category of extremely difficult geotechnical conditions. The tunnel lining is made of reinforced concrete MB-30 with thickness d = 0.50 m for three-lane and d = 0.65 m for four-lane tunnel (Kuželički and Ružić 2008). Cavern in Pećine Tunnel at km 1 + 412 (Figs. 7.55, 7.56, 7.57 and 7.58) contains sea water. Water is 21 m deep, and cave is 37 m long.

7.9 Rožman Brdo Tunnel

Rožman Brdo Tunnel is located on A-6 highway near Vrbovsko and has two tunnel tubes with length 523 m, in that tunnel were 17 caverns. The longest cavern is at km 67 + 368, and it is long 119 m.

Important caverns in "Rožman Brdo" Tunnel are at km 66 + 926 (Fig. 7.59), km 67 + 208 (Fig. 7.60), km 67 + 368 (Figs. 7.61, 7.62 and 7.63).

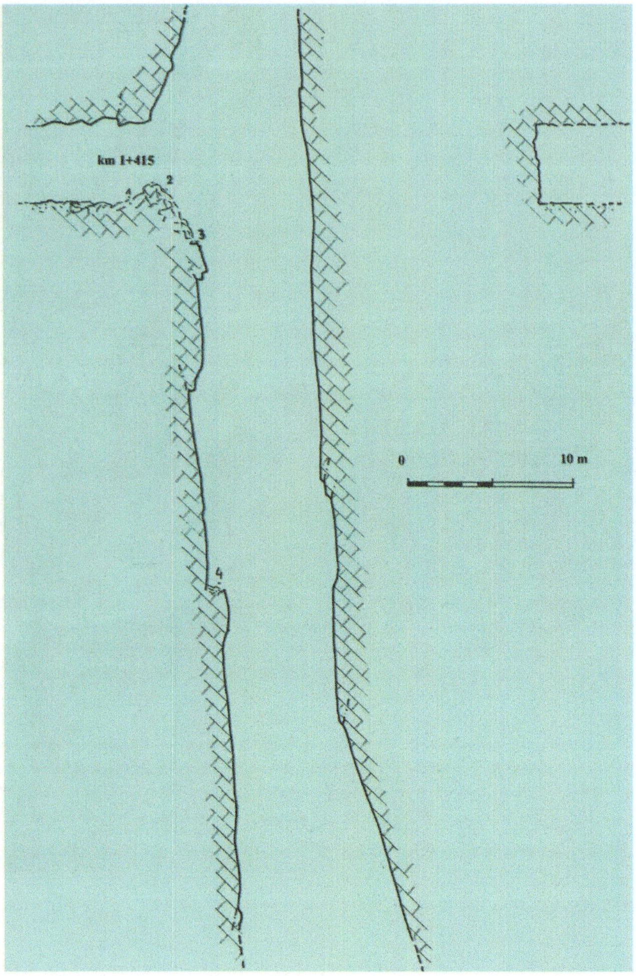

Fig. 7.50 Profile of cavern at km 1 + 415 in Sveti Ilija Tunnel

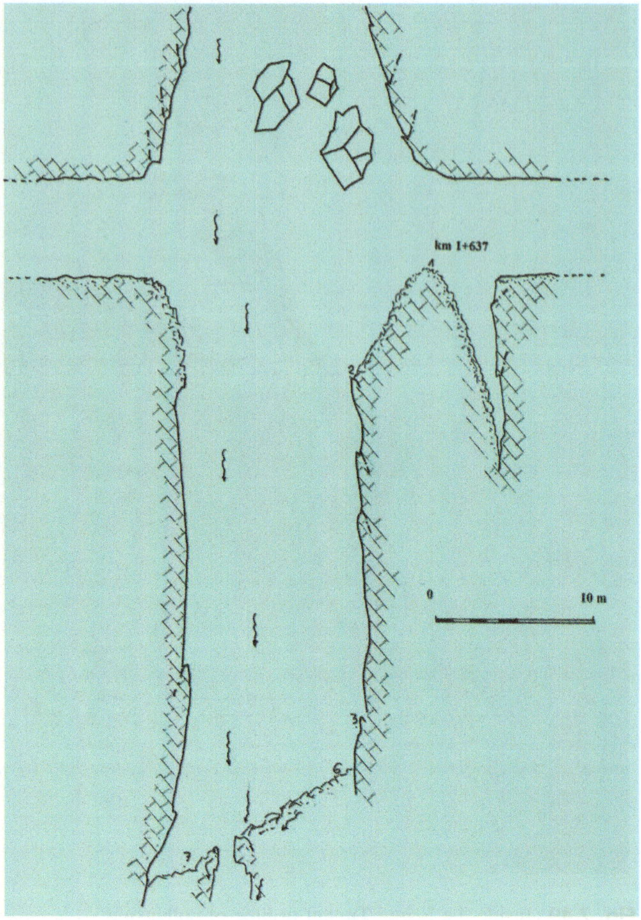

Fig. 7.51 Profile of cavern at km 1 + 637 in Sveti Ilija Tunnel

7.10 Puljani Tunnel

Tunnel "Puljani" is a 417 m long double-tube road tunnel located on highway A-10. Cavern at km 1 + 965 (Figs. 7.64, 7.65, 7.66 and 7.67) is mentioned here.

7.11 Hydrotechnical Tunnel for Drainage of the Konavle Field

The hydrotechnical tunnel for drainage of the Konavle field was drilled in 1958 in the length of 1961 m. Its role is to direct the waters over the rainy season to the edge of the Pasjača Bay during the rainy season. This prevented the

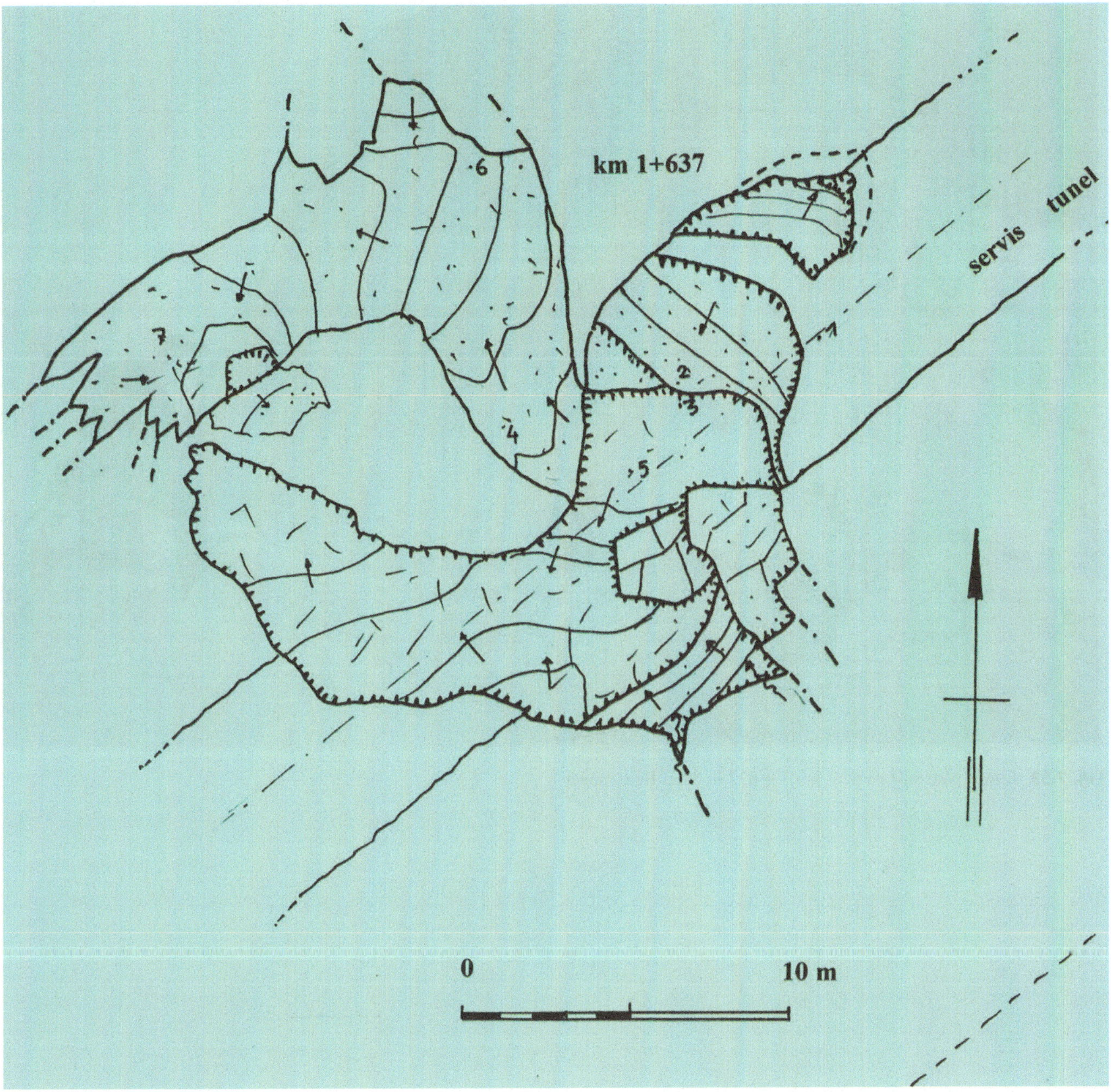

Fig. 7.52 Plan of cavern at km 1 + 637 in Sveti Ilija Tunnel

Fig. 7.53 Lower parts of cavern at km 1 + 637 in Sveti Ilija Tunnel

Fig. 7.54 Final points reached during the first explorations of lower parts of the cavern at km 1 + 637 in Sveti Ilija Tunnel

Fig. 7.55 Cavern at km 1 + 412 in Pećine Tunnel

Fig. 7.56 Descending to cavern at km 1 + 412 in Pećine Tunnel

Fig. 7.57 Lake in cavern at km 1 + 412 in Pećine Tunnel

Fig. 7.58 Cave diving in lake in cavern at km 1 + 412 in Pećine Tunnel

Fig. 7.59 Cave karren in cavern at km 66 + 926 in Rožman Brdo Tunnel

Fig. 7.60 Cavern at km 67 + 208 in Rožman Brdo Tunnel

Fig. 7.61 Entrance from tunnel to cavern at km 67 + 368 in Rožman Brdo Tunnel

Fig. 7.62 View from Cavern at km 67 + 368 to Rožman Brdo Tunnel

Fig. 7.63 Longest cavern at km 67 + 368 in Rožman Brdo Tunnel

irrigation of farmland in the field. The tunnel pierces the coastal ridge below Popovići with an elevation of the bottom at +41.90 m above sea level and at the exit +11.33 m above sea level (average decrease of about 1.4%). Water enters the tunnel after it reaches a height of 43.70 m.a.s.l. because a threshold with an upper angle of +43.70 m.a.s.l. is constructed in the entrance structure. After reconstruction (1972–1977), the tunnelling capacity increased from 40 to 60 m^3/s (Roglić and Baučić 1958). The new environmental impact study was made in 2017, in that tunnel are tens of caverns.

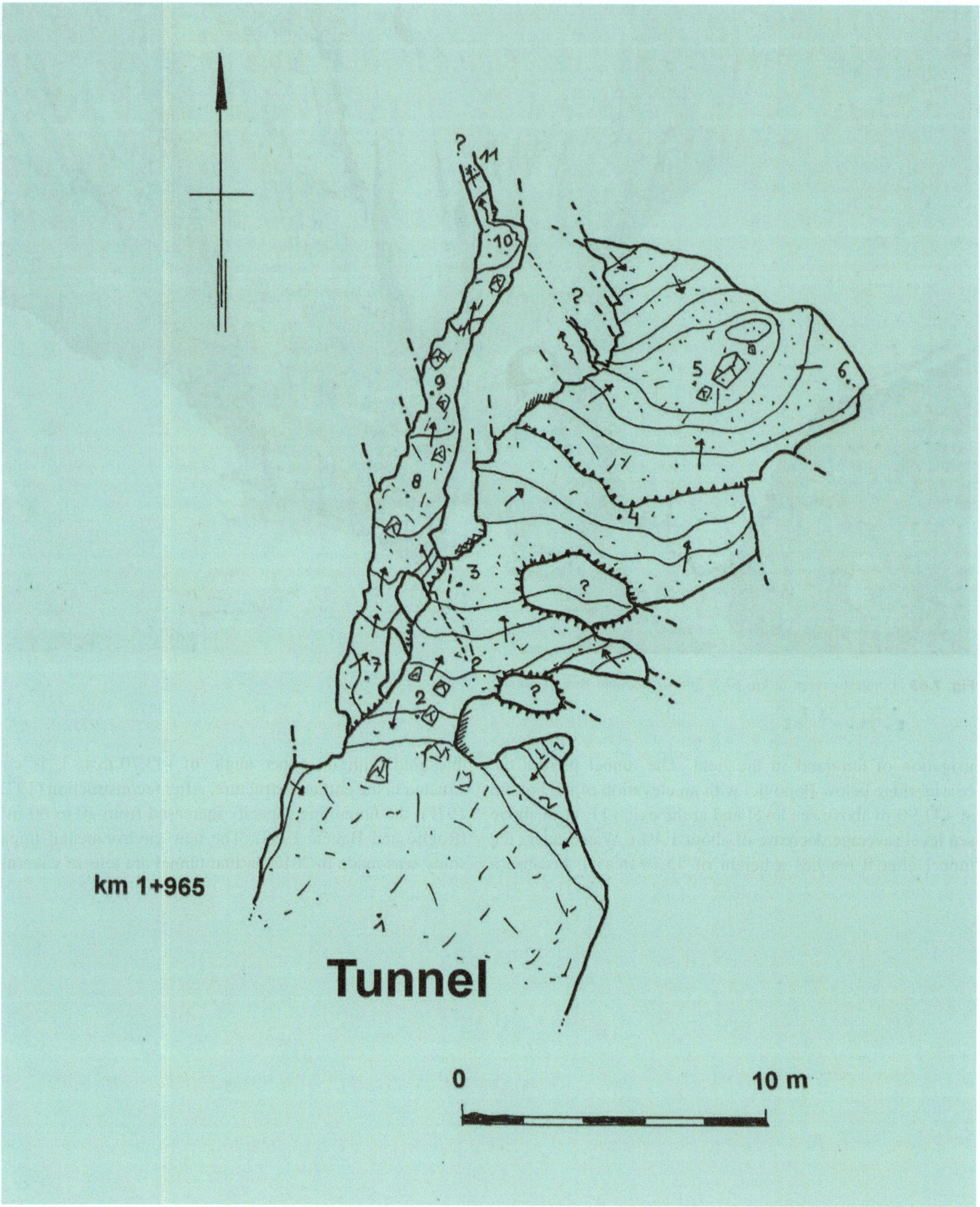

Fig. 7.64 Profile of cavern at km 1 + 965 in Puljani Tunnel

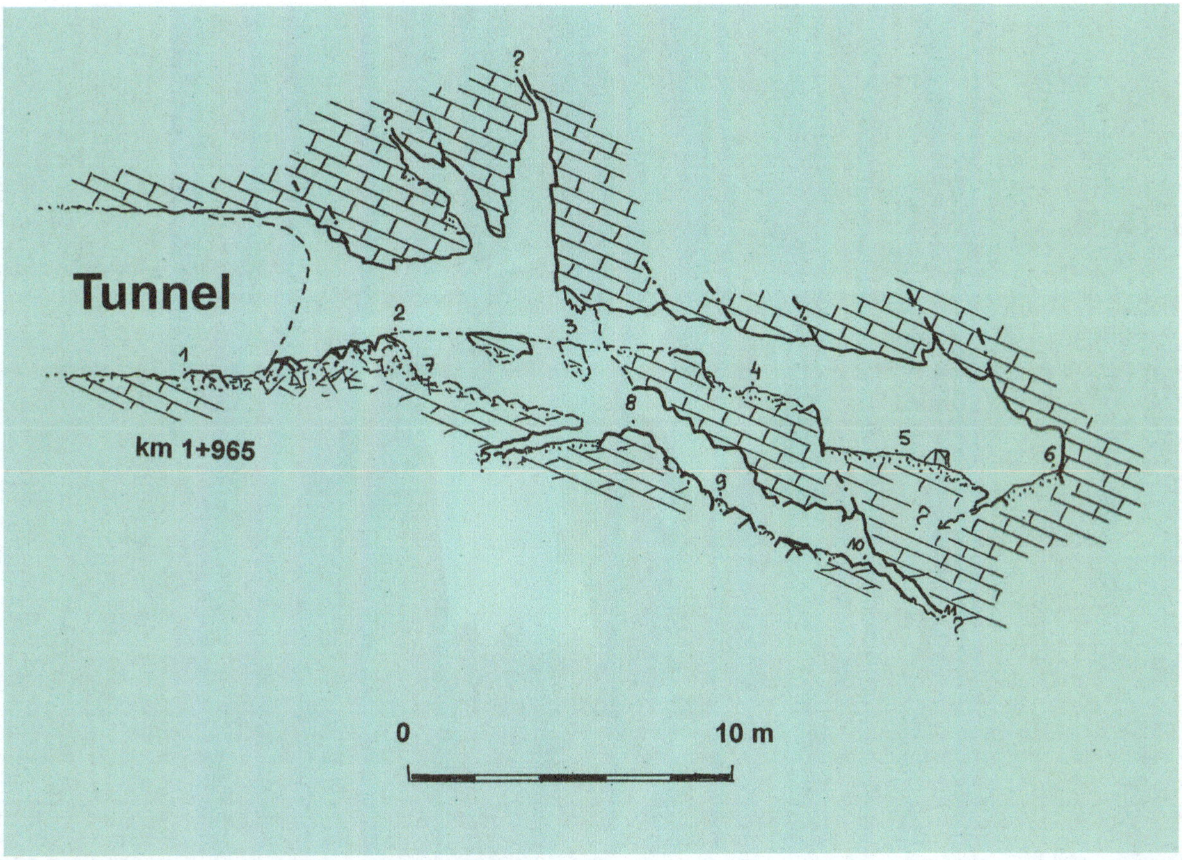

Fig. 7.65 Plan of cavern at km 1 + 965 in Puljani Tunnel

Fig. 7.66 Cavern at km 1 + 965 in Puljani Tunnel (view from tunnel to cavern)

Fig. 7.67 Cavern at km 1 + 965 in Puljani Tunnel (view from cavern to tunnel)

8 Caverns Found and Explored During the Construction of Bridges, Viaducts, Roads and Railroads

Abstract

Caverns found in the foundations of bridges and viaducts have always posed a threat to the safety of the building. The area of the bridges over the Karst Rivers Dobra and Kamačnik on the A-6 motorway belongs to the Black Sea Hydrological Basin. Numerous caverns were discovered in the foundations of the viaducts "Zečeve Drage", "Osojnik" and "Jezerane" but the most interesting is the discovery of a large hall under the "Kačjak 2" viaduct in the city of Rijeka. During the construction of roads and railways, numerous caverns were found and explored. Over 60 caverns were found at the junction of the A-1 and A-6 highways "Bosiljevo 2". Due to the instability of the terrain, sometimes the cavern entrances had to be secured by installation of concrete or metal pipes. Hundreds of previously unknown caverns were explored and documented along the roads in the Dinaric Karst in Croatia.

8.1 Bridges and Viaducts

8.1.1 Dobra River Bridge

A detailed speleological survey of the cave in the area of the Dobra Bridge, in the foundation of the U-6 pillar, at km 66 + 406, on the Karlovac-Rijeka highway, Kupjak-Vrbovsko section, provided information on the location, dimensions and morphology of the speleological object. It is a pit of simple morphological type, with a vertical difference of over 14.30 m. The entrance (1.05 × 2.75 m) is followed by a vertical jump of 12.20 m leading to a smaller room (2.0 × 3.50 m) with extensions to the south-east and north-east. This is the so-called false bottoms and because of the narrowness and dimensions of the further passage of cavers is not possible. Speleothems and corrosion forms are present in this cave. The lowest point of the cavern is 14.30 m below the entrance level. The total volume of the surveyed part of the pit is estimated to be about 90–100 m^3, but it can be assumed that the volume of the cavern is much larger. No steady active water flow was observed. Intense drip water created speleothems and corroded the cavern. The pit was developed in a north–south direction, corresponding to a fault plains that played a significant role in its genesis.

The speleological object extends below the foundations of the U-6 pillar of the Dobra Bridge, and the special remediation was suggested.

The rocks in which the pit was formed belong to the Middle Jurassic, Dogger limestone J_2, whose thickness of the layers, if visible, varies between 40 and 200 cm.

The deposits of Dogger are represented here by limestones, which are characterized by a large thickness of layers from 40 to 200 cm, good layering and the rare appearance of layers of dolomite. Light grey to grey, brownish, and rarely darker in colour. They are mainly represented by mudstone and wackestone, packestone, grainstone and rudstone (intrabasic breccias). Grainstones-type oolites also occur, from textural forms to millimetre lamination (centimetre bands of darker and lighter mudstone limestones) as well as wavy to sloped lamination.

8.1.2 Kamačnik Bridge

During the construction of the bridge (2002) for the A-6 motorway across the Kamačnik River near Vrbovsko in Gorski Kotar, a cavern opened in the foundation of the S-1 pillar of the right road lane of the Kamačnik Bridge. It was thought to be small in size, but speleologists noted that the nearby karst spring of the Kamačnik River was dived to a depth of 86 m and continues further into the depths. This indicated intense rock karstification of the area. Subsequently, over 400 m^3 of concrete was injected into the foundation of right the pillar S-1 of the Kamačnik Bridge at km 65 + 000 (Fig. 8.1), and prior to speleological research it

M. Garasic, *The Dinaric Karst System of Croatia*, Cave and Karst Systems of the World,
https://doi.org/10.1007/978-3-030-80587-6_8

was thought that 5–7 m^3 would be too much. Specifically, concrete and grout passed through a narrow surface of cracks in the rocks and filled the cracks only in a deeper area. This enabled a solid foundation of the bridge, and the speleologists once again confirmed their knowledge of karst and caves in the Dinaric Karst of Croatia. The rocks in which the pit was formed belong to the Upper Jurassic, Lower Malm dolomites $J_3^{1,2}$, whose layer thickness, if noticeable, varies between 25 and 45 cm. The Lower Malm deposits are represented here by dolomites. Limestones, i.e. dolomitic limestones or limestone dolomites, rarely occur. The entry vertical is only 3.70 m. Here, the cave extends into an underground room measuring 4.50 × 1.0 m. The continuation of the cave is north (330°), i.e. towards the main fault and south (150°), i.e. towards the river Kamačnik (Fig. 8.2). Here, the pit continues with a vertical jump of 7.50 m, but it is a crack measuring 0.35 × 0.30 m through which speleologists could not go any further. The bottom of the pit is covered with a rocky embankment, and most likely it was the so-called. "False bottom", which turned out to be the right conclusion. Here, a slight airflow in the direction of the cavern was measured.

8.1.3 Zečeve Drage Viaduct

Caverns were found and explored during the first and second stages of road construction (1993–1994 and 2004–2005) in the foundations of the pillars of the Zečeve Drage viaduct on the Rijeka—Zagreb A-6 highway near Vrbosko in Gorski Kotar. Some caverns found in the second phase extended below the viaduct columns built in the first phase. Some of them were reinforced with additional works and some did not pose any danger to the existing road. More important caverns are at km 70 + 714, km 70 + 758, km 70 + 208, km 70 + 139, km 70 + 398—pillar S-5 left, km 70 + 506—pillar S-7 left, km 70 + 510—pillar S-7 right (Figs. 8.3 and 8.4).

8.1.4 Osojnik Viaduct

In the foundation of S-5 pillar of viaduct Osojnik two caverns exist. At km 76 + 925 and at km 76 + 918, pillar S-5 left. Other caverns were found at km 76 + 800, pillar S-2 right and at km 76 + 850, pillar S-3 right, km 76 + 955, pillar S-6 right (Fig. 8.5) (Garašić 2005).

8.1.5 Kačjak II Viaduct

As part of the construction of the Rijeka bypass road, the Kačjak II Viaduct (2004) was designed. During the construction of the Pillar S-4 foundation, a large cavern appeared (Figs. 8.6 and 8.7). Previous exploratory drilling and geophysical testing did not determine the cavern. Kačjak II is a particularly long highway viaduct that crosses the important Rijeka-Zagreb railway line in the notch "Sveta Ana" (known for its archeological findings in the caves), located on the A-7 motorway (Fig. 8.8). The dimensions of the cavern below the viaduct are 56 × 32 m with a height difference of 31.60 m (Fig. 8.9). Fortunately, the positions of discontinuities, layers and cavern do not compromise the viaduct and highway stability. The occasional underground stream flows through a cavern in a large hall towards the Adriatic Sea and Martinščica Bay. In some places, the underground hall is over 25 m high. This is the largest underground hall in the city of Rijeka. Just above half the hall, there is a pedestrian passage below the viaduct (Garašić 2004).

8.1.6 Jezerane Viaduct

In the area of the foundations of the pillar S-3 (left) viaduct "Jezerane", at km 2 + 020 on the route of the Bregana—Zagreb—Dubrovnik highway, section Jezerane—Žuta Lokva, near Jezerane, a pit with a knee or columnar morphological type was explored. The speleological object has a vertical difference of over 10.50 m, which is likely to extend deeper, but the passage of the speleologists was prevented by a narrow passage (Fig. 8.10). The total volume of the pit is estimated at 45–50 m^3. The rocks in which the pit was formed belong to the Lower Cretaceous Neocomian limestones K_1^{1-3}, whose layer thickness varies between 30 and 50 cm. They are characterized by grey and light grey colour and good stratification.

8.1.7 Crna Draga Viaduct

The entrance to the speleological object (cavern) is located on the north-east side of the foundations of pillar S-1 (right), about 17 m away from the left pillar S-1, at km 204 + 550, highway Bregana—Zagreb—Dubrovnik, section Tunnel Sveti Rok—Bridge, near Baricevic in Dalmatia, at an altitude of approximately 451 m.

The entrance to the cavern at km 204 + 550 (Fig. 8.11) was opened after blasting to excavate the foundation on 1 February 2003. The vertical entrance to the cavern was 2.80 × 2.95 m. The entrance to the speleological object was irregular, round in shape, with a vertical pit jump. The rocks in which the cavern is formed belong to the Lower Cretaceous limestones and limestones of the Cenomanian part, $K_{1,2}$, whose layer thickness varies between 20 and 120 cm. They are distinguished by grey, light grey and grey-brown in

Fig. 8.1 Cavern at km 65 + 000 under the pillar S-1 of the Kamačnik Bridge

Fig. 8.2 Caves in Canyon of Kamačnik River are very near "Kamačnik Bridge" at km 65 + 000

Fig. 8.3 Entrance to cavern at km 70 + 510 under pillar S-7 of Zečeve Drage Viaduct

colour and good layering. These deposits are built by the south-western slopes of the Velebit Mountains, with Tertiary clastic deposits ("Jelar deposits"). The tertiary cover was torn apart by denudation.

8.2 Roads and Railroads

8.2.1 Žuta Lokva Road Junction

A detailed speleological survey of the speleological object, in the embankment of the road at km 18 + 130 (Fig. 8.12), on the highway Bregana–Zagreb–Dubrovnik, near Rapajin Klanac in Lika, provided information on the position, dimensions and morphology of the speleological object. A cavern is a vertical speleological object (pit), of a knee-shaped morphological type with a total height difference of over 38.50 m, a horizontal projection of 47.30 m, which probably extends deeper but the passage for speleologists is impossible due to the narrow passages and the so-called "False bottom". The total volume of the investigated part of the cavern is estimated from 1200 to 1500 m^3. It is undoubtedly established that there is an occasional water course in the pit that during the rainy season. A larger amount of water recharges the Gacka River and belongs to the Adriatic Basin. In the geological past and today, groundwater has aggressively corroded the rocks. There are speleothems in the entire cavern in the form of stalactites, cortices, corals, botroids and curtains. There are also crystalline forms. Underground scabs in the lower parts of the pit indicate the considerable paleohydrogeological activity. The cavern was generally developed in a north-west–south-east direction, corresponding to the position of the fault plain that played a significant role in its genesis. On Žuta Lokva road junction has explored 16 caves (Fig. 8.13).

8.2.2 Martina Jama in Front of Krka Bridge

The entrance to the speleological object is in the right lane of the motorway in the embankment at km 71 + 754, on the highway Bregana—Zagreb—Dubrovnik, section Pirovac—Skradin, near Skradin in Dalmatia, about 500 m from the bridge "Krka". Although the construction workers only discovered the entrance during the construction work, the locals had previously known of its existence. They called it

Fig. 8.4 Cavern at km 70 + 510 under pillar S-7 of Zečeve Drage Viaduct

Martina Jama. The first entrances to the underground space of Martina Jama were made by the inhabitants of the village of Bičine, using ropes and ladders. At that time, part of the calcite formations that the first "explorers" pulled from the cave as souvenirs were devastated. The first speleological survey was undertaken by members of the Croatian Speleological Society (HSS) in 1961, followed by the SD "Ursus spelaeus" from Zagreb in 1980. The following investigations were carried out on several occasions in 1988 and 1989 by members of the SO PD "Kamenjar" from Šibenik. In 1989, a speleological team from the Croatian Museum of Natural History in Zagreb was also exploring this speleological object. They made a sketch of the cave and photographs of the underground space. However, until this latest research, this speleological facility has not been fully explored and documented for safe highway construction.

The vertical entrance to the cavern is 1.50 × 0.70 m (Fig. 8.14). The entrance to the speleological object is irregular, ellipsoidal in shape, first with a vertical cave jump, and then the branched cave channel mainly in the north-west–south-east direction. The rocks of the cavern entrance, due to intensive processes of karstification, construction work and blasting, are intensely fractured and cracked, and care should be taken when anchoring ropes. In the entrance part of the cavern, a primary clay of a hard, consistent state and medium plasticity appears in the cracks. The width of the cracks in this part of the pit is up to a maximum of 10 cm.

The rocks in which the cavern is formed belong to the Upper Cretaceous Senonian Rudist limestones, K_2^3, whose thickness varies between 20 and 120 cm. They are distinguished by grey, light grey and grey-brown in colour and pronounced lamination.

It is an anticlinal form at the core of which are the Senonian rudist limestones, which continue continuously from the Turonian deposits. The stratigraphic affiliation of these limestones to the Senonian has been established on the basis of numerous analysed rudist specimens.

According to the existing and recognized classifications of speleological objects, a speleological object at km 71 + 754 is a combined speleological object with a vertical entrance (pit), of branched morphological type (Figs. 8.15 and 8.16).

The dimensions of this cave are over 26.50 m of total hight and a length of 172.20 m. The cavern probably extends deeper, but the passage of the speleologists is

Fig. 8.5 Cavern at km 76 + 955 under pillar S-6, right of Osojnik Viaduct

prevented by a narrow passage. The aforementioned speleological object mostly extends below the embankment of the road and represented a certain problem of changing stability or solidity of rocks in the embankment of the road.

8.2.3 Bosiljevo 2 Interchange

Hundreds of caverns were found and documented during the construction of roads, highways and railroads. The highest concentration of their occurrence during surface excavation was found during construction works at the "Bosiljevo" junction, i.e. the junction of motorways A-6 (Zagreb—Rijeka) and A-1 (Zagreb—Split—Dubrovnik). Over 60 smaller caverns (Fig. 8.17) that posed problems when constructing the highway were explored. Here will be presented only five of caves from A-1 highway, part Bosiljevo—Josipdol. Caverns at km 2 + 235 (Figs. 8.18, 8.19 and 8.20), at km 8 + 560 (Figs. 8.21, 8.22 and 8.23), at km 10 + 795 (Figs. 8.24 and 8.25), at km 17 + 004 (Figs. 8.26, 8.27, 8.28 and 8.29) and at km 17 + 174 (Figs. 8.30 and 8.31).

8.2.4 Debelo Brdo Cutting (Slopes)

In the intersection of the new section of the D-1 Zagreb—Split road, Korenica—Udbina section, near Bjelopolje in Lika, a dozen caverns appeared and were investigated. Most of them were vertical and four were horizontal. The most interesting of these were found at km 1 + 860, on the left side of the cutting. On the right side of the Debelo Brdo slope, three caverns were found at km 1 + 880, km 1 + 870 (Fig. 8.32) and km 1 + 895. They are all part of a cave system that is cut off by a newly constructed road. These caves are ones of the most beautiful and saved for future visiting. Two more caves are on km 2 + 860 and km 2 + 770.

The rocks where the caverns are formed belong to limestones, dolomitic limestones and limestone dolomites. Rarely do they belong to the dolomites from the transition between the Lower and Upper Cretaceous, most commonly the Late Cretaceous (Barremian–Aptian–Albian) K_1^{3-5} and the Upper Cretaceous-Turonian limestones $K_2^{1,2}$, whose layer thickness varies between 30 and 70 cm. The deposits are characterized by grey and light grey colour and good stratification.

Lithologically, these are mudstone-to-wackestone-type limestones, then skeletal packstone and grainstone, and floatstone ridge and reef facies of limestone with rudists.

The lower part of the cenomanian is approximately 40 m thick and consists of mudstone and wackestone limestone. They are well stratified (0.10–0.60 m), light brown in colour. Wackestones contain various skeletal particles, foraminifera, algae, shellfish and echinoderms. Non-skeletal particles are common algal lumps and pellets with sludge bases. Bioclastic margins are often micritized, and traces of bioerosion are visible on larger shell fragments.

In the immediate vicinity and in the cave itself (as well as in nearby speleological objects), the slope of carbonate deposits (limestones) is between 37° and 40°, and the direction of their slope is south or south-west (180–220°). The fractures observed in the object are up to ten centimetres wide, flat and smooth. This is the main stage of speleogenesis. In the vicinity of the cavern (on the surface), the cracks are mostly filled with clay material, with a difficult to digest consistency. Even on the surface (in a notch of the road), sometimes in crystalline or amorphous form, in addition to the clay and carbonate calcite binder. There are speleothems that occur in the form of crusts, stalagmites, stalactites and cave corals (Figs. 8.33, 8.34, 8.35 and 8.36). At the location

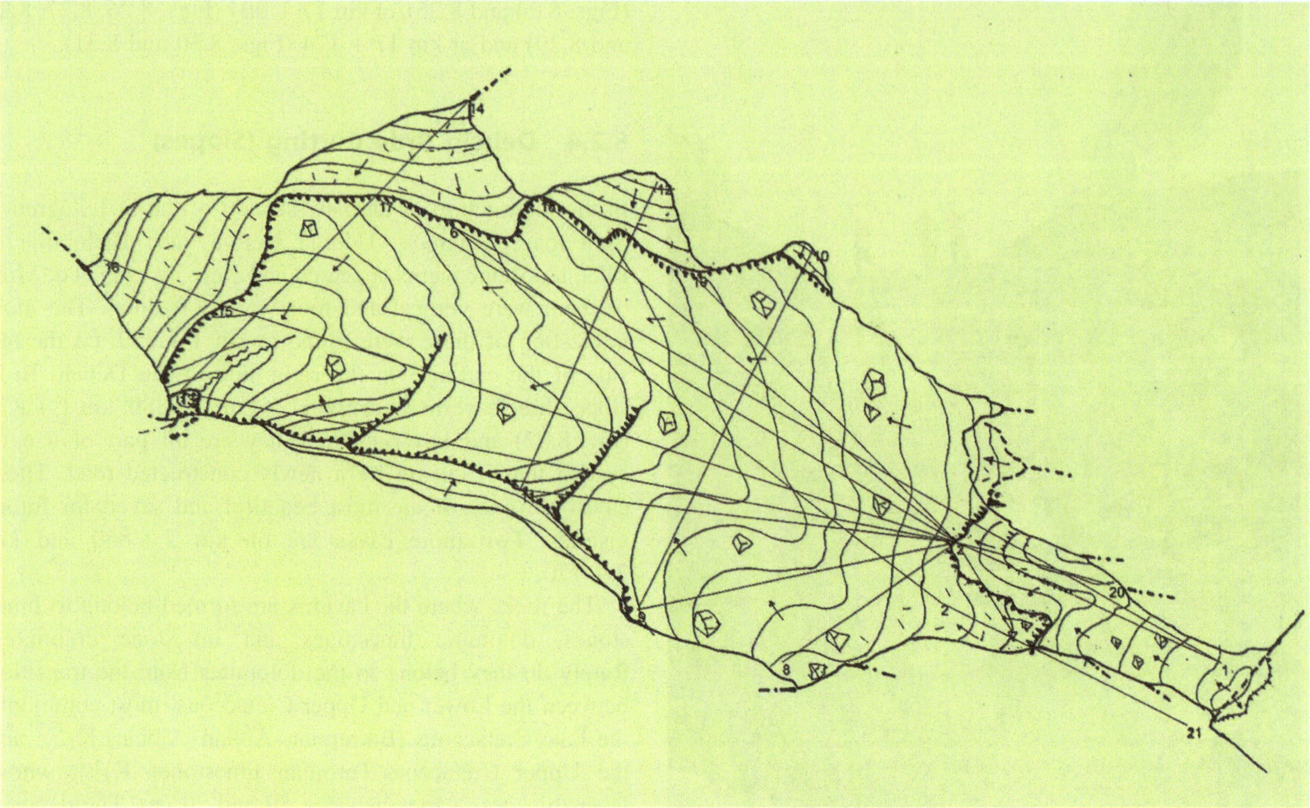

Fig. 8.6 Plan of cavern at km 1 + 281 under pillar S-4 of Kačjak 2 Viaduct

of the speleological object, there is a more intense fissure and fault zone of width of about 2.0–3.0 m, which, in addition to lithological and hydrogeological conditions, occupied a significant place in the genesis (occurrence) of the main parts of the studied cavern and some surrounding speleological objects. No milonitic fault zones were observed in the speleological object. The evaluation of the speleological survey is that it is a relatively young but now inactive fault, as evidenced by the thin coatings of the speleothems that cover the fault plain in the cavern without being broken or deformed. There are also old deposits of speleothems that have been broken by the action of paleoseismic forces. Such formations have been broken by the action of neotectonic movements in the last 30,000 to 50,000 years. A number of speleothems broken by the forces of recent mining in their immediate vicinity have also been detected. The main fault has a direction of 24–204°, with a very steep incline (from 86° to 88° to the west). In addition to this fault, there is also a significant fault in the direction of providing approximately 105–285° in the entrance part of the cavern, which transversely touches the previously mentioned fault inside the cavern below the steep and vertical parts. The fracture system is approximately perpendicular to the main fault and has an orientation of 95–275° with a slope of about 63°. Corrosion of the surrounding rocks is very intense.

The whole caverns are filled with speleothem, and there are many clay sediments in the interior. Their position, shape and stratification indicate occasional groundwater flow, especially at the end of the main channel. Depth of karstification in the area of this speleological object is estimated at 250–300 m, and the zone of vertical circulation about 100–150 m. This is followed by a zone of horizontal circulation, which transports groundwater to the Krbava River basin, i.e. partly to the Black Sea drainage line (towards Lika Pljesevica) and partly to the Adriatic Sea drainage line (Lika middle ground). Without additional hydrogeological research, it is difficult to say for sure what the drainage is. Since the Lower and Upper Cretaceous limestones are in hydrogeological terms, due to their intense secondary (fracture) porosity and permeable, it is expected that there are several similar structures in the continuation of this fault zone, which do not have access from the surface, but further work in notch could come across them. These forecasts were confirmed during the execution of the works, and these speleological objects were also investigated.

Length of this cave system is long about 650 m, and depth is around 25 m.

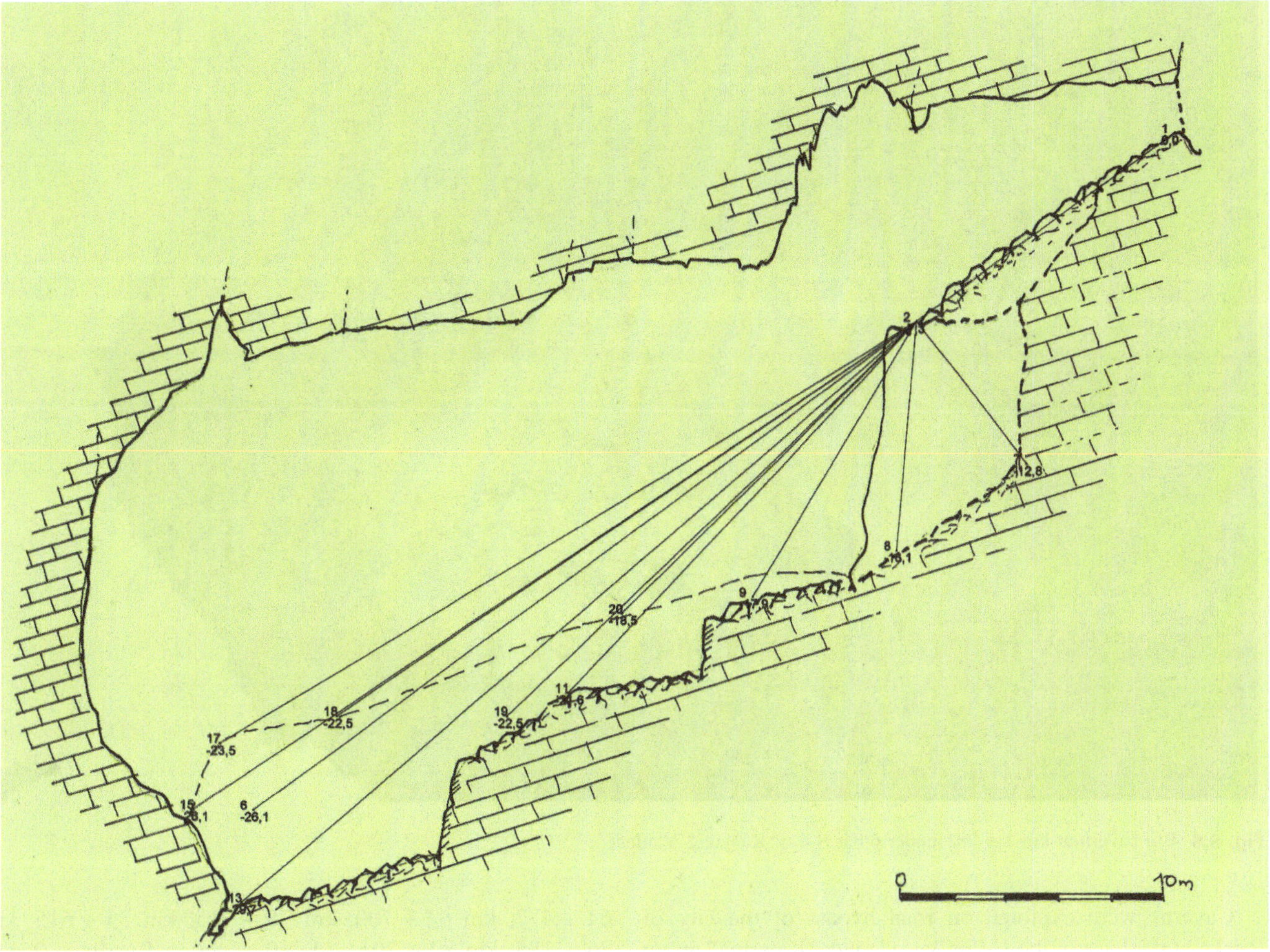

Fig. 8.7 Profile of cavern at km 1 + 281 under pillar S-4 of Kačjak 2 Viaduct

8.2.5 Pipe in the Embankment

While using the A-6 motorway near Jablan in Gorski Kotar, in front of the Čardak Tunnel on the Zagreb side, in May 2006 a cavern opened in the southern part of the road embankment at km 22 + 950 (Figs. 8.37, 8.38 and 8.39), section Vrbovsko—Ravna Gora. Since the embankment was made of loose material and the cavern extended well below the embankment, it was decided to install concrete pipes reaching the solid carbonate rocks for the safety of the speleologists. The pit was explored to a depth of 28.50 m and later successfully rehabilitated.

A similar situation was present when exploring the cavern at the base of the S-9 (right) pillar in the "Zečeve drage" viaduct near Vrbovsko on the A-6 motorway. Impact drilling for the foundations of the S-9 pillar (Figs. 8.40 and 8.41) revealed a cavern, which was later expanded and secured with metal tubes that allowed the entry of speleologists who explored it up to 21.50 m deep.

8.2.6 Pit in the Green Grass Area

In the fall of 2013, about 500 m in front of the tunnel "Veliki Gložac" on the Zagreb side of the A-6 highway in the green grass area between two lanes, at km 9 + 000, a cavern with an opening size of approximately 4 × 3 m and a depth of 5.50 m opened in minutes (Fig. 8.42). It is fortunate that the cavern only opened between the two lanes of the road because the highway was operational. During the construction of the highway, there was no knowledge of any speleological object in the area. This was an additional challenge to investigate the phenomenon more thoroughly. In addition to detailed speleological exploration, geophysical methods (Fig. 8.43) (seismic, geoelectrics, GPR), exploratory drilling, and laboratory geotechnical analyses and testing of core samples were used. The cavern was found to extend up to 14.50 m deep, and its extension was found some fifty metres south also in the green belt. The caverns were successfully rehabilitated.

Fig. 8.8 Big cavern at km 1 + 281 under pillar S-4 of Kačjak 2 Viaduct

Caverns were explored on road bypass of the city of Rijeka at the base of the U-2 pillar, viaduct "Katarina" at km 0 + 393 (Figs. 8.44 and 8.45), at junction "Rujevica" (Fig. 8.46), ramp 5, km 0 + 106 and km 0 + 160, at junction "Diračje" (Figs. 8.47 and 8.48) between profile 375 and 375, km 3 + 660 on D404.

Following is a list of important caves on slopes of the highways (roads)/the order on each highway is by exploring dates from 1990 to 2010/:

On highway "Istrian Y"—A-9: cavern at km 30 + 550, etc.

On highway "Istrian Y"—A-8: caverns at km 13 + 070, km 16 + 080, km + 16 + 100, etc.

On highway A-7 (Rupa—Rijeka—Žuta Lokva): Caverns at km 18 + 884, km 18 + 922, km 18 + 927, km 24 + 567, km 25 + 954, km 27 + 925, km 0 + 243 and km 245 on junction Šapjane, km 14 + 306, km 8 + 740, km 1 + 000, km 6 + 305, junction Sveti Kuzam, etc.

On highway A-6 Rijeka—Bosiljevo—Karlovac—Zagreb): caverns at km 43 + 520, km 58 + 790, km 62 + 570, km 63 + 378, km 63 + 760, km 56 + 260, km 56 + 652, km 62 + 560, km 62 + 698, km 57 + 940, km 58 + 050, km 57 + 340, km 1 + 993, km 65 + 000, km 2 + 240, junction Bosiljevo 2, km 2 + 180, junction Bosiljevo 2, km 61 + 477, km 62 + 700, km 75 + 715, km 75 + 615, km 76 + 185, km 82 + 291, side 10, junction Bosiljevo 2, km 82 + 376, side 10, junction Bosiljevo 2, km 82 + 685, side 10, junction Bosiljevo 2, km 82 + 676, side 10, junction Bosiljevo 2, km 0 + 340, side 4, junction Bosiljevo 2, km 0 + 280, side 4, junction Bosiljevo 2, km 0 + 290, side 4, junction Bosiljevo 2, km 0 + 178, passage Hrsine, junction Bosiljevo 2, km 0 + 160, passage Hrsine, junction Bosiljevo 2, km 0 + 147, passage Hrsine, junction Bosiljevo 2, km 82 + 342, side 10, junction Bosiljevo 2, km 82 + 383, side 10, junction Bosiljevo 2, km 82 + 307, side 7, junction Bosiljevo 2, km 0 + 424, junction Vrbovsko, km 83 + 367, side 10, junction Bosiljevo 2, km 82 + 789, side 10, junction Bosiljevo 2, km 82 + 784, side 10, čvor Bosiljevo 2, km 82 + 682, side 10, junction Bosiljevo 2, km 79 + 339, km 77 + 537, km 80 + 485, km 77 + 520, km 77 + 500, km 77 + 475, km 68 + 040, km 76 + 450, km 76 + 420, km 79 + 325, km 61 + 800, km 75 + 185, km 72 + 360, km 72 + 305, km 72 + 305, km 72 + 360, km 69 + 292, km 69 + 274, km 11 + 260, km 17 + 793, km 17 + 240, km 77 + 537, km 80 + 485, km 77 + 520, km 77 + 500, km 77 + 475, km 68 + 040, km 76 + 450, km 76 + 420, km 77 + 630, km 23 + 100, km 39 + 658, km 39 + 815, km 41 + 198, km 38 + 180, km 38 + 800, km 37 + 230, km

Fig. 8.9 Entrance to cavern at km 1 + 281 under pillar S-4 of Kačjak 2 Viaduct

Fig. 8.10 Cavern at km 2 + 020 under pillar S-3 (left) Jezerane Viaduct

Fig. 8.11 Cavern at km 204 + 550 under the left pillar S-1 Crna Draga Viaduct

Fig. 8.12 Cavern at km 18 + 130 in Žuta Lokva junction

Fig. 8.13 Sixteen caves were explored in Žuta Lokva junction

Fig. 8.14 Cave in the embankment at km 71 + 754—Martina Jama near Krka Bridge

37 + 180, km 41 + 262, km 22 + 950, km 52 + 750, km 42 + 030, km 43 + 195, km 43 + 270, km 63 + 760 (Figs. 8.49 and 8.50), km 63 + 378 (Figs. 8.51 and 8.52), km 62 + 570 (Figs. 8.53 and 8.54), etc.

On highway A-1 (Zagreb—Bosiljevo—Split—Dubrovnik):

Caverns at km 17 + 004, km 17 + 234,15, km 17 + 174.60, km 16 + 387, km 8 + 560, km 2 + 235, km 10 + 840, km 9 + 100, km 14 + 960, km 10 + 795, km 10 + 512, km 10 + 835, km 10 + 985, etc. *on part Bosiljevo —Josipdol.*

Caverns at km 6 + 340, km 1 + 305, km 1 + 341, km 6 + 424, km 6 + 286, km 14 + 229 (Fig. 8.55), Km 0 + 470, etc. *on part Josipdol—Tunnel Mala Kapela.*

Caverns at km 0 + 920 (Fig. 8.56), km 7 + 220, km 7 + 500, km 13 + 150, km 18 + 130, km 18 + 285, junction Žuta Lokva, km 2 + 570, junction Žuta Lokva, km 0 + 424, junction Žuta Lokva, km 18 + 240, km 4 + 528 (Fig. 8.57), km 3 + 870, km 3 + 168 (Fig. 8.58), km 3 + 315, km 3 + 320, km 3 + 300, km 3 + 865, km 7 + 225, km 4 + 750 (Fig. 8.59), km 4 + 795, etc. *on part Tunnel Mala Kapela—Žuta Lokva.*

Caverns at km 4 + 940, km 5 + 530, km 4 + 840, km 18 + 880 (Fig. 8.60), km 18 + 500, km 23 + 340, km 10 + 220, km 3 + 105, km 3 + 180, km 4 + 860, km 11 + 260, km 17 + 793, km 17 + 240, km 9 + 100, junction Otočac, km 11 + 735, km 13 + 625 (Fig. 8.61), km 7 + 330, km 7 + 870, km 7 + 890, km 7 + 774, km

Fig. 8.15 Speleothems in cave in the embankment at km 71 + 754

0 + 090, junction Otočac, side 2, km 11 + 040, km 11 + 200, km 22 + 560, km 18 + 350 (Fig. 8.62), etc. *on part Žuta Lokva—Ličko Lešće*.

Caverns at km 19 + 660, km 28 + 760 (Fig. 8.63), km 18 + 765, etc. *on part Ličko Lešće—Lički Osik*.

Caverns at km 20 + 755, junction Zadar 2, km 0 + 170, junction Maslenica, km 24 + 703, km 25 + 050, km 48 + 060, km 0 + 100, side 3, junction Pirovac, km 0 + 103, side 3, junction Pirovac, km 71 + 754, km 47 + 975, km 71 + 870, km 67 + 925, km 80 + 640, km 80 + 400, km 65 + 510, etc. *on part Maslenica—Zadar2—Šibenik—Dugopololje*.

Caverns at km 22 + 065, km 21 + 190, km 21 + 090, km 3 + 077, km 10 + 830, junction Bisko, km 22 + 925, PUO Mosor (Figs. 8.64, 8.65 and 8.66), etc. *on part Dugopololje—Šestanovac—Zagvozd*.

Along the Istrian Y highway.

Primary research for the "Istrian Y" (A-8 and A-9 highways) began in 1989 with the engineering-geological mapping of tracks. At that time, 15 speleological objects that were on the route of the future highway were reconvened and partially explored.

In 1996, six more caverns on the A-8 were thoroughly explored. A detailed speleological survey of the pit at the km 13 + 070 east of the "Istrian Y" road, Kanfanar—Rogovići (Pazin) section on A-8, in April 1996 contributed information on the location, dimensions and morphology of the object. It is a pit with a depth of 41.60 m, whose total volume is estimated at about 250 m^3. The cave continues, but due to the dumped garbage, further passage to cavers was prevented.

Fig. 8.16 Martina Jama or cavern at km 71 + 754

Fig. 8.17 Map of Bosiljevo 2—the junction of A-6 and A-1 motorways with positions of explored caves

A detailed speleological survey of the cave at km 16 + 080 road west of the "Istrian Y" road, section Kanfanar—Rogovići (Pazin) on A-8, in April and May 1996 contributed information on the location, dimensions and morphology of the object. The pit is 86.40 m deep, with a horizontal projection of 33.70 m, with a total volume estimated at around 2300 m^3. The cave continues, but it has been estimated that for the purpose of this research it is not necessary to cross the underground lake at a depth of 86 m. Its further dimensions are unknown so far, and most likely it is the so-called hanging water in the groundwater, which means that the level of permanent groundwater is even deeper. An occasional active stream of water was observed in the pit, with a lake about 1.0 m deep. The rocks in which the pit was formed were very intensively eroded and corroded by water drip, and speleothems were formed in many places. The pit was developed in a north-west–south-east direction, corresponding to a fault plain that played a significant role in its genesis.

During 2010 and 2012, four more caverns on the A-9 were explored.

Detailed speleological survey of the speleological object, in the drainage canal, at the km 30 + 550 station on the Istrian Y highway, section Nova Vas—Višnjan, at the Višnjan junction near Višnjan in Istria, contributed information on the position, dimensions and morphology of the

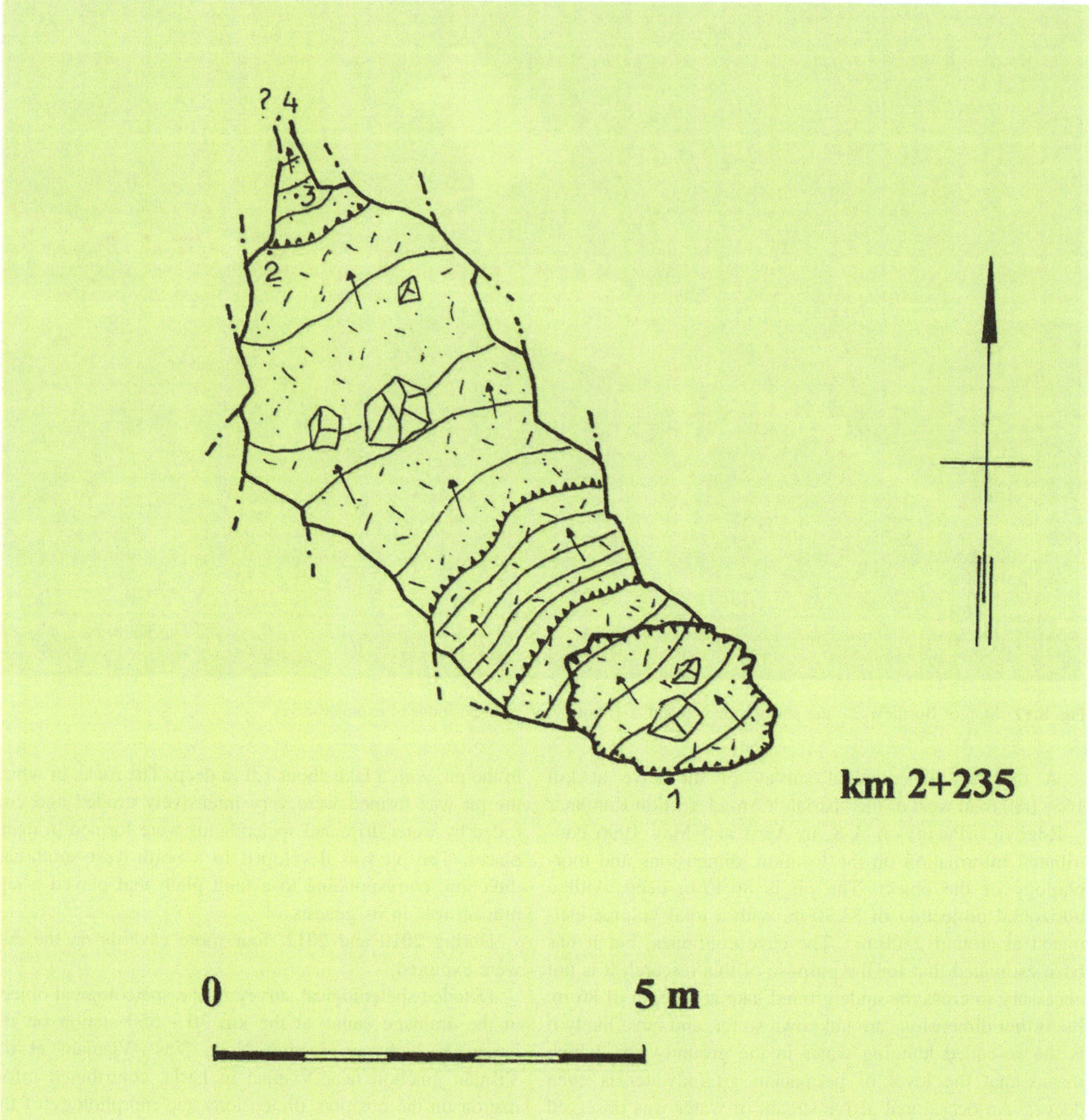

Fig. 8.18 Plan of cavern at km 2 + 235 on Bosiljevo—Josipdol section

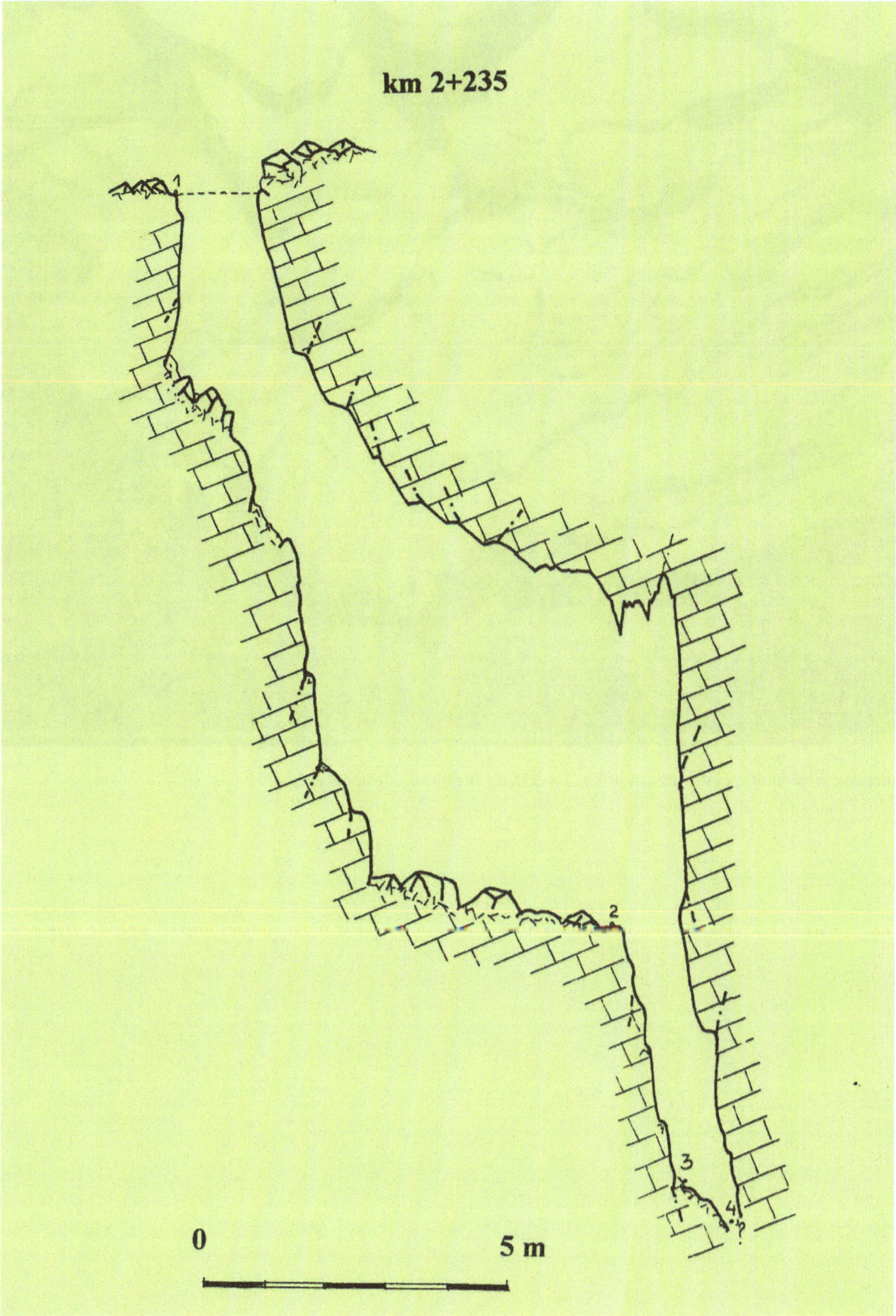

Fig. 8.19 Profile of cavern at km 2 + 235 on Bosiljevo—Josipdol section

Fig. 8.20 Intensive erosion of rocks in cavern at km 2 + 235 on Bosiljevo—Josipdol section

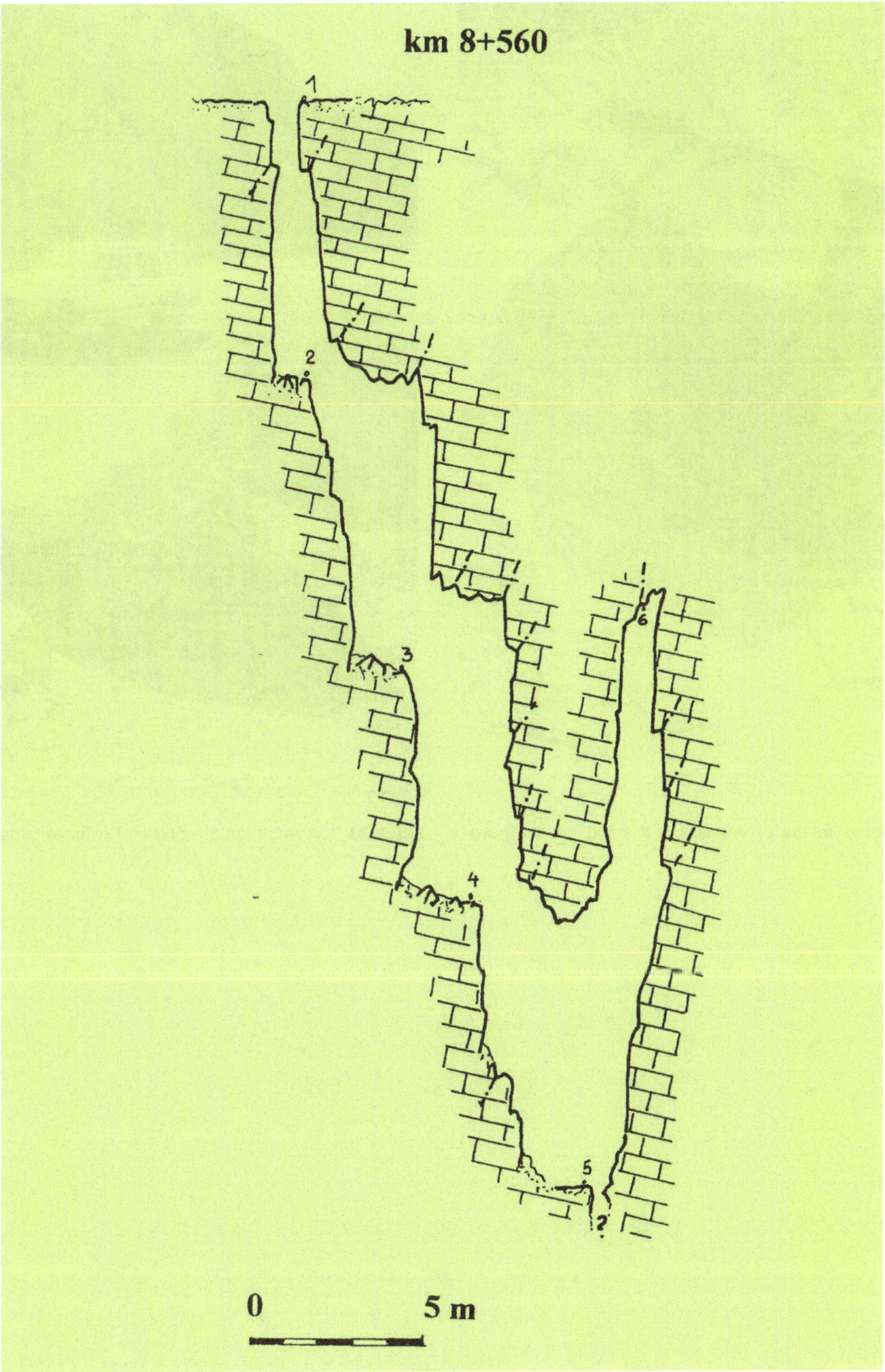

Fig. 8.21 Profile of cavern at km 8 + 560 on Bosiljevo—Josipdol section

Fig. 8.22 Entrance to the cavern at km 8 + 560 on Bosiljevo—Josipdol section

Fig. 8.23 Cavern at km 8 + 560 on Bosiljevo—Josipdol section

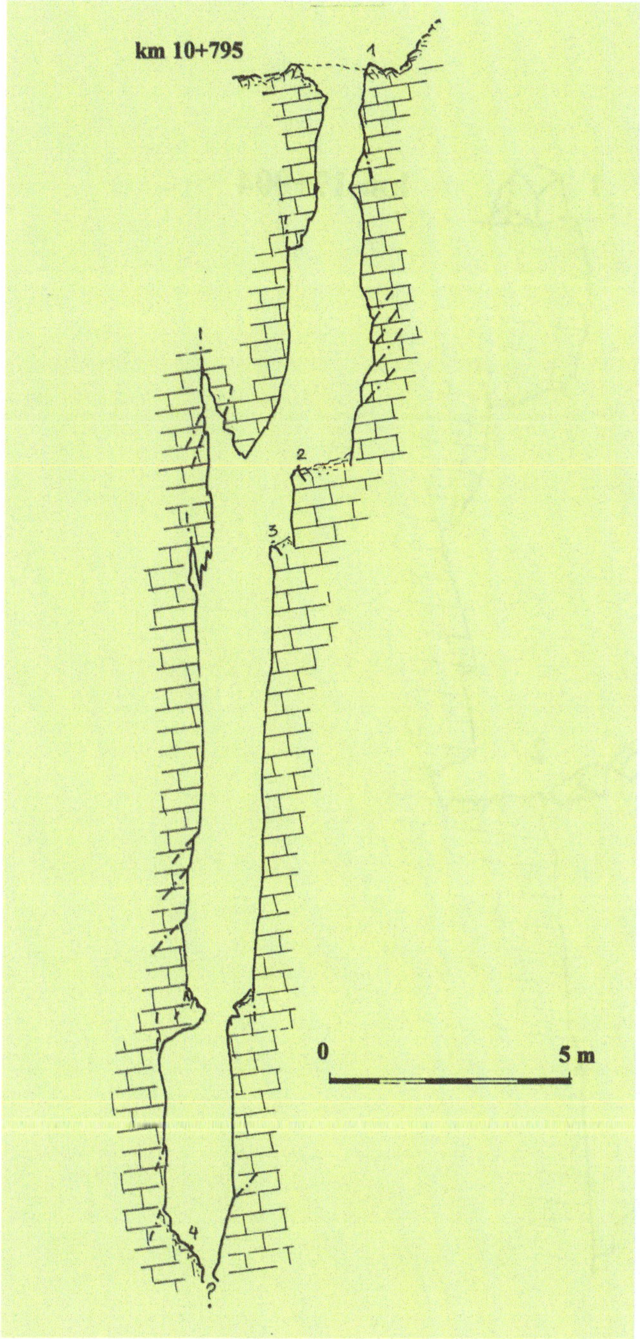

Fig. 8.24 Profile of cavern at km 10 + 795 on Bosiljevo—Josipdol section

Fig. 8.25 Entrance to cavern at km 10 + 795 on Bosiljevo—Josipdol section

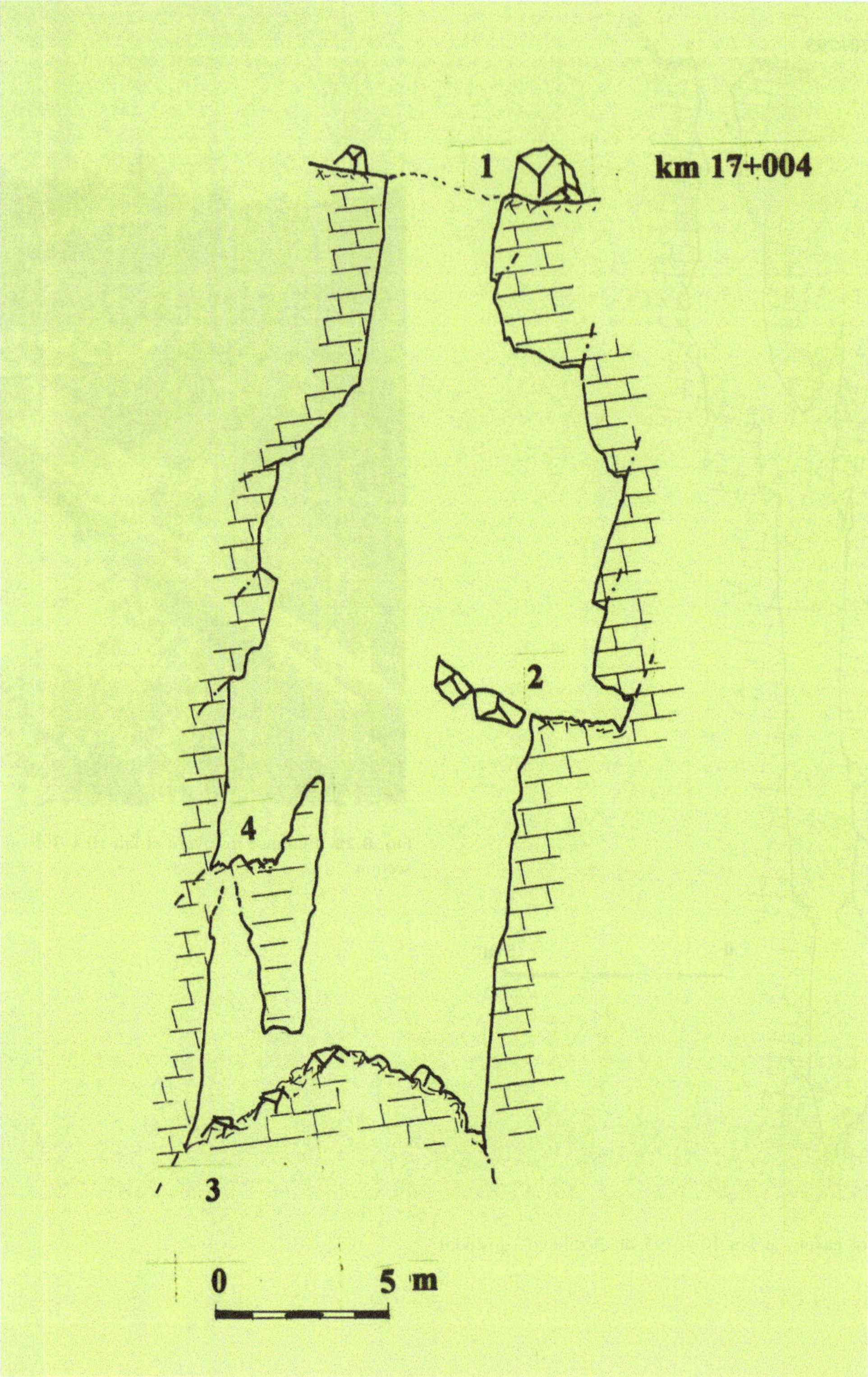

Fig. 8.26 Profile of cavern at km 17 + 004 on Bosiljevo—Josipdol section

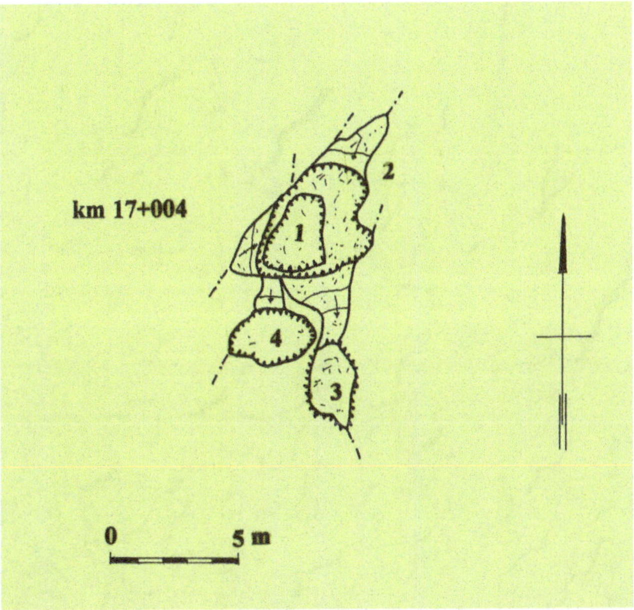

Fig. 8.27 Plan of cavern at km 17 + 004 on Bosiljevo—Josipdol section

Fig. 8.28 Entrance to cavern at km 17 + 004 on Bosiljevo—Josipdol section

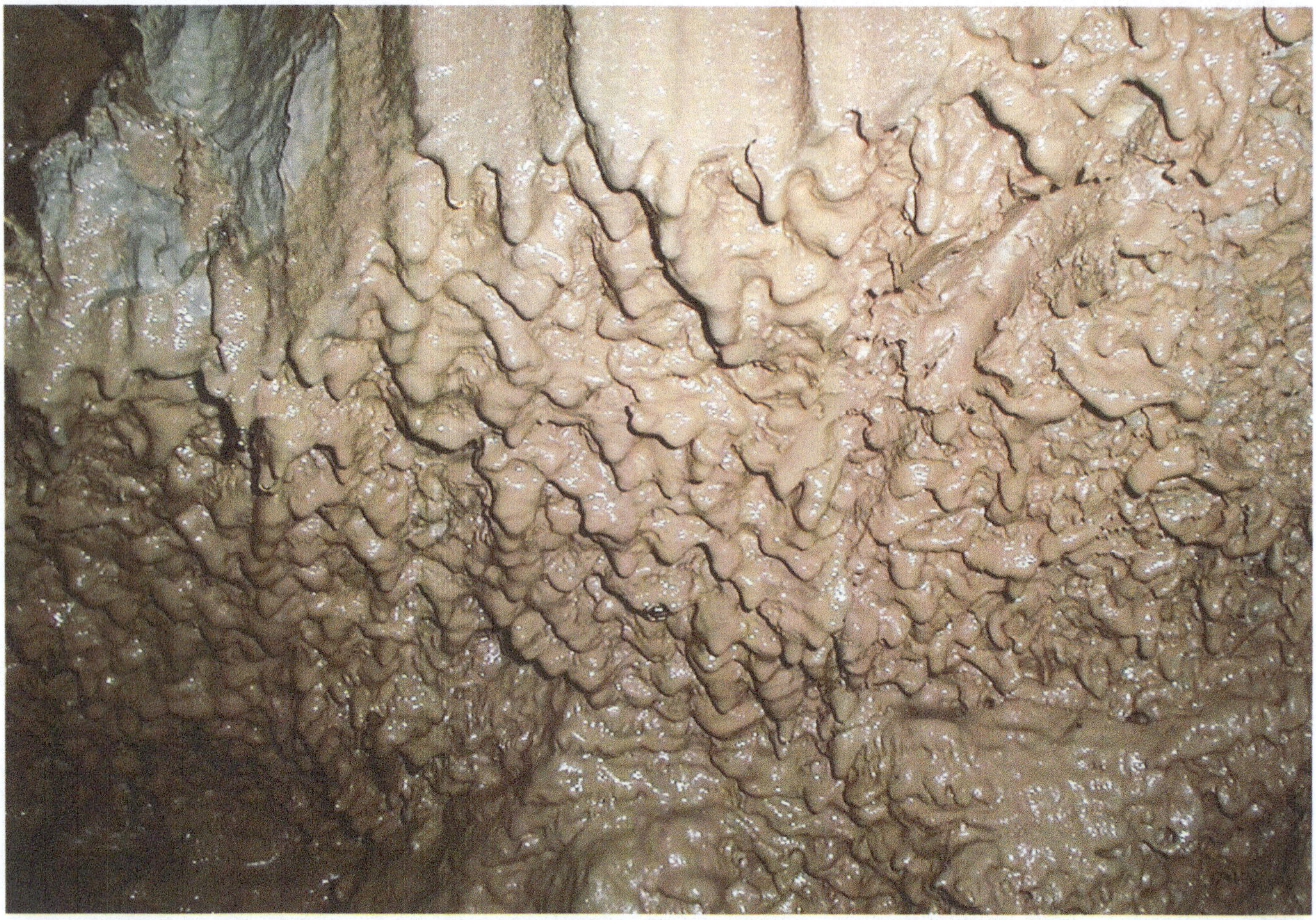

Fig. 8.29 Mud speleothems in cavern at km 17 + 004 on Bosiljevo—Josipdol section

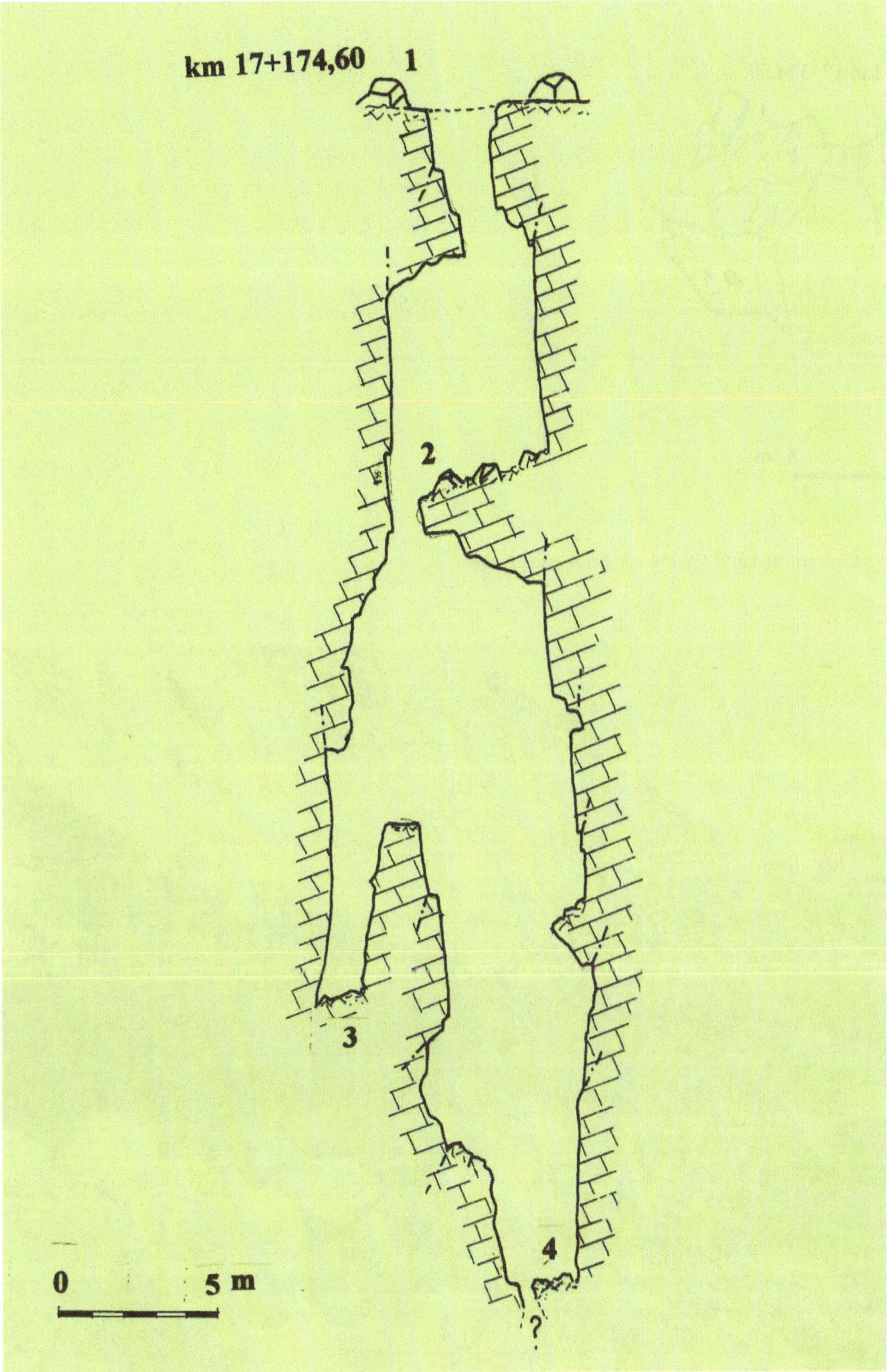

Fig. 8.30 Profile of cavern at km 17 + 174 on Bosiljevo—Josipdol section

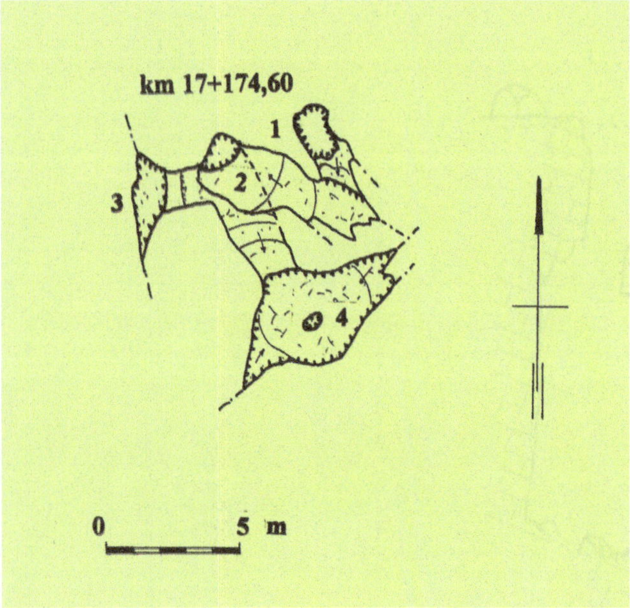

Fig. 8.31 Plan of cavern at km 17 + 174 on Bosiljevo—Josipdol section

Fig. 8.32 Cavern called Debelo brdo near Bjelopolje in Lika on Korenica—Udbina section of the D-1 road

Fig. 8.33 Cavern at km 1 + 870 on Debelo brdo section

Fig. 8.34 Speleothems in cavern at km 1 + 870 on Debelo brdo section

Fig. 8.35 Total length of caverns on Debelo brdo section is about 650 m

Fig. 8.36 Cavern on Debelo brdo section

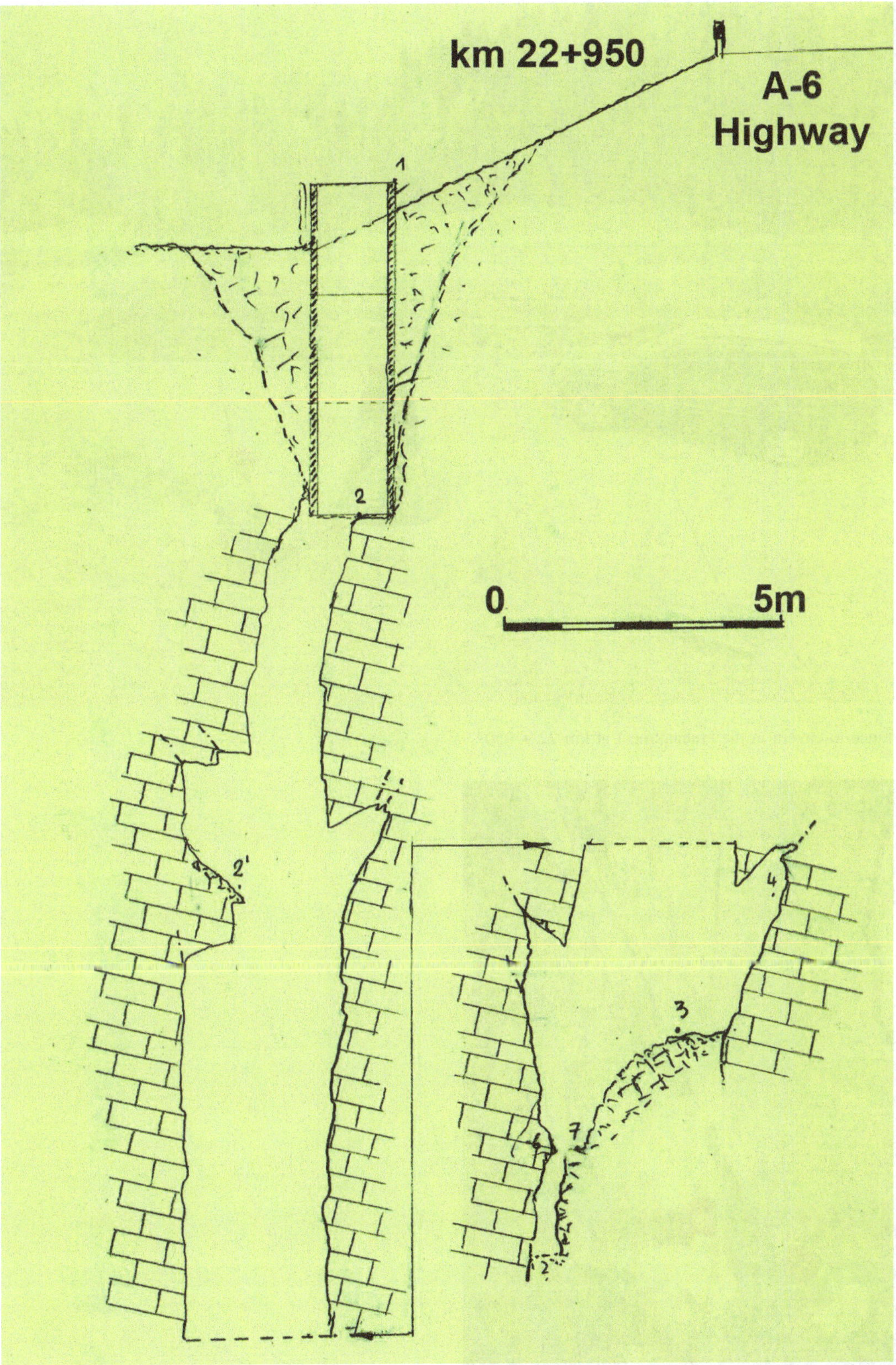

Fig. 8.37 Profile of cavern in the embankment at km 22 + 950 near Čardak Tunnel in Gorski Kotar

Fig. 8.38 Entrance to cavern in the embankment at km 22 + 950

Fig. 8.39 Cavern in the embankment at km 22 + 950

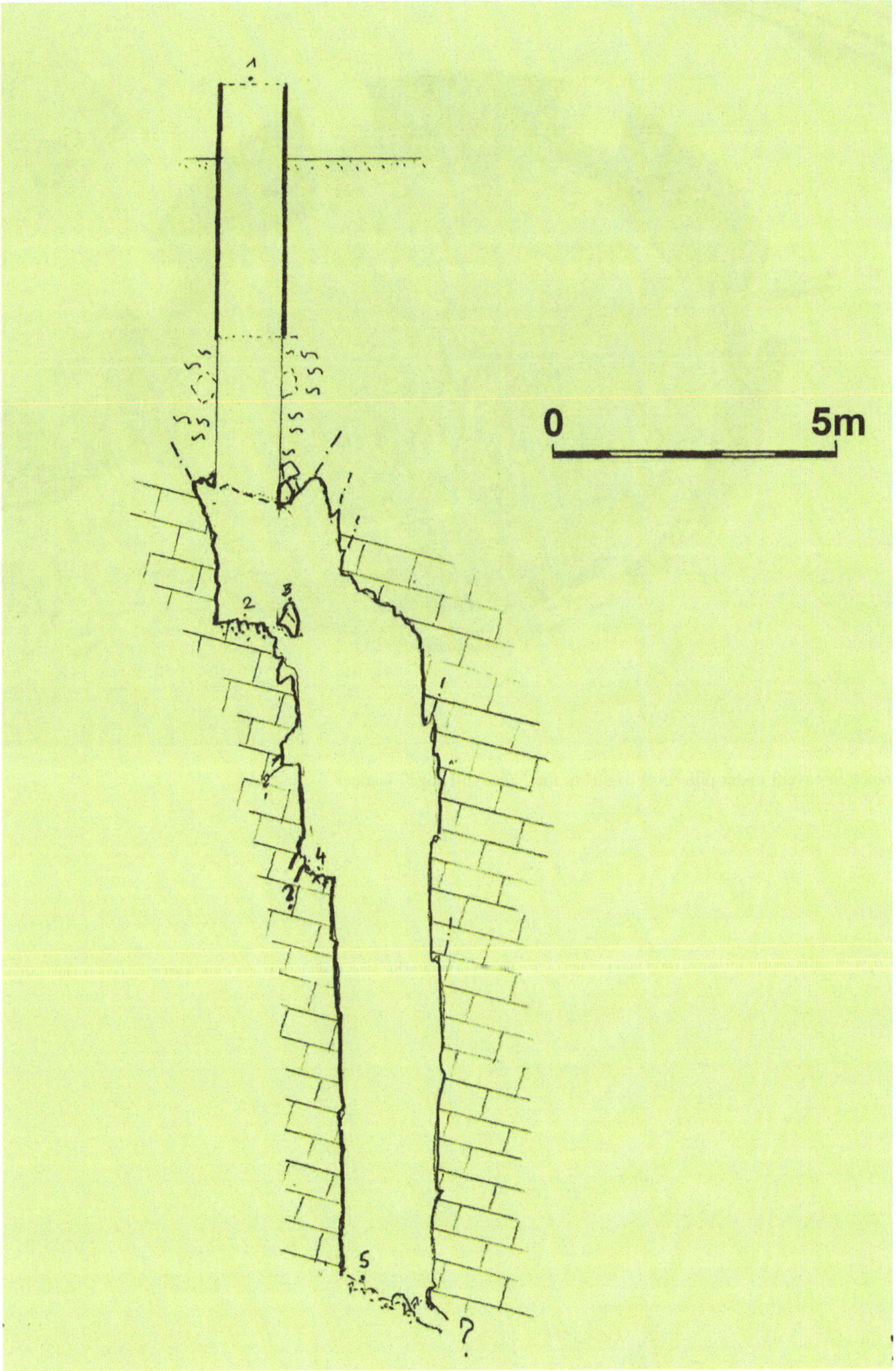

Fig. 8.40 Profile of cavern under pillar S-9 (right) in the "Zečeve Drage" viaduct near Vrbovsko

Fig. 8.41 Entrance in cavern under pillar S-9 (right) in the "Zečeve Drage" viaduct

Fig. 8.42 Cavern at km 9 + 000 in the green grass area between two lanes,

Fig. 8.43 Use of geophysical methods in cavern at 9 + 000

Fig. 8.44 Cavern at km 0 + 393 under pillar on U-2, Katarina viaduct

Fig. 8.45 Cavern at km 0 + 393 under U-2 pillar, Katarina viaduct

Fig. 8.46 Cavern at km 0 + 160 on Rujevica road junction, ramp 5, in Rijeka

Fig. 8.47 Entrance to cavern at km 3 + 660 in Diračje road junction in Rijeka

Fig. 8.48 Lower part of cavern at km 3 + 660 in Diračje road junction

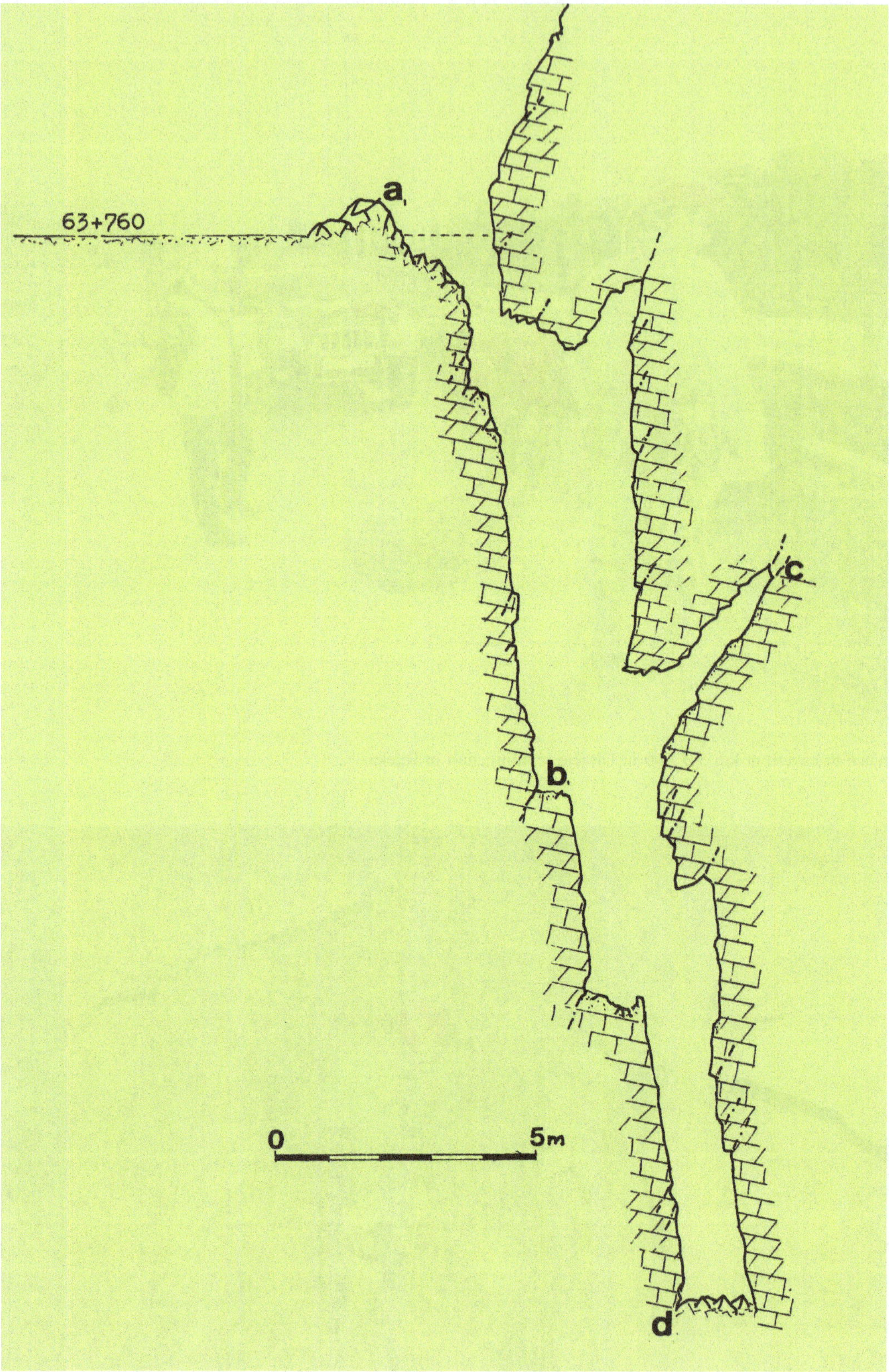

Fig. 8.49 Profile of cavern at km 63 + 760 on A-6 highway, Kupjak—Vrbovsko section

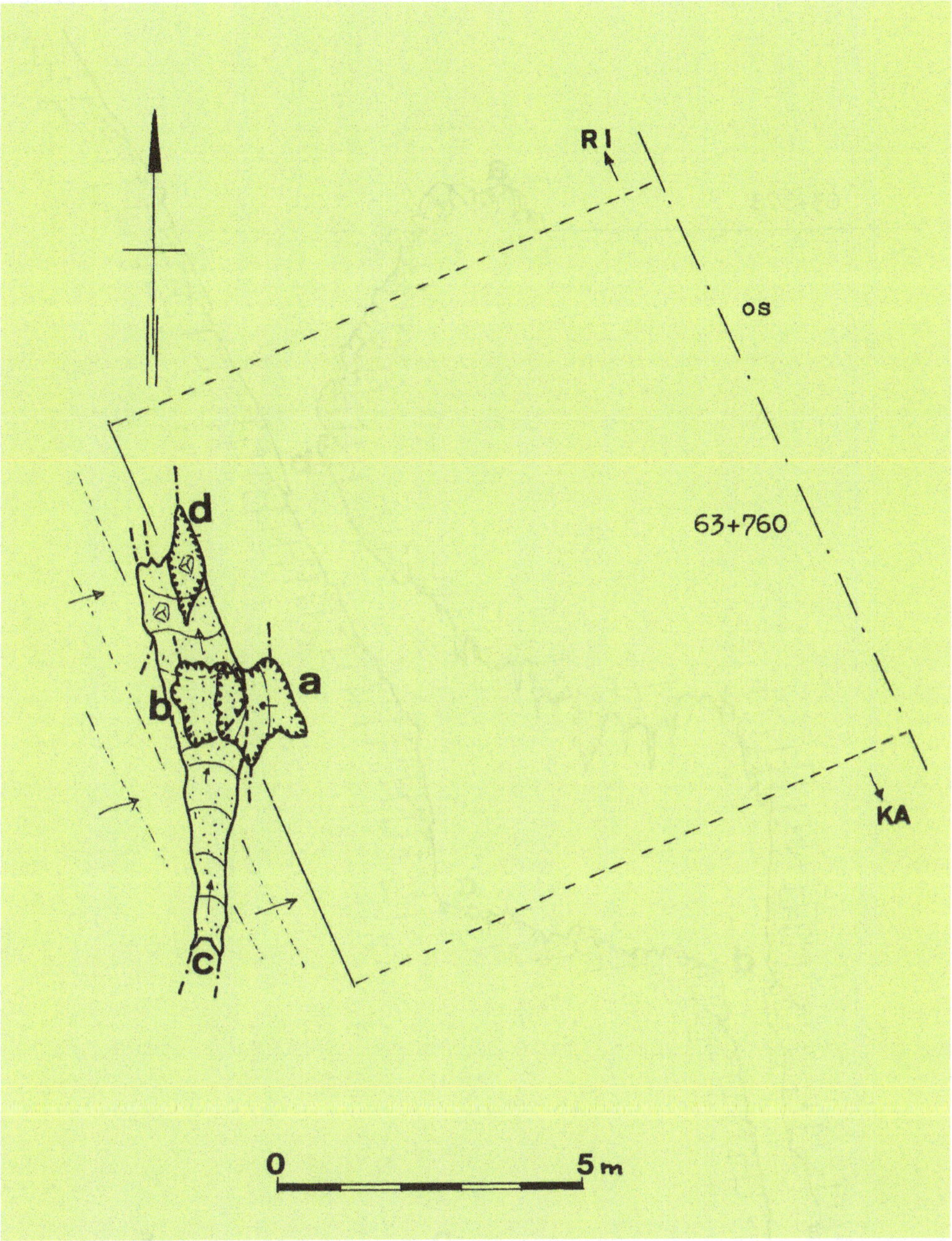

Fig. 8.50 Plan of cavern at km 63 + 760 on A-6 highway, Kupjak—Vrbovsko section

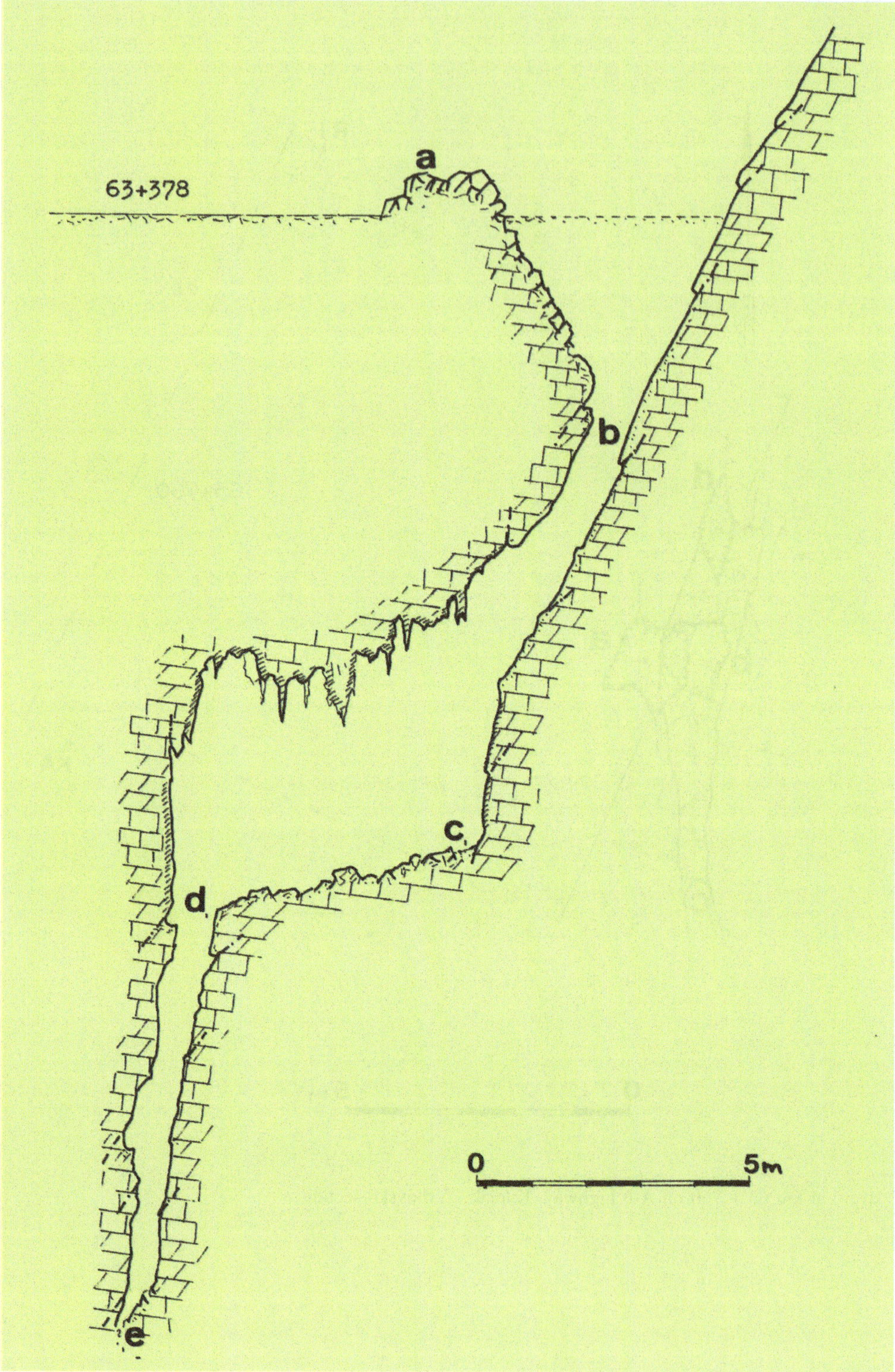

Fig. 8.51 Profile of cavern at km 63 + 378 on A-6 highway, Kupjak—Vrbovsko section

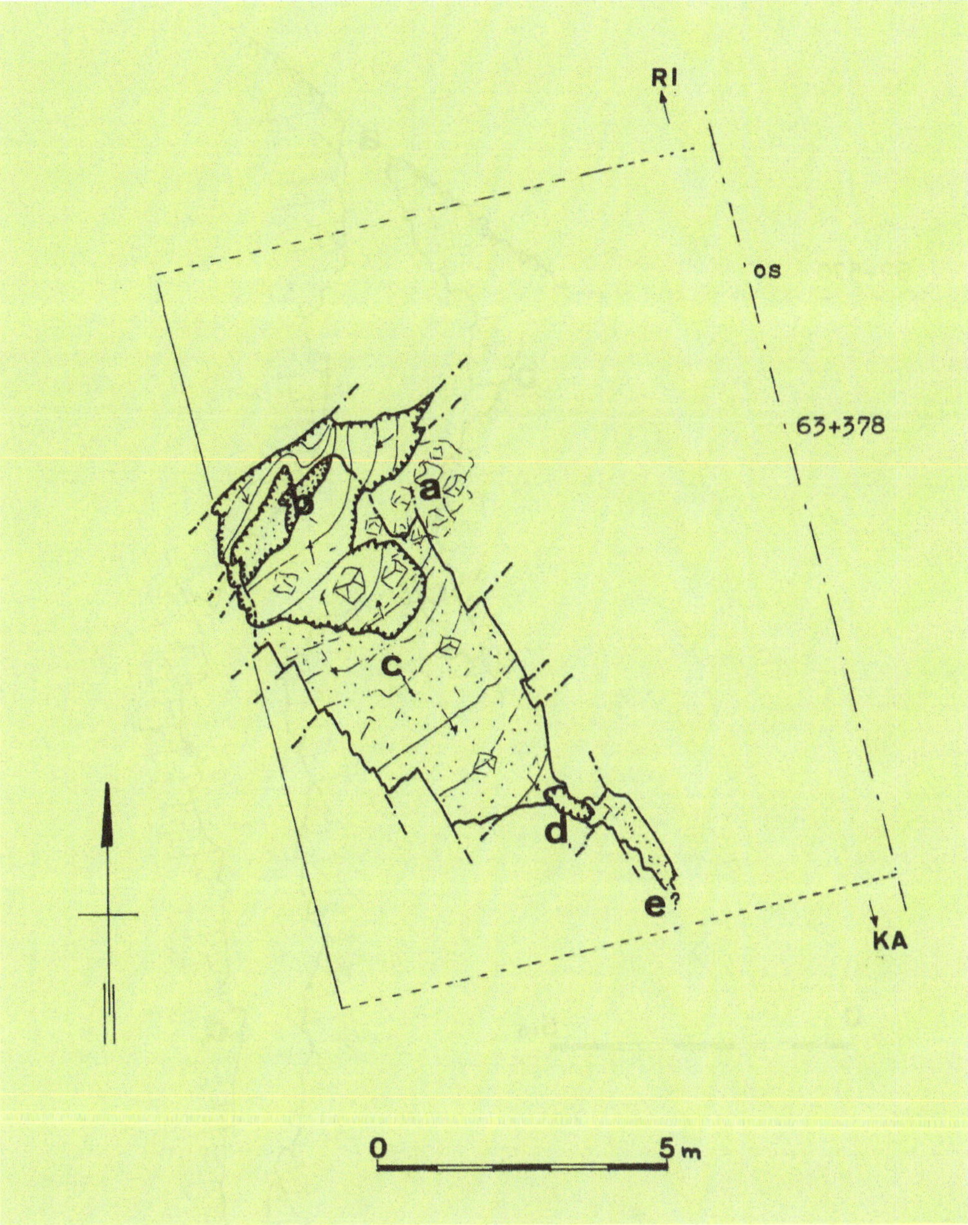

Fig. 8.52 Plan of cavern at km 63 + 378 on A-6 highway, Kupjak—Vrbovsko section

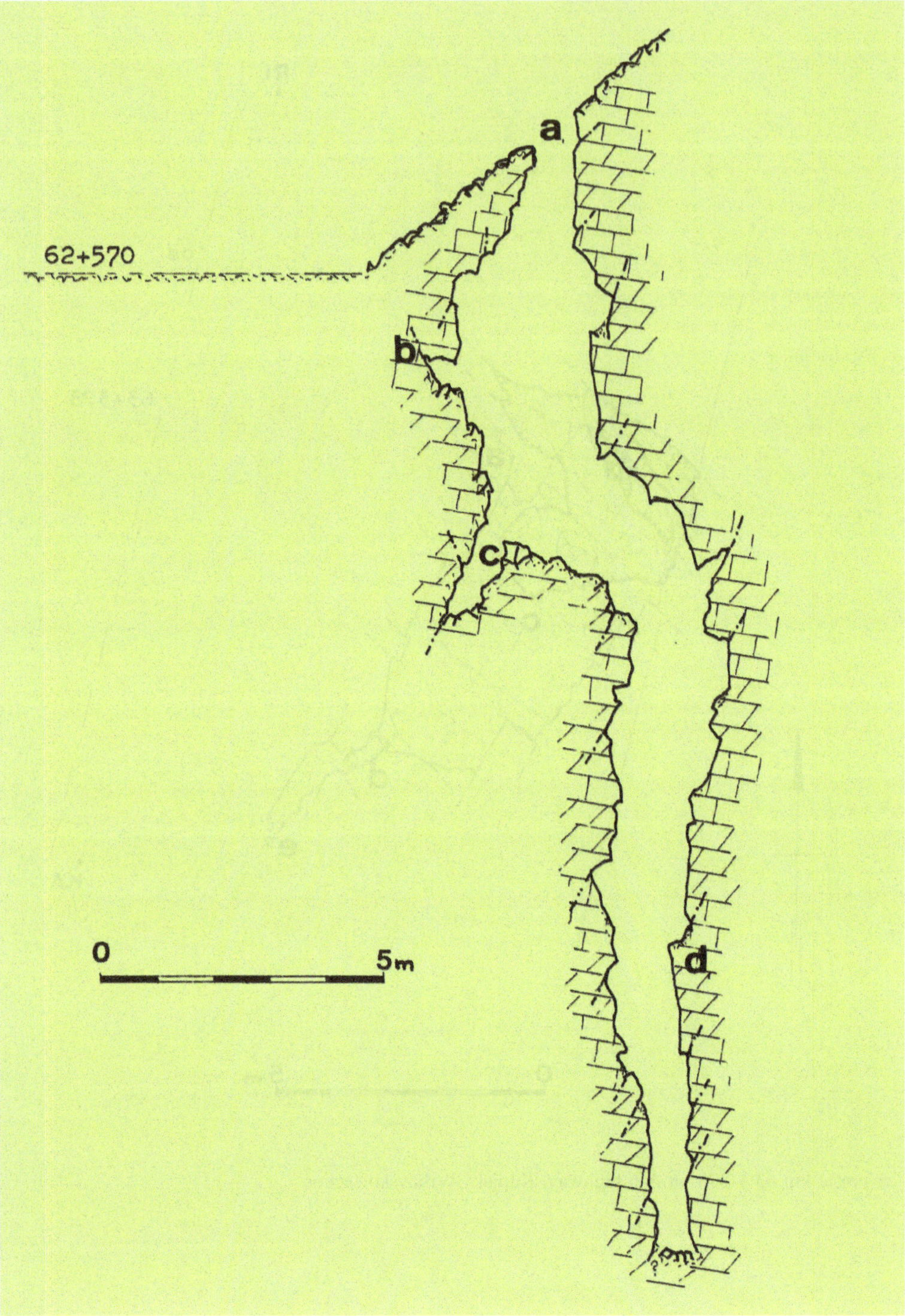

Fig. 8.53 Profile of cavern at km 62 + 570 on A-6 highway, Kupjak—Vrbovsko section

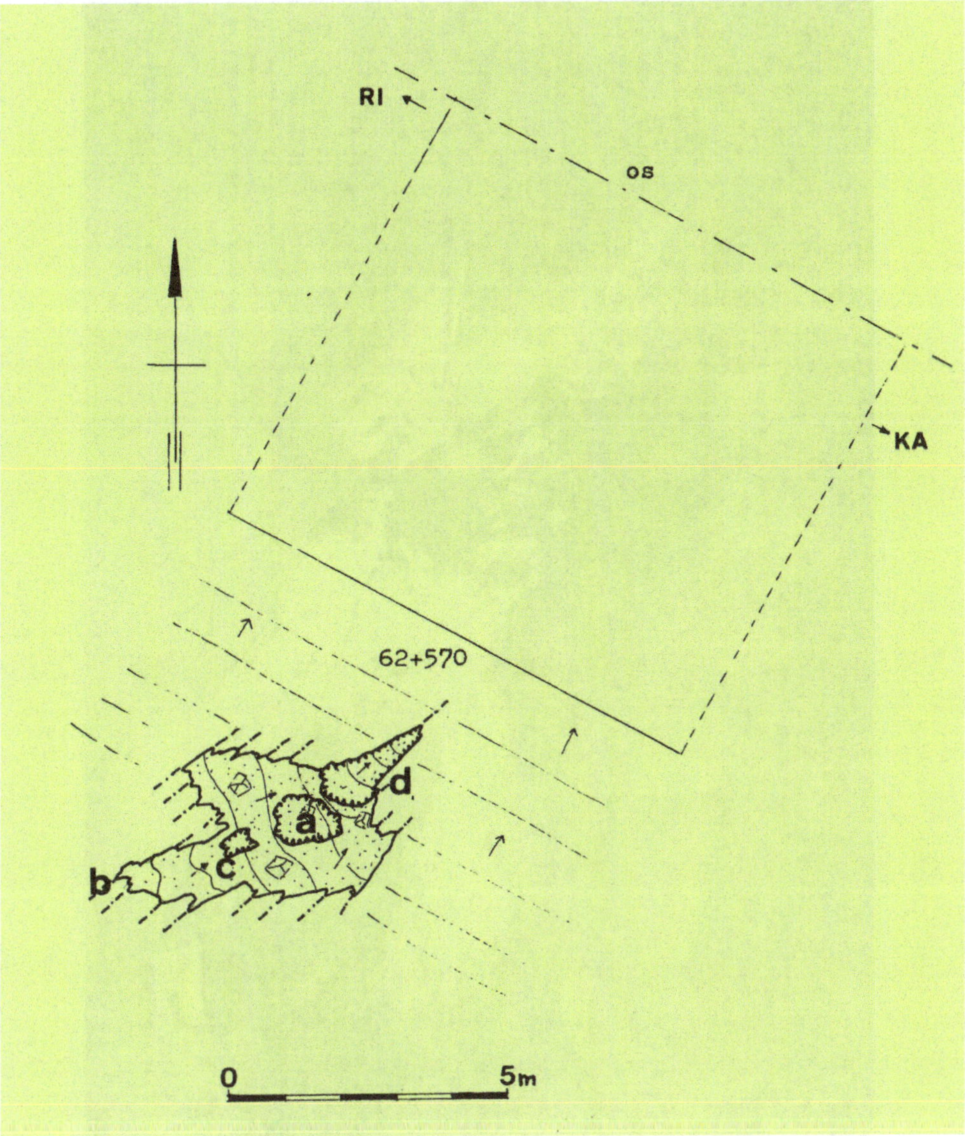

Fig. 8.54 Plan of cavern at km 62 + 570 on A-6 highway, Kupjak—Vrbovsko section

Fig. 8.55 Cavern at km 14 + 229 on A-1 highway, Josipdol—Mala Kapela Tunnel section

Fig. 8.56 Cavern at km 0 + 920 on A-1 highway, Mala Kapela—Žuta Lokva section

Fig. 8.57 Cavern at km 4 + 528 on A-1 highway, Mala Kapela—Žuta Lokva section

Fig. 8.58 Cavern at km 3 + 168 on A-1 highway, Mala Kapela—Žuta Lokva section

Fig. 8.59 Cavern at km 4 + 750 on A-1 highway, Mala Kapela—Žuta Lokva section

Fig. 8.60 Cavern at km 18 + 880 on A-1 highway, Žuta Lokva—Ličko Lešće section

Fig. 8.61 Cavern at km 13 + 625 on A-1 highway, Žuta Lokva—Ličko Lešće section

Fig. 8.62 Cavern at km 18 + 350 on A-1 highway, Žuta Lokva—Ličko Lešće section

Fig. 8.63 Cavern at km 28 + 760 on A-1 highway, Ličko Lešće—Lički Osik section

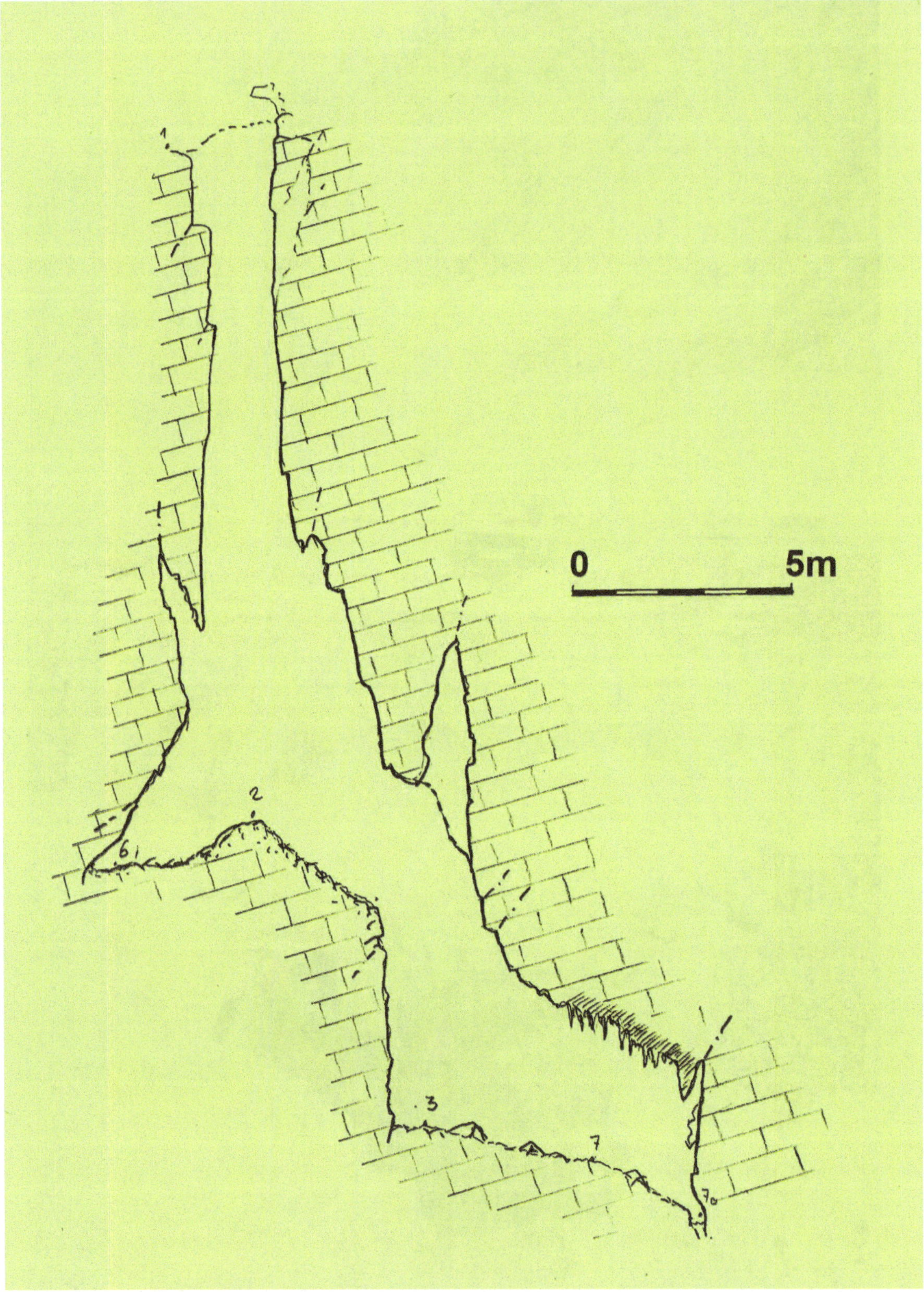

Fig. 8.64 Profile of cavern at km 22 + 925 on Mosor rest station

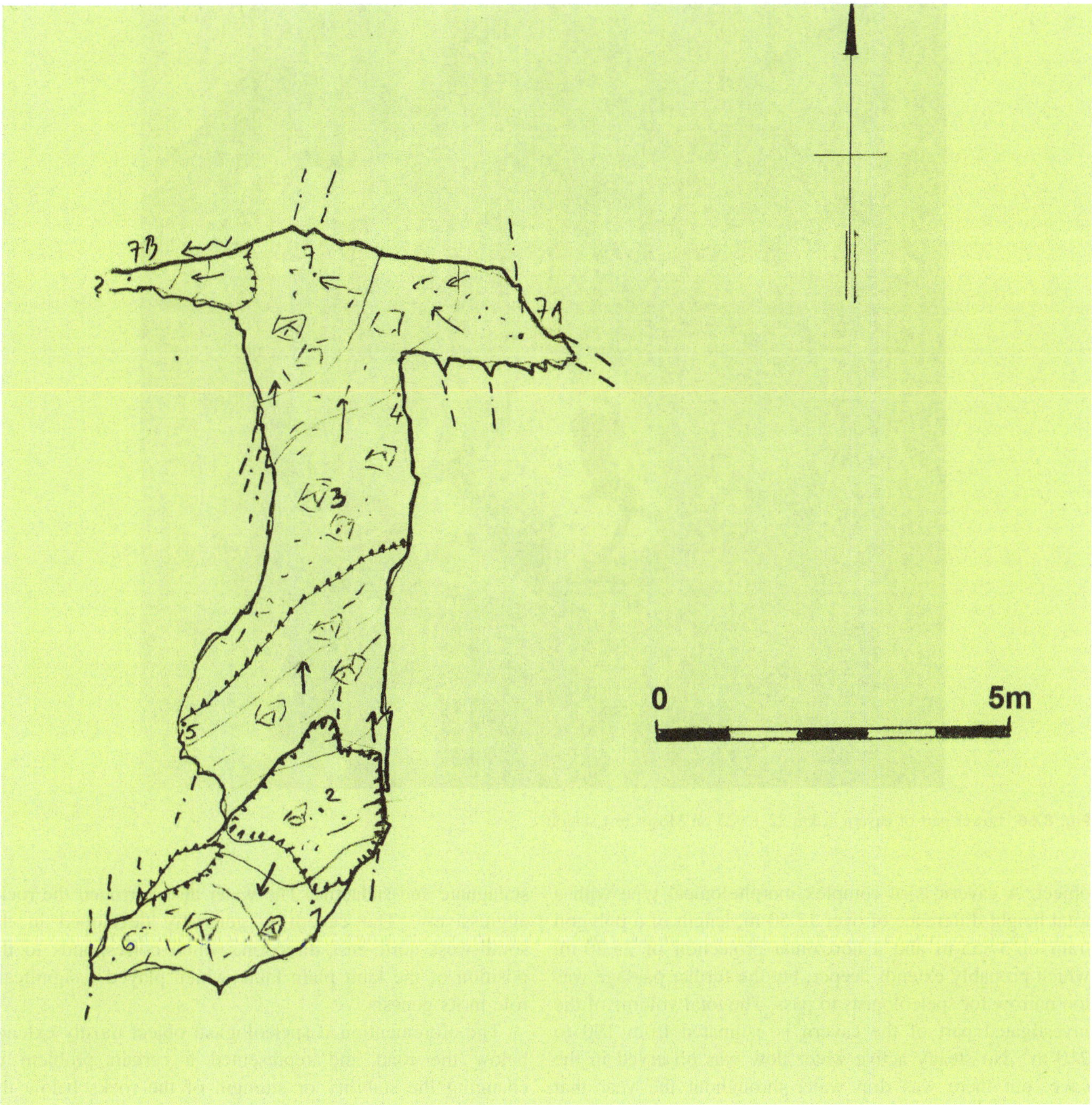

Fig. 8.65 Plan of cavern at km 22 + 925 on Mosor rest station

Fig. 8.66 Lower part of cavern at km 22 + 925 on Mosor rest station

object. A cavern is of complex morphological type with a total height difference of over 12.50 m, length of a polygon train of 37.55 m and a horizontal projection of 25.20 m, which probably extends deeper, but the further passage was too narrow for speleologists to pass. The total volume of the investigated part of the cavern is estimated from 190 to 210 m^3. No steady active water flow was observed in the cave, but there was drip water throughout the year that created speleothems in the form of crust, stalagmite, stalagnate and stalactite. The water also corroded the rocks aggressively. The cavern is generally developed in the south-west–north-east direction, which corresponds to the position of the fault plain fault, which played a significant role in its genesis.

The aforementioned speleological object mostly extends below the road and represented a certain problem of changing the stability or strength of the rocks below the future road.

9 Caverns Found and Explored During the Construction of Airports and in Quarries

Abstract

During the construction of airports, quarries and their facilities in the Dinaric Karst, hundreds of caverns were found, explored and documented. Most of these caverns are less known to the public because of their strategic and military significance. During the construction of the runways of Dubrovnik Airport, about twenty caverns were found, one of which was turned into a show cave. Caverns in quarries are common, but information about them depends on the users and owners of these quarries. Unfortunately, their existence is often discovered only when they are destroyed or buried without prior research. Some very significant and large speleological systems have been discovered in the quarries. Groundwater and its condition are always taken under consideration, and caverns were also found during karst mining, for example, Labin, Raša, Omiš, Kistanje, Drniš, etc.

9.1 Airports

Caves were discovered during the construction of Croatian Karst airports. Unfortunately, until 1990, these caverns were only known through personal contacts, since papers about them were not published. It is known that large cavities were found during the construction of military airports such as Željava near Bihać, in the Croatian terrain with several caves where the halls were larger than 60 × 50 m were found between 1955 and 1965, in the Ćilipi Airport near Dubrovnik in 1960 and 1962 where 4 pits about 20 m deep and one cave about 150 m long were found, in Krk Island airport one pit about 15 m long was found, and in military airport in Divulje near Trogir several pits of unknown dimensions were found.

Caverns in Dubrovnik Airport.

Ever since the construction of the first civilian airport in Dubrovnik in 1961, it has been known that in the period from 1950 to 1960, the Yugoslav army found several caves and pits on the site of the military airport. These speleological objects posed a threat to the stability of the control tower, administrative buildings, parking lot and airport runway. Nothing was known about these caves in public until recent research in May 2012. Then, during the works for the expansion of the runway, several new pits were found and explored. (Garašić and Cvetković 2012). On that occasion, information was collected about several previously known pits and caves that were explored and rehabilitated during the period from 1950 to 1960. The first detailed description of all new pits was given by Garašić (2013). Five vertical speleological objects were found in the Upper Cretaceous limestones. In them, the water moves vertically towards the sea, creating vrulja (submarine springs) along the coast in the vicinity of Molunat and Ćilipi. Fleury et al. (2007) wrote about such submarine freshwater springs. We should also mention the Đurović Cave, which is located inside the airport. Its length is 197 m (Kovačević 2006; Buzjak 2006). To date, it is the only cave in the world that is arranged for tourist visits and is located below an international airport. Subsequent research for the expansion of building "C" at Dubrovnik Airport (Fig. 9.1) was conducted from May to October 2014. In that period, six new vertical caves were explored, of which K-5 (Fig. 9.2) is the deepest. The pit was explored to a depth of 58 m, where the channels get too narrow for the person to pass (Garašić and Garašić 2017).

M. Garasic, *The Dinaric Karst System of Croatia*, Cave and Karst Systems of the World,
https://doi.org/10.1007/978-3-030-80587-6_9

Fig. 9.1 Building “C” at Dubrovnik Airport

For the widening purposes of the terminal building “C” of Dubrovnik Airport (Fig. 9.1), detailed speleological survey was conducted in May and October 2014. Five new caverns were added to the list of six previously known caves under the Dubrovnik Airport which sums up to total of eleven caves. One of the caves is open for tourists which is unique example in the world. The longest cave under Dubrovnik Airport measures 197 m, and the deepest one measures 58 m in depth. In explored caves (K-1a, K-1b, K-2, K-3, K-4, K-5) (Fig. 9.2) show signs of gravitational karstification in their upper parts, and in K-2, the lower parts of the cave show traces of regressive karstification, i.e. aggressive progression process of karstification towards the surface. The situation is similar in the nearby Đurovic Cave. Traces of groundwater oscillations were found, but the corrosion forms speleothems were seen.

During the construction of the foundations of the building “C” at Dubrovnik Airport, a deterioration of the stone material in the underground indicated the existence of caves. A number of such areas was determined (from K-1a to K-5) (Figs. 9.3, 9.4, 9.5 and 9.6). Those caves were not known or discovered in the earlier researches. Detailed cave explorations determined that all of the caverns lie below the future building “C” and therefore must be carefully treated. No active groundwater flow has been detected, and the dimensions of caves do not indicate to the possibility of groundwater level rising. Occasional groundwater level might be at least seventy or more metres below the lowest point of those caverns, i.e. maximum of 40–50 m above the sea level. Caverns are not in any relationship with the nearby Đurović Cave. All of the explored caves were formed in layers of carbonate rocks of the Upper Cretaceous (Garašić and Garašić 2017) which are slightly inclined (between 10° and 20°) to the northeast. But the basic predisposition of their genesis is related to fracture and fault systems that are almost perpendicular to the so-called Dinaric direction. The accuracy of these assumptions can be verified by comparing the position of layers with tectonic elements within the nearby Đurović Cave which was done during the last two research expeditions. Besides the famous Đurovic Cave in the

Fig. 9.2 Cavern K-4 at Dubrovnik Airport

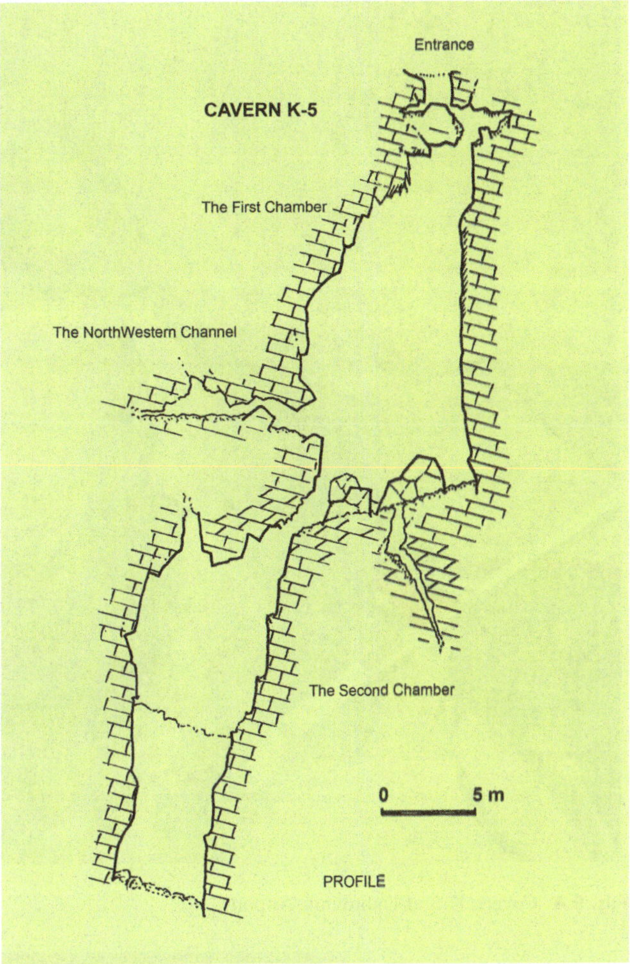

Fig. 9.3 Profile of cavern K-5 at Dubrovnik Airport

vicinity of the mentioned caverns, there are two more caves ("Velika jama" with the depth of 31 m and "Mala jama" with the depth of 10.5 m). They are documented during the speleological research for the needs of control tower and terminal building of Dubrovnik Airport construction back in 1961. These data indicates the specific intensive vertical karstification in this area. The altitude of this area is about 155 m above the sea level. There are turning axis of tectonic movements. Detailed speleological research of new caves (caverns K-1 to K-5) in the area of Dubrovnik Airport's building "C" next to Ćilipi, Konavle, in Dalmatia, contributed to the information about the position, dimensions and morphology of the objects. They are mostly simple pits with the vertical difference of 11 m (K-4) and 58 m (K-2) which most likely extends even deeper, but it has not been proved yet. Total volumes of the newly researched caverns are estimated to be from 20 to 3000 cubic metres. As mentioned before, no active water flows were detected, but the existence of speleothems indicates the presence of dripping water. Speleothems are found in the forms of crust, stalactites, salagmites and corrosion forms. Caves are formed mainly in the northeast–southwest direction which corresponds to the fault plane system which played a great role in the speleogenesis of this area. It is likely that the future construction works will encounter more caverns with the

Fig. 9.4 Cavern K-5 at Dubrovnik Airport

Fig. 9.5 Lower parts of cavern K-5 at Dubrovnik Airport

Fig. 9.6 Cavern K-2 at Dubrovnik Airport

similar position along the same or the parallel fault planes that are in the same tectonic movement (with a rotating axis) caused by general subduction.

During the construction (1965–1975) of the Željava military airport in the Croatian part of the Lička Plješevica Mountain, several cavities were encountered in several places. Some of them were used for the underground storage areas of the airport, which was no longer in use after the Croatian War of Independence (1991–1995). It was completely devastated and partially collapsed. Near the Željava Airport, in Hodak Cave, one of the larger underground halls is located. As well as within Baričeva cave, which connect the areas of Krbavica, Klokot and Kuruzovića pećine—interesting hydrogeological phenomena on the border with Bosnia and Herzegovina. Unfortunately, scientists and cavers do not know all the details about the speleological objects that were found during the construction of the airport.

9.2 Quarries

9.2.1 Vrelo Quarry

The Vrelo Cave was discovered in a quarry near Fužine in Gorski Kotar in 1950. Cave is about 350 m long (Fig. 9.7).

9.2.2 Tounj Quarry

Cave in Tounj quarry (8639 m long—Lacković 1989) is connected (in 2019) with Tounjčica Cave through submerged channels. Together they make Tounjčica Cave system with the length of 9104 m and vertical difference of 124 m (Figs. 9.8, 9.9 and 9.10).

Speleological explorations of caves in the Tounj quarry are of recent date, but records of Tounjčica Cave date back

Fig. 9.7 Vrelo Cave near Fužine in Gorski Kotar

to the seventeenth century. The earliest information about the Tounjčica Cave dates from 1689 when Ivan Vajkard Valvazor, a Slovenian nobleman researcher and scientist, wrote the work "Slava vojvodine Kranjske" about the town of Tounj and gave a pictorial representation of the town where the entrance to the Tounjčica Cave is clearly visible. The first real speleological survey of the Tounjčica Cave was published in 1935 by geologist and speleologist Josip Poljak. More recent cave exploration of this cave has continued since 1956 when it was topographically recorded for the first time. Research is ongoing today. Mining at the Tounj quarry has repeatedly opened different caverns, but most of them have been completely excavated or buried in waste rock.

Here are some rare photos (Fig. 9.11) from the first cave dive organized there. On that occasion, 9 June 1973, cave divers in the syphon reached the depth of −33 m and length of 60 m. After this action, many cave divers (Croatian and international) tried to find the connection between this two caves but without results. Cave diving in Tounjčica Cave in 1973 was the beginning of modern cave diving in Croatia (Figs. 9.12 and 9.13).

Fig. 9.8 Cave channel in Tounjčica Cave

Fig. 9.9 Lake in Tounjčica Cave

Fig. 9.10 Syphon connecting two caves into Tounjčica Cave system

9.2.3 Cave in Debeljača Quarry

For the construction purposes of A-1 motorway in the karst part of Lika, a quarry on Debeljača hill near Gornja Ploča was opened. During the exploitation years, between 2002 and 2004, many caverns were detected. Some of them were unfortunately buried. A large cavern was explored only partially. The owners and users of the quarry did not give permission for longer detailed exploration. The cavern was first surveyed in 2002. It was established that after the vertical part, the cave continues horizontally. The last research was carried out in 2012 finding that the cave is 1082 m long and 67 m deep. The original entrance was 2 × 3 m, and today, it is 10 × 20 m (Bajo 2006).

9.2.4 Caves in Srinjina and Podoštra Quarries

In "Srinjine" quarry, several vertical caverns were opened at the northeast border of the exploitation field, at altitudes of 215 and 221 m.a.s.l. The quarry was located on the north side of Perun hill in Dalmatia and was operational between 2000 and 2004. The rocks in which the caverns are formed belong to the Upper Cretaceous, Senonian K_2^3 limestones. Their layer thickness varies between 20 and 70 cm. The deposits are well stratified, white, sometimes grey and light grey in colour. The deposits are part of the overturned anticline, and the layers are tilted towards northeast.

The cavern at 221 m a.s.l. is of a knee-shaped morphological type with a total height difference of 27.30 m and a

Fig. 9.11 First cave dive organized on 9 June 1973 in Tounjčica Cave

total horizontal projection (length) of 16.80 m. The cavern is likely to continue, but the passage for speleologists was prevented by the so-called false bottom at the entrance part of the cave and the large amount of speleothems in the interior of the cavern. The total volume of the investigated part of the cavern is estimated to be between 750 and 800 m^3. The cavern at 215 m a.s.l. is 39 m deep, and its volume is estimated to be between 1,400 and 1,450 m^3.

Podoštra quarry is located near Gospić in Lika. Cave discovered in that quarry has two entrances. One is vertical (16 m deep), and the other is horizontal. Total length of the cave is 176 m (Figs. 9.14 and 9.15). An underground lake exists in the cave.

9.3 Groundwater Protection When Discovering New Caves

In Croatian Karst, groundwater is constantly present in about 18–20% of all caves. Research shows that, due to the global climate change, groundwater levels are changing more intensively in some places. According to the data, between 1980 and 1985, groundwater was found in about 35% of all known speleological sites (5,263) (Garašić 1986). This means that groundwater is present in about 1/3 of all caves (Figs. 9.16 and 9.17). If caverns were added, the percentage of water containing caves would be about 50%. Particular attention is paid to the protection of groundwater (Biondić et al. 1998). Substantial amounts of groundwater were found in some caverns. The water is transported by special pipelines to the surface for the purpose of water supply, for example, Učka Tunnel, Sveti Ilija Tunnel, etc. Many towns and villages in the Dinaric Karst are supplied or were supplied with water extracted directly from the springs (Rijeka–Rječina, Split–Jadro, Dubrovnik–Ombla, Delnice–Kupica spring, Prezid–Trbuhovica, Fužine–Ličanka spring, etc.) or caves (Vrgorac–Betina Cave, Zadar–Golubinka Cave, Skrad–Zeleni Vir, Opatija–Učka cavern, etc.).

In the study of caverns along roads, particular attention was paid to the protection of groundwater. A good example is the construction of a bypass around Vransko Lake on the island of Cres (2006–2008), where the greatest accumulation of fresh drinking water in the Adriatic islands is located. Its maximum depth is -84 m, with a volume of about 220 million m^3 of water. The entire road was isolated from substrates with special foils. Speleological objects (caverns) are specially protected and left in their primary form.

During construction works in the Učka Tunnel, a watercourse that is now used for water supply of the tourist town Opatija was found in a cavern. During the summer months, the water from cavern in Učka Tunnel is used as an additional water supply, because of the large consumption by tourists. The entrance to the cavern is located at the altitude of approximately 500 m. The watercourse was discovered in 1977, partially explored in 1978 and 1980, and is in use since 1982. Prior to that, potential contaminants on the surface (former INA resort on Poklon, Učka Mountain) had to be repaired. Subsequent surveys in 2018 and 2020 extended the cave system by several kilometres.

Fig. 9.12 Tounjčica Cave system is longer than nine kilometres

Fig. 9.13 Underground lake in the Tounj quarry

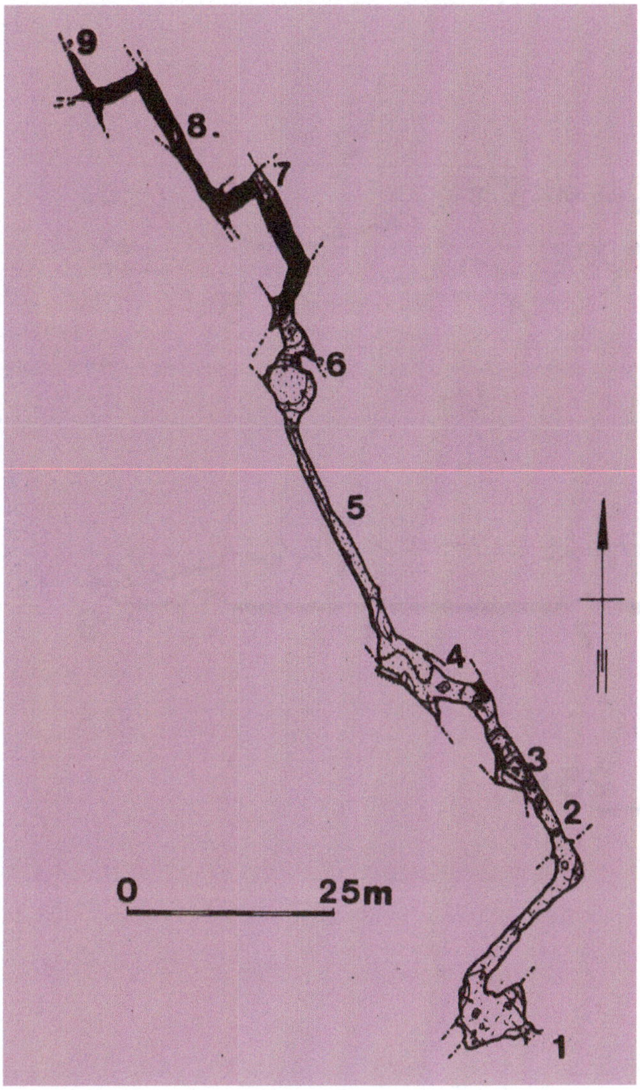

Fig. 9.14 Plan of cavern in Podoštra quarry

Great care for the protection of groundwater was also taken during the excavations of Tuhobić Tunnel, Sveti Rok Tunnel (Velebit Mountain) and Sveti Ilija Tunnel (Biokovo Mountain). The caverns found along the highways and roads were protected from additional contamination by special foils. The original groundwater flow remained unchanged in most of them.

9.4 Caverns Found in Mines and Exploratory Wells

In the long tradition of mining in Croatia, caverns were occasionally found in the underground mines. Caverns were not particularly interesting to miners unless the water spilled out from them or they were filled with speleothems and minerals. For example, in the Istrian coal mines, several specimens of Proteus anguinus appeared in the mines from the caverns with groundwater. After mining in the Istrian coal mines stopped, groundwater filled the large areas of abandoned mines. In Ravni Kotari, alumina, an aluminium ore was dug from paleo caverns, and marl was used for cement production in Omiš. Iron ore was exploited in underground mines in Kotlenice. On the island of Brač, bitumen was extracted from the underground at Škrip. In Gorski Kotar near Mrzle Vodice and in Lika near Štikada, barite was extracted. In the marl mine near Omiš, a few caverns were discovered (Fig. 9.18). In all these mines, caverns were found, which unfortunately were not explored during the mining operations, but some of them provide a wealth of information today. In Drniš, in some places abandoned lignite mines were not properly rehabilitated. They now represent a threat of collapsing. Groundwater from the caverns seeped into some of the mines.

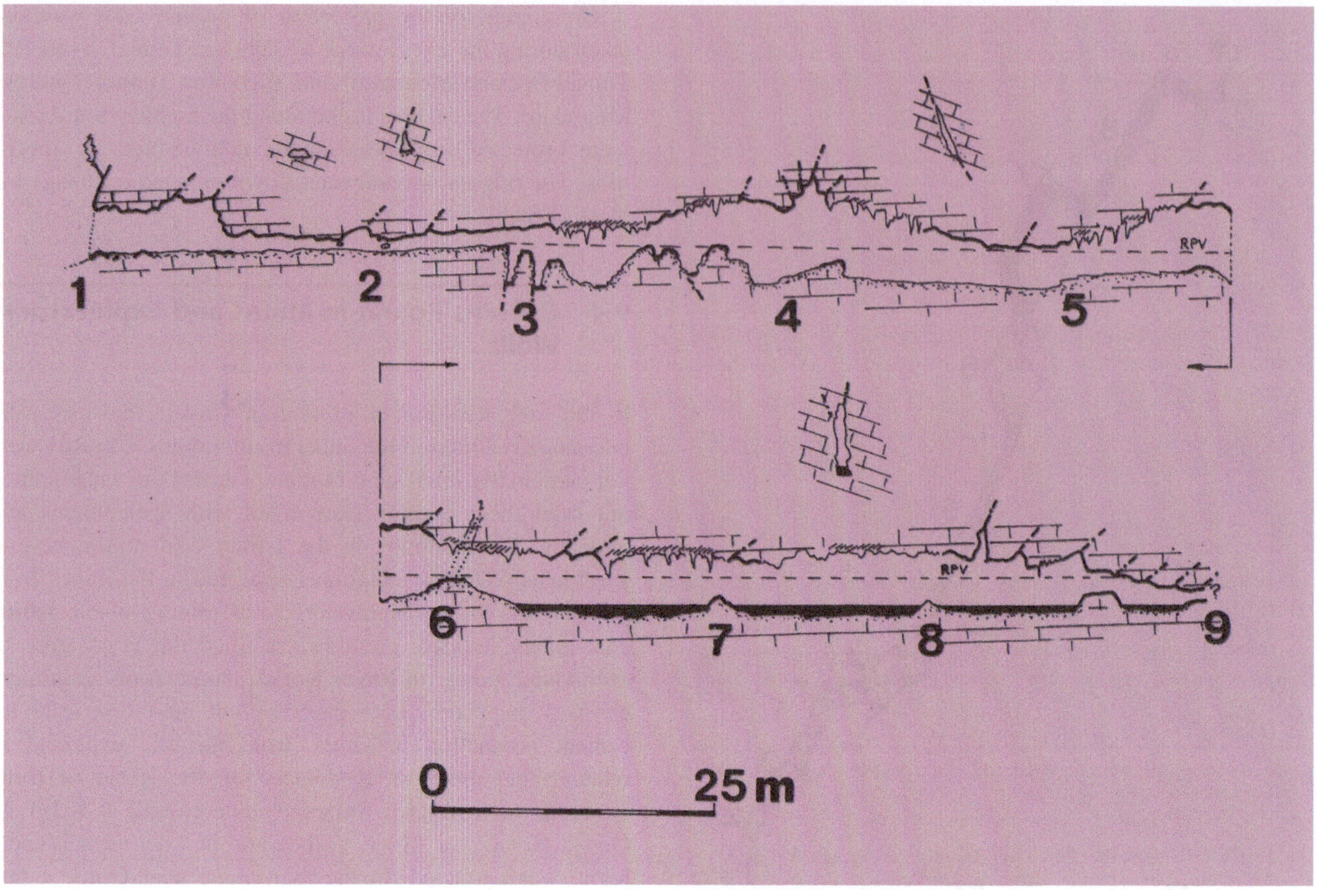

Fig. 9.15 Profile of cavern in Podoštra quarry

Fig. 9.16 Groundwater is present in about 1/3 of all caves (Zala Cave)

Fig. 9.17 Groundwater in Zala Cave

Fig. 9.18 Caverns in marl mine near Omiš

During exploratory drilling for mineral resources, geotechnical testing or hydrocarbons (gas, oil, bitumen, asphalt), the so-called sinking of drilling equipment is often observed, which means the appearance of caverns. Sometimes the caverns are several inches wide, sometimes several metres even tens of metres. The deepest discoveries of caverns in the Dinaric Karst in Croatia were found at a depth of 3,150 m. This was detected in a deep well in the Adriatic when a cavern filled with speleothem was discovered at that depth. This information represents the evidence that karstification in this part of the world is several kilometres deep. Croatia has 133 deep oil wells in its area of the Adriatic Sea. In the deepest drilled exploration well Vlasta-1, 5,400 m deep, indicators of deep karstification appear below 4 km deep. Cavities were found below 4,500 m in JJ-1a well which is 4,750 m deep. It is estimated that carbonates (and thus potentially karstification) are also present at 6100 m as it is mentioned in some INA (Croatian Oil Company) reports (Labaš 1987; Grandić et al. 2002).

Both exploration wells and mining operations in the Dinaric Karst in Croatia almost always, if more extensive, come into contact with the caverns in some way. This information is evidence of the abundance of caverns in this area.

10 Particularly Interesting and Rare Occurrences in Caves of the Dinaric Karst in Croatia

Abstract

Some phenomena that occur frequently in the Dinaric Karst of Croatia are rarely seen in other parts of the World, so they are mentioned here. For instance, vruljas (underwater freshwater springs under the surface of the Adriatic Sea) are sometimes impressive with the amount of water discharge and the length of submerged channels. They are always present under the large mountain massifs of Učka, Velebit, Mosor, Biokovo, etc. Estavelas—caves located on the edges of karst fields have a dual hydrogeological function: spring and ponor. The largest karst fields in the world, which are located in the Dinaric Karst, are also mentioned here. Large vertical shafts are a special feature of this karst area. Rare occurrences of caves in clay, marl and tufa, special and rare types of speleothems, caves in ice, underground bifurcations and measurements of radon gas concentration in the caves of the Dinaric Karst in Croatia are also mentioned.

The peculiarity of the Dinaric Karst in Croatia in relation to other areas is striking. Only in Croatia Dinaric Karst is observed below the surface of the Adriatic Sea. Many vruljas or submarine freshwater springs (Benac et al. 2003, 2004) were found and explored there, while they are not found in other parts of Dinaric Karst. Vruljas are deeper than 100 m and longer than 3000 m. It is similar to the occurrence of anchial caves and caves in the coast and islands.

The area of Dinaric Karst in Croatia hosts many rare and unique karst phenomena, such as one of the strongest estavelas (Velika pećina), a unique clay cave, caves in which snow and ice are present throughout the year although they are only ten or less kilometres away from the Adriatic coast (Gorski Kotar, Biokovo, Velebit, etc.), caves formed in travertine barriers which are several hundred m long and caves formed in marls (longer than 1000 m). Deep karst springs (explored by diving to the depths of over 200 m), vertical shafts deeper than 500 m, which are among the deepest in the world, very rare or unique speleothems, etc., are found in this area.

10.1 Vruljas

Vruljas are submarine freshwater springs (Fig. 10.1). They are exclusive karst phenomena occurring in the sea beneath high and significant mountain ranges, for instance, beneath the mountains of Učka, Velebit, Mosor, Biokovo, etc. Vruljas were described in the scientific literature over a hundred years ago. The first list of vruljas in Dinaric karst was given by Alfirević (1969a). At that time, cave diving was not yet developed. Therefore, many vruljas from that list were later proved to be significantly smaller or significantly larger.

Hydrogeological function of most vruljas began about 18,000–20,000 years ago when the level of the Adriatic Sea rose about 86–106 m (Šegota 1968). It is clear that almost all speleological objects slightly changed since that time. They were created much earlier, i.e. millions or more years ago. Analysis of the age of the speleothems found in vruljas on the eastern coast of the Adriatic Sea undoubtedly confirms that the sea level did not rise continuously, but in several stages with different intensities. The age, size and shape of speleothems in vruljas helped determinating the speleogenesis, but also the (neo) tectonics and (paleo) hydrogeology of the area in which they are formed (Garašić 1991).

The longest vruljas in the Dinaric Karst in Croatia were explored near the villages Modrič and Zečica (Petricioli et al. 1995; Buzjak et al. 2013). Modrič vrulja is over 3 km long (Figs. 10.2, 10.3 and 10.4), Velika Vrulja (1554 m long) and Zečica vrulja is over 1.5 km long (Vasseur 2019).

Their position and morphology show that they are the part the same cave system that transports water from the Ričica area in Lika to the Adriatic Sea.

Another particularly interesting area of vruljas is located along the Adriatic coast beneath the Biokovo Mountain, in

M. Garasic, *The Dinaric Karst System of Croatia*, Cave and Karst Systems of the World,
https://doi.org/10.1007/978-3-030-80587-6_10

Fig. 10.1 Vruljas are submarine freshwater springs (Lovran)

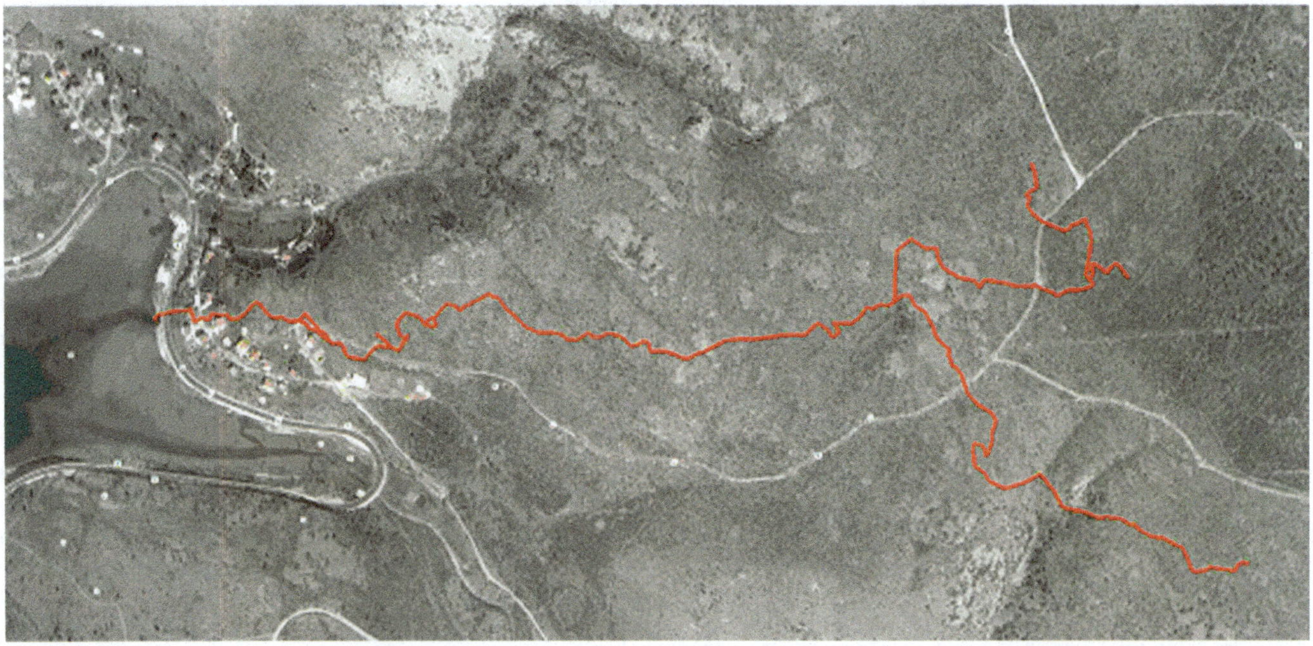

Fig. 10.2 Position of Modrič vrulja

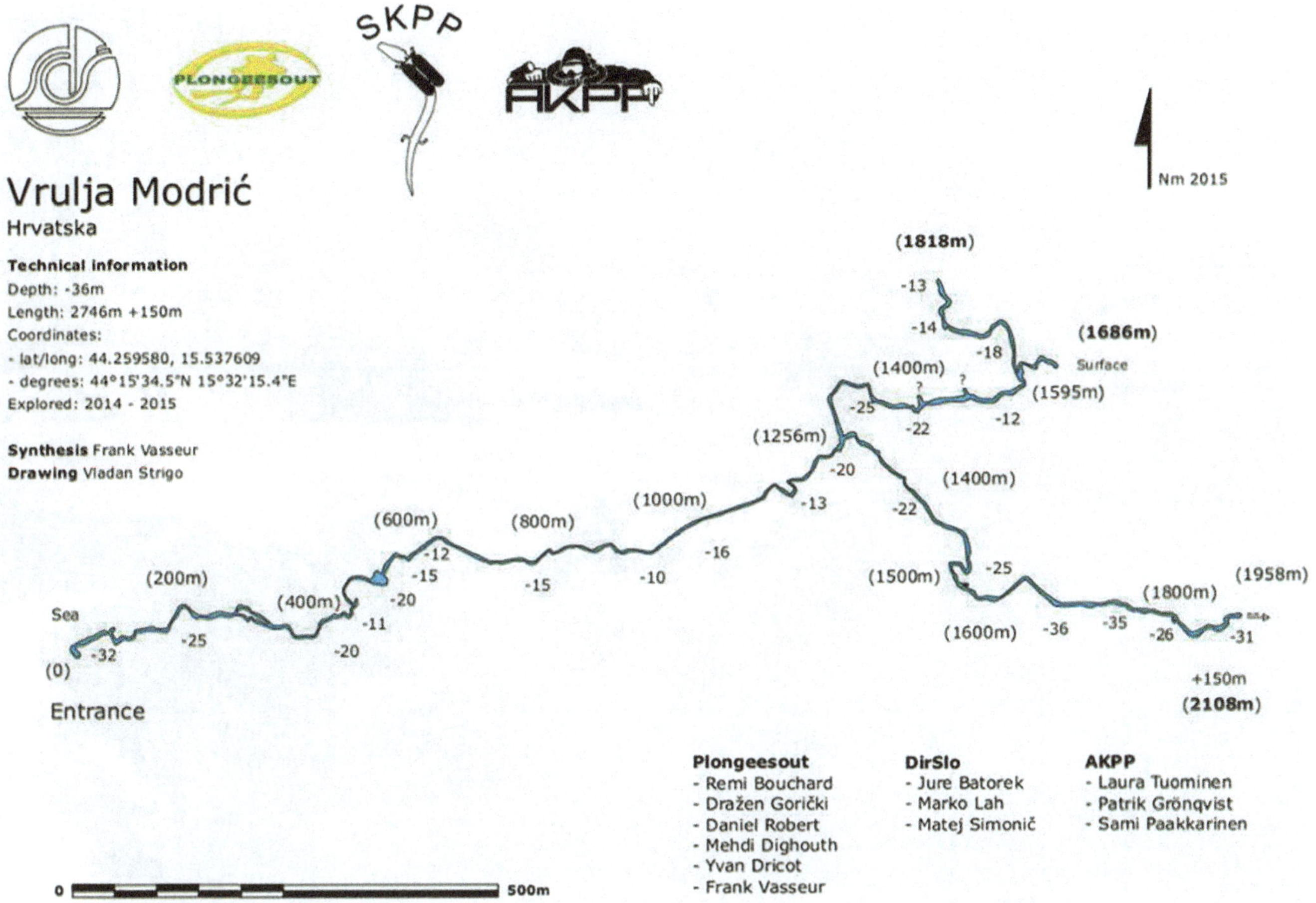

Fig. 10.3 Plan of Modrič vrulja

the area around Dubci, Brela, etc. The most famous vrulja in this area is located in Dubci. During the rainy months the flow is between 20 and 50 m^3/s, which is the highest recorded flow in this part of Dinaric Karst. These numbers are just an estimation because it is not possible to measure the exact flow rate with simple methods. However, this flow rate is far stronger than any known individual vrulja flow on the east coast of the Adriatic Sea. The flow rate was estimated based on the data from the entrance (at −50 m depth) and at depths of −40, −20, −10 and 0 m. Polish cave divers were the first to descend to −122 m in Vrulja Dubci (Fig. 10.5). According to most recent research, there are several outlets of this vrulja system. One is approximately 25 m from the coast at a depth of about 30 m, and the other one is approximately 50 m from the coast at a depth of about 50 m. There were several attempts of cave diving in this vrulja. Divers reached −100 m of depth through the first entrance in 2000. In 2002, the divers reached the depth of −125 m through the second entrance, but further diving was prevented by strong water currents. Several attempts were made to dive deeper into this vrulja, but cave divers experienced safety problems at the very entrance. The danger of tangling the safety leash was great due to the strong water current. Two years later (1 May 2004), one of participating divers, the leading Polish diver Wiktor Bolek died during preparatory deep dives in Wildschutz (Germany). After his death, no one continued diving in Dubci Vrulja. According to Wiktor, there are indications that the entrance part of Vrulja in Dubci is at least 150–160 m deep which could be foreseen from the first dives.

Although there is no direct colour water tracing evidence, it can be assumed that some underground flows from the Imotski polje (Slišković and Ivičić 2000), i.e. from the Red Lake, Blue Lake and other ten very deep valleys, could have their vruljas in the Adriatic Sea. The distance is approximately 20 km. Tectonically and lithostratigraphically, the vrulja area in Dubci could be a part of a large underground hydrogeological system. This system spreads from Herzegovina and is perpendicular to the spreading direction of the Dinarides. Some parts of this hydrogeological system were found in the Biokovo Mountain during the excavation of Sveti Ilija tunnel (2008–2010) when groundwater was found in several caverns.

Fig. 10.4 Modrić and Zečica vruljas

Anchialine caves

Speleological objects that are close to the sea and have underground water connections with the sea are called anchialine caves. They contain brackish water, i.e. mixed fresh and seawater. In fact, these are submerged caves or Vrulja that are genetically the same as those that have remained above the surface of the sea in the immediate vicinity today. Special conditions for the salinization of groundwater or the concentration of metals (Žic et al. 2011; Kwokal et al. 2014) in sediments may, in biological terms, create special habitat conditions called anhcialine caves. Such special caves are known in the karst of Croatia on the islands (e.g. the island of Mljet) or along the eastern coast of the Adriatic Sea. In the karst of Croatia over 200 are known (Surić 2006; Bakran-Petricioli and Petricioli 1999, 2008). Inorganic iodine speciation study described the waters of seven Croatian coastal caves: Urinjka Cave, Cave on Punta Ert, Vrtare Male Cave, Špilja pod Orljakom, Živa voda Cave, Lenga Pit and Bjejajka Cave. These are anchialine caves as they are connected hydraulically with Adriatic Sea surface water. According to the study of Žic et al. (2007), the tide inside the cave rise and fall as it does outside, but replenishment of the water is restricted by the karst rock. In effect, the water in the cave probably acts more like a piston and, although moving slightly vertically, has a long residence time compared to a fully flushing cave. Anchialine environments display a number of unusual features, e.g. a well-developed pycnocline, hypoxia and endemic fauna. Iodate and iodide were determined by differential pulse voltammetry and cathodic stripping square wave voltammetry, respectively. Low iodide concentrations have been consistently identified in the bottom water of the caves where concentrations of 90–100 nM would ordinarily be expected from intrusion of Eastern Adriatic surface seawater where total inorganic iodine concentrations behave conservatively with salinity the loss of the iodide implies oxidation to iodate. The respiration mechanism is favoured because of enhanced alkalinity found in the near surface waters of the caves (Žic et al. 2011). Traces of metals in the column of

Fig. 10.5 Dubci vrulja

Urinjska Cave (Cukrov et al. 2006). A cavern with an underground lake which is connected to the sea was in Pećine tunnel reported by Garašić in 2006. In the narrower area of Pećine Tunnel, there are deposits of Upper Cretaceous (Turonian–Senonian) and a small part of the breccia, which are believed to be of the Upper Eocene and Lower Oligocene ages. Upper Cretaceous deposits are represented by rudist limestones, which are usually microcrystalline and cryptocrystalline, white, white–grey or light brown, and may also be light pink in their youngest parts. Sometimes, they are represented with breccia with white fragments and reddish calcite binder. It is rarely observed on the surface that limestones are poorly dolomitic, but it can be assumed that they contain thinner layers of dolomitization breccias locally. The breccias, which are assumed to be of Upper Eocene to the Lower Oligocene age, are equivalent to "Jelar"—deposits. They are usually oligomictic, but sometimes polymictic. They are made up of angular, rarely sub-angular sections of white, light greyish or brownish upper Cretaceous Turonian–Senonian limestones, and a binder of porous calcite, coloured by admixtures of a reddish-brown ferrous substance. These breccias lie irregularly and transgressively on the Cretaceous limestones (Kuželički and Ružić, 2008). Along the coast of Rijeka Dubrovačka near Dubrovnik, three caves with a high concentration of sulphur in combination with seawater were found (Cukrov et al. 2012). The largest cave (Mokošička Cave) is flooded today due to the construction of the road.

10.2 Estavelas

One of the peculiarities of the Dinaric Karst in Croatia is the appearance of estavelas, i.e. caves with dual hydrogeological function. Usually, during the rainy season, they have a hydrogeological function of springs, and during dry season, they have a hydrogeological function of ponors (swallow-holes). They occur along the edges of karst fields. Examples are Velika Pećina near Blata in Lika, Tović near Fužine in Gorski kotar (Figs. 10.6 and 10.7), etc.

Between 1920 and 1933, Josip Poljak visited and explored Velika pećina Cave (about a hundred metres from the entrance) and in its immediate vicinity (Poljak 1935). He specifically described Blata as periodic lake with caves. The

Fig. 10.6 Tović estavela near Fužine in Gorski kotar

Fig. 10.7 Tović estavela

surrounding speleological objects that Poljak assumed were hydrogeologically connected (e.g. Sinjac) were described, in speleological terms, by Garašić (1986).

After Josip Poljak, the hydrogeological approach was taken by Bahun (1968, 1970) without going into deeper speleological issues. The functioning mechanism of Velika pećina estavela, as well as the catchment area with possible water connections, remains an open question. In addition to the speleomorphological approach, speleological studies of Mala and Velika Kapela massifs in the period from 1960 to 1962 as well as from 1975 to 1989 (Garašić and Kovačević 1990) were aimed to establish possible water connections with springs of Dretulja, Vrnjika, Zagorska Mrežnica, Siljevac, Suvaja, Studenac, Velika pećina, etc.

Neotectonic research and direct measurements of neotectonic activity within speleological objects were carried out during 1980 and 1981 (Garašić 1980).

Estavela Velika pećina is located in the far south-western part of the karst periodic lake Blata on the southeast slopes of the Mala Kapela massif in Lika. The entrance to the speleological object is located approximately 2200 m air distance from the railway station Blata, on the railway line Ogulin-Knin, in the southeast direction. The closest settlement is Vrcelji Begovacki (about 350 m from the entrance). The size of the entrance to the cave is 7 × 10 m and is very easy to find at the end of the meadow below the vertical rocks.

According to the morphological type, this estavela is a branched cave with constant water flow. Velika pećina is formed in the Lower Cretaceous (Neocomian) limestones of K_1^{1+2} in close proximity to fault contact with Jurassic (upper Malm) dolomites J_3^3 (normal fault). There is also an indication of overthrust contact near the field. The geological structure of the Mala Kapela massif is a Cretaceous syncline. The oldest rocks in this section are Neocomian micrite limestones. The limestones of shallow-water lagoon sedimentation (allochemical and cryptoalgal limestones) follow. Allochemical limestones contain predominantly micrite bases with intraclasts and micritized fossils. Favrein pelmicrites and biomicrites are common. Sparitic cement is present. The limestone layers are 30–120 cm thick and are sloped mostly southwest. Their inclination is between 20° and 30°. The speleological object was created along the fracture system perpendicular to stratification, which approximately follows the slope direction. Local faults were observed in the cave. More pronounced faults and fractures are of the Dinaric direction of spreading, and fractures are rarely filled with clay or calcite binder. These are fractures in syncline that are very wide (of metre dimensions), long and smooth. Due to the erosion and corrosion effects of groundwater, fractures in the syncline (vertical), as well as interlayer fractures in the lower Cretaceous limestones, were expanded, thus creating this speleological object. Due to its two-character effect (springs and swallow-hole), this estavela deposited the secondary sediments of the genesis in front of the object in the Quaternary deposits of Blata and returned to the object itself. This is understandable when it is known that the surface of Lake Blata is a Jurassic dolomite, which is generally water-impermeable, and the water must practically return to Velika pećina from which it burst to the surface, once the hydraulic balance has been reached. As a result, the bottom of the cave is filled with clay and calcareous material, which is repeatedly transported and sedimented, in the inlet and in the dry channels. The effect of water (aggressiveness) on the rocks has been observed in several places and is manifested by the formation of hieroglyphs, veneers (streams) and underground scabs. The very idea of exploring the estavela Velika pećina near Blata, which in the rainy season creates a periodic lake Blata, approximately 2.5 km × 1.5 km deep, several metres deep, is attention-grabbing. Namely, such a quantity of water (several million m^3) must create a large reservoir in the underground. According to observations and measurements, it takes 3 to 5 rainy days in the rainy season for the estavela to work (i.e. the outbreak of water begins). This was confirmed by studies. The size of the canals (dimensions of width and height) and the amount of water in them indicate that such canals continue into the interior of Mala Kapela. The high chimneys, which are most likely the result of inverse karstification (Maucci 1952), bring in water that sinks from surface to crevices, cracks and pits in the area of Bukov vrh (661 m) and Klek (1,124 m). This water reaches the object relatively quickly (vertical circulation), but it takes longer for the large cavities to fully saturate with water. Most of the water arrives via a syphon path (horizontal circulation) from the Glibodol field and the surrounding catchment area. No direct links (open entrances) have been established so far, but this is likely to be achieved indirectly through one of the deep pits at Mala Kapela near Samar. The water temperature in the cave was about 10° C at the time of the survey.

10.3 Large Verticals

Large or deep vertical underground channels of speleological objects are the result of tectonics of the area in which the speleological object is located. The largest verticals in the Dinaric Karst in Croatia are found in the Velebit and Biokovo mountains. The deepest pits of more than 1,400 m (4 deeper than 1,000 m) are also present and vertical jumps deeper than 500 m (two deeper than 500, three deeper than 3,00 m, etc.).

10.3.1 Patkov Gušt

Patkov gušt Pit at North Velebit is −553 m deep. Entrance altitude is 1,450 m, and maximum vertical length is 553 m. This was the second deepest vertical entrance known in the world at the time of the survey in 1997 up until 2017 when it became the third (Slovenia, Iran, Croatia, China and so on). The cave is located in the strict nature reserve Hajdučki and Rožanski kukovi in northern Velebit, 150 m north of Gornji kuk. The cave was named in honour of the speleologist Zoranu Stipetić–Patak, who dived in a syphon at the bottom of Lukina Jama pit in 1994 and died diving in the sea in 1996 (Bakšić 1998).

10.3.2 Jama Velebita

Velebita Cave system is −1,026 m deep and 3,346 m long. Velebita is connected in one cave system with Dva javora cave. The greatest vertical in the world inside a cave (−513 m) is found here. Cave is located in National Park Sjeverni Velebit (Hajdučki kukovi) (Bakšić 2004a, b).

10.4 Caves in Clay

Caves in clay are very rare. The author is aware of only one such phenomenon in Croatian karst, which was discovered during the construction of Zagreb–Rijeka (A-6) highway near Vukova Gorica, about 30 m away from the southbound lane of the highway, at km 59 + 940 (Figs. 10.8 and 10.9). A highway drainage canal and an M-6 separator were designed and constructed in the area. After heavy rainfall, landslide appeared in the lower parts of the drainage canal. The entrance to the cave appeared in that place. The cave was 27 m long with two rooms. It was completely formed in clay. The clay covered the Lower Triassic dolomites, and its age is estimated to be from older Pliocene. The thickness of the clay in this area is between 20 and 50 m. The dimensions of the cave channel was 4 × 5 m, except for one narrowing. The stability of this unique cave in clay was provided by the clay column that prevented collapsing (Figs. 10.10, 10.11 and 10.12). The overburden (clay top above the cave) is between 15 and 25 m thick. Water flowed from the surface throughout the cave, partly coming from the drainage canal. In the back hall of the cave, water from the surface dipped into a 25 × 35 cm swallow-hole. The opening of the swallow-hole was too small for further passage of speleologists, and only at this point did the carbonate (dolomitic) rock substrate appear. The ceiling and the walls of this cave seemed very unstable, and, because of the danger of collapsing, speleologists did not stay and explore it thoroughly. The entrance opening of this cave was unchanged for about 3 months until the landslide remediation was carried out. After the landslide was rehabilitated and the retaining wall was built, the cave collapsed naturally. It is not known for sure whether the cave collapsed completely or if some of its part still exist underground. There are no any marks of depression or suffosion on the surface. In the world, the longest caves in clay are up to about 100 m long and collapse relatively quickly (over a maximum of one decade).

10.5 Caves in Marl of the Dinaric Karst

Cave in Šušnjar was found and explored in 1989. This is the longest cave in Banovina region, totally formed in marl. It is 712 m long and 23 m deep (Garašić and Kovačević 1991). It is a typical dendritic type of cave. Today, the entrance to this cave is completely filled with soil due to erosion (Štefanac 2011).

In Istria, near the town of Gologorica, there is an active occasional swallow-hole formed in the Eocene marls. It is Piskovica Cave, named after the local name for marl (pisak) that shaped the surrounding landscape. The marl erosion between the two limestone slabs in the flysch facies created a 1036 m long cave channel (Figs. 10.13, 10.14 and 10.15). The depth of the cave is −38 m. The cave was locally known since the nineteenth century. The cave drew attention of geologists during geological mapping of that area in 1958 by

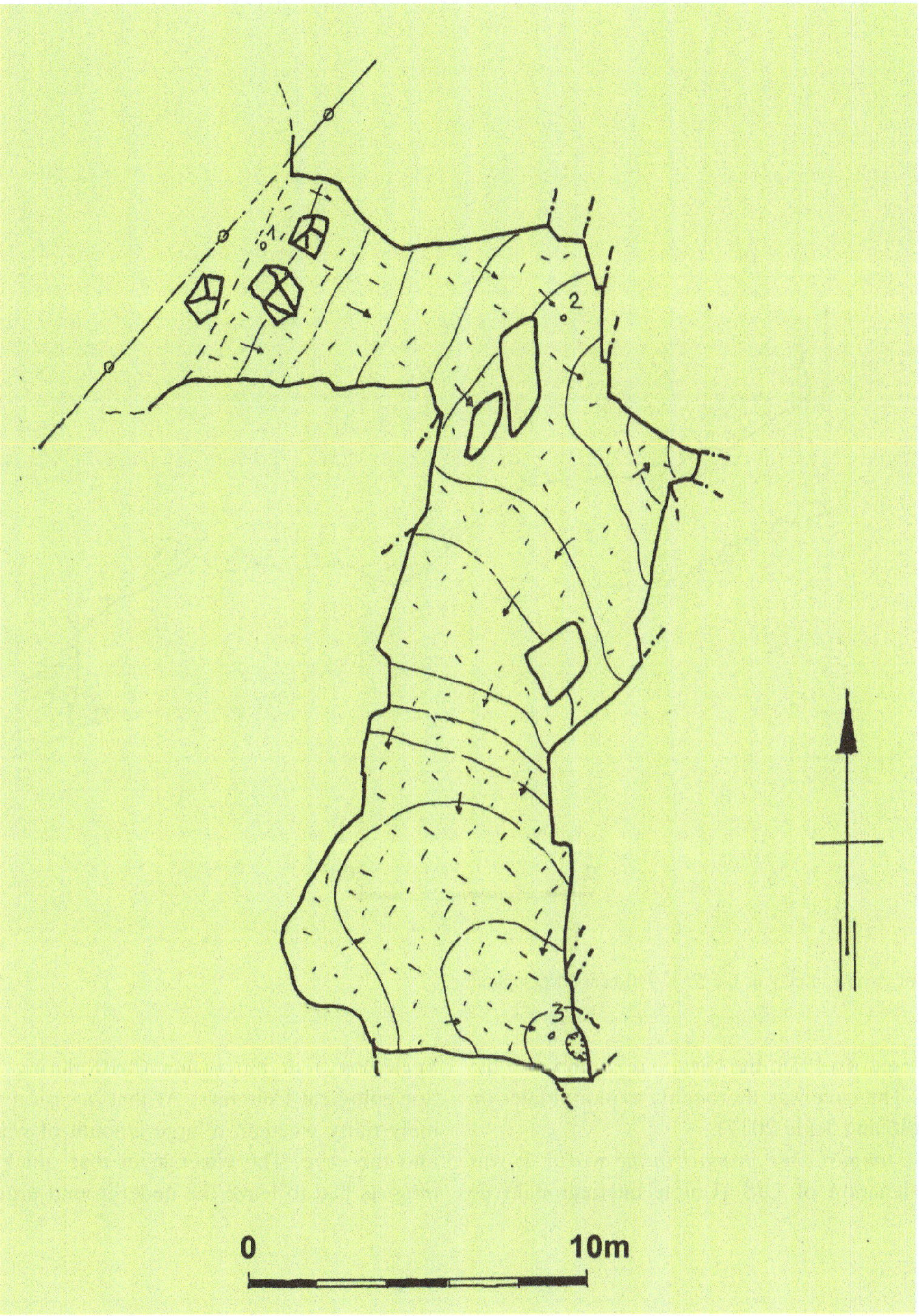

Fig. 10.8 Plan of cavern in clay at km 59 + 940 near M-6 separator

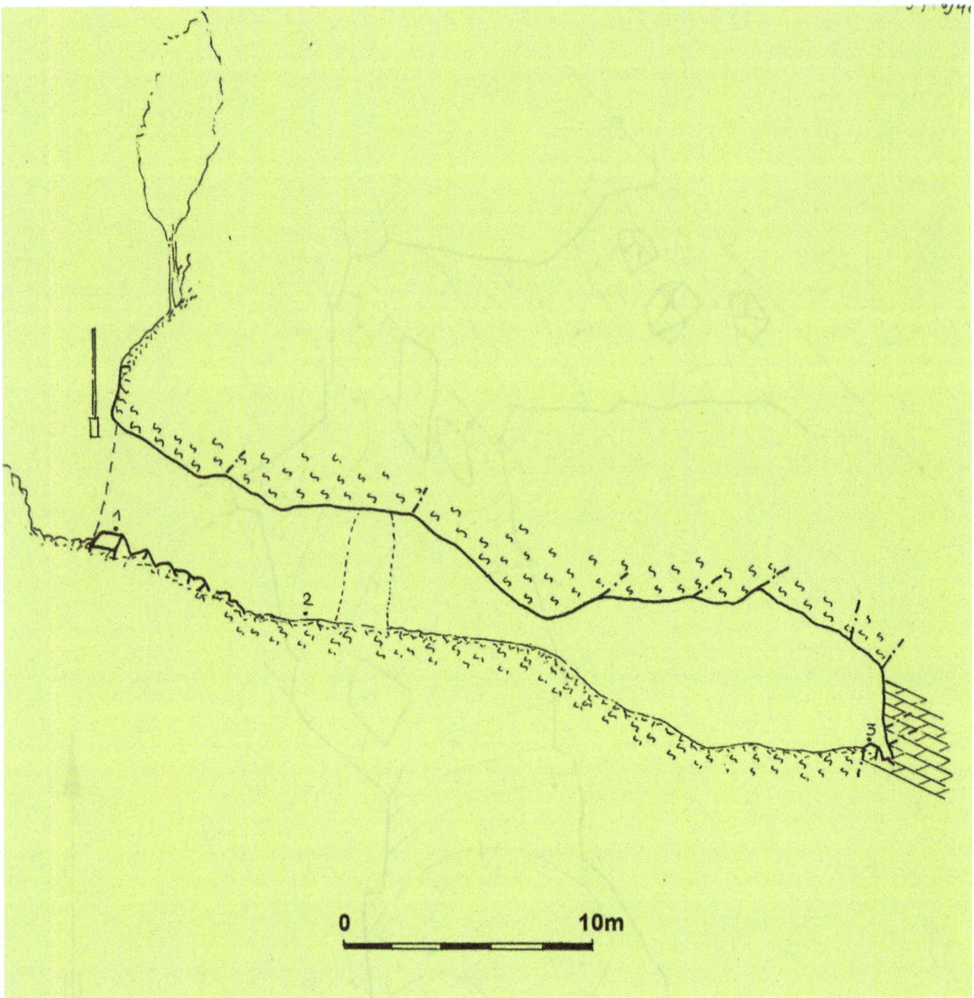

Fig. 10.9 Profile of cavern in clay at km 59 + 940 near M-6 separator

Prof. A. Magdalenić. In 1958, the cave was explored to the depth of 350 m. The cave was thoroughly explored later on (Zlokolica-Mandić and Jekić 2017).

Since it is the *longest cave in marl in the world*, it was visited by a delegation of UIS (Union Internationale de Speleology) in November 2010 during the 1st Croatian Speleological Congress. At that occasion, due to the extremely rainy weather, a large amount of water began to flow into the cave. The water level rose quickly and the speleologists had to leave the underground urgently.

Fig. 10.10 Cavern in clay at km 59 + 940

Fig. 10.11 Cavern in clay at km 59 + 940

Fig. 10.12 Cavern in clay at km 59 + 940

Fig. 10.13 Piskovica Cave near Gologorica in Istria

Fig. 10.14 Piskovica Cave is formed in marl

Fig. 10.15 Piskovica Cave

Fig. 10.16 Krčić Cave is formed in tufa

10.6 Cave in Tufa

Krčić is a waterfall on the river of the same name, 22 m high, located on the north side of the Knin field. It is also called "Veliki Buk" and "Topoljski Buk". Below the waterfall, from the Krčić Cave, which is formed in tufa rocks, the spring of karst River Krka is located. The length of the explored cave is over 250 m. In-depth and long-term investigations of the Krčić Cave (Figs. 10.16 and 10.17) were carried out to design the Krčić HPP project (Božičević 1982).

Tufa covers are found in the swallow-hole Perinka (one of the ponors of Gacka River) and at the bottom of a large vertical in Tučić swallow-hole near Gračac (Božičević 1966).

A large number of caves in tufa were found in canyons and along the waterfalls of rivers Zrmanja, Krupa, Krka, Una, etc.(Cukrov et al. 2003). The most famous tufa barriers and caves found in them are located in the area of Plitvice Lakes (Figs. 10.18 and 10.19). They are smaller than Cave below the Krčić waterfall, which is among the longest ones in the world completely formed in tufa.

10.7 Micro-Waterfall

During January 1976, while exploring new channels in Bezdanjača under Vatinovac or Horvatova Cave (length 1176 m, depth 202 m) near Vrhovine, at the depth of 175 m, the author noticed a hole of millimetre dimensions in one stalagmite. The water spewed under pressure from the hole in stalagmite up to twenty cm in height. When the hole was clogged, the water broke out through the second, previously unnoticed hole, and then the third one above the first one. The author never saw such phenomenon anywhere else in the world (Figs. 10.20, 10.21 and 10.22). He visited that place several times between 1976 and 1978 and photographed the phenomena in IR and UV in order to find some explanation. At the International Seminar on Photodocumentation in Speleological Research held in Postojna, IR photographs of stalagmite from which the aforementioned waterfall emerges were presented which drew a lot of attention of the audience (Garašić 1978). To date, no evidence of any assumptions made by the author nor any other explanation was found.

Fig. 10.17 Tufa in Krčić Cave

Fig. 10.18 Tufa barriers of Plitvice Lakes (winter)

Fig. 10.19 Tufa barriers of Plitvice Lakes (winter)

10.8 Circular Anemolites

The Vrelo Cave was discovered in a quarry near Fužine in 1950. Many anemolites were found there, but their genesis was not explained until 2016. The genesis of anemolites or stalactics that are bent due to the influence of air currents has been explained in many papers and publications in the last hundred years. These are stalactites that are bent in the direction of air flow. However, the stalactites in the Vrelo Cave are bent and rotated in different directions (Figs. 10.23, 10.24 and 10.25). One is circularly bent to the left and the other one, near it, is circularly bent to the right. The question was how can air flow change at such a short distances. The causes are changes in watercourse height, proximity to waterfalls, rapid whirlpool in the underground stream and circular air flow that lasted for a long time, occurring each time at high waters and intense stalactite deposition. These stalactites were very rarely submerged, but the water level was in their immediate vicinity. This happened and is still happening today because of the special positions of the cave channels, their shape, amount of groundwater, the differences in temperatures and the flow of air (Fig. 10.26). These special anemolites are called circular or circular anemolites and were not described or found anywhere else in the karst underground.

Due to the splitting of the cave channels and the different flow rates of the underground water flow, air currents appear at certain levels, which are sometimes in the opposite direction, but circular in shape. This is the root cause of the emergence of these unique speleothems (Garašić and Getvaj 2020).

10.9 Other Rare Phenomena (Speleothems with Tiles, Anti-acoustics of Botroids, Speleothems Under the Sea, the Appearance of Gases, Mud Speleothems, etc.)

Very rare speleothems were found in some caves of the Dinaric Karst in Croatia, horizontal tile speleothems, for example. This shows that speleogenesis in this part of the world is sometimes similar or identical to other parts of the karst, but is sometimes completely unique. Specifically, in quiet separate sedimentation basins within the formation of

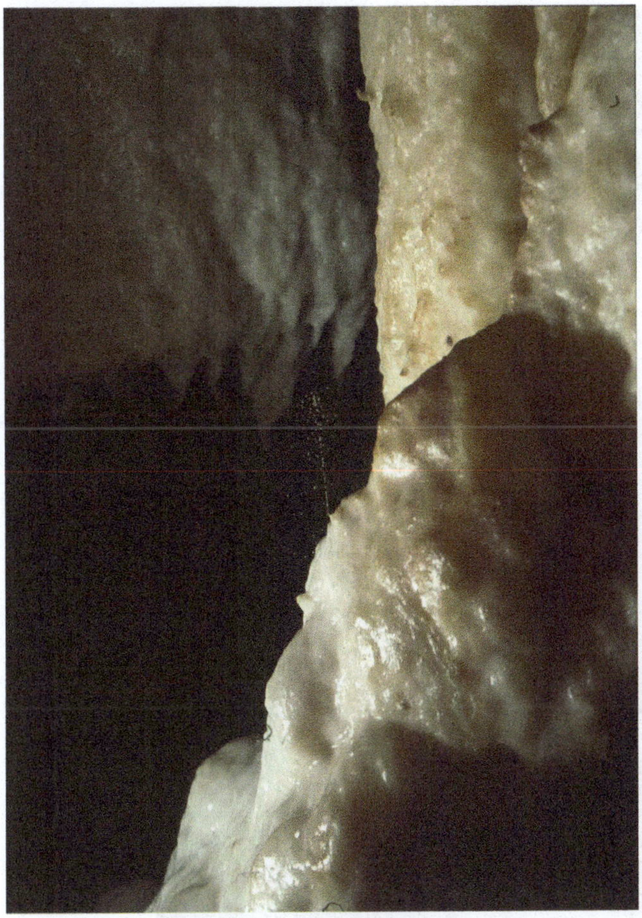

Fig. 10.20 Micro-waterfall in Bezdanjača pod Vatinovcem or Horvatova Cave

speleothems in a calm environment with strong mineralization, special forms of conoliths (stalactites with tiles that sometimes have their "counterpart" on stalagmites) were created, for example at Kojina Cave in Kordun. A similar phenomenon was found in Modrič Cave (Fig. 10.27) near Rovanjska in Dalmatia (Malez 1987; Kuhta et al. 1999; Miko et al. 2002).

Botroids are a special type of speleothems that crystallize in calm closed sub-lake conditions. Sometimes, they are mistaken for pizolites, but they have a different genesis. Unlike botroids, they are not attached to the substrate or surrounding rocks.

Sometimes in larger halls completely crystallized by botroids, due to their surface and shape, the sounds are muffled. This is where the "damping" of sound occurs, and some call it "deaf halls" because sometimes the sound is absorbed so much that communication between individuals is very difficult (Fig. 10.28). The hall in Bezdanjača under Vatinovac at a depth of about 185 m is a perfect example of a deaf hall.

In some speleological sites in Istria, Ravni kotari and Dalmatinska Zagora, high concentrations of carbon dioxide (CO_2) were measured. Locally, even in quantities that are hazardous to humans. In some caves these concentrations can be variable. In some parts of the year, they become normal, but in some places (Istria area), they remain dangerous throughout the year. The cause may be related to the morphology of the speleological object, i.e. the small sizes of air openings, air flow caused by changes in atmospheric pressure, type and amount of groundwater, type and amount of speleothems, rotting of organic material, proximity of mines, etc.

In some speleological objects near the sea, paleo- and neoeffects of the sea on speleothems (anchialine caves) can be observed. Sometimes, such objects are hundreds of m away from the sea, and in the example of Vrulja Modrič, we can see the influence of the sea several kilometres inside the karst massif.

The optical effect of the so-called Blue caves is created when sun rays bounce off light rocks at the bottom of the sea and return at a certain angle through the aquatic entrance to the cave. It can be seen in dozens of caves on islands and on the east coast of the Adriatic Sea. The most famous Blue Cave is Modra špilja on the island of Biševo (Fig. 10.29) which is accessible to tourists. It is a well-known cave that served as an inspiration for the artist Ramsonet to make the first painting (oil on canvas) from a speleological object in Croatia.

Similar occurrences on the islands of Krk, Cres and Hvar are less known. One such phenomenon was described as far back as 1920, but unfortunately, it was partially buried by irresponsible actions. This is the Pijavica Cave near Novi Vinodolski. On the island of Cres is cave Škuja va Žanji, Medova Buža on island Rab, etc.

Kojina Jama belongs to the inner karst belt, located on Mašvina hill between Slunj and Rakovica. It has a vertical entrance with horizontal and vertical extensions. It is extremely rich in speleothem. Numerous speleothems of plate and crystalline forms were detected in it. It is 753 m long and 178.5 m deep, which is the deepest known speleological object in Kordun area. It is interesting that today it is located about 250 m above the karstification zone of the Korana River, but the Mašvina hill is in intense uplift. Garašić (1984) described the geological relationships and neotectonic activities in this area and predicted the appearance of deep pits in addition to long caves (Panjkov ponor, Varićak Cave, Crno Vrelo, etc.) (Fig. 10.30). This was later proved to be correct. The pits in this area are often deeper than 200 m.

The "mud speleothems" were first described in 1967 after the widening of the narrow passage in the lower Cerovac

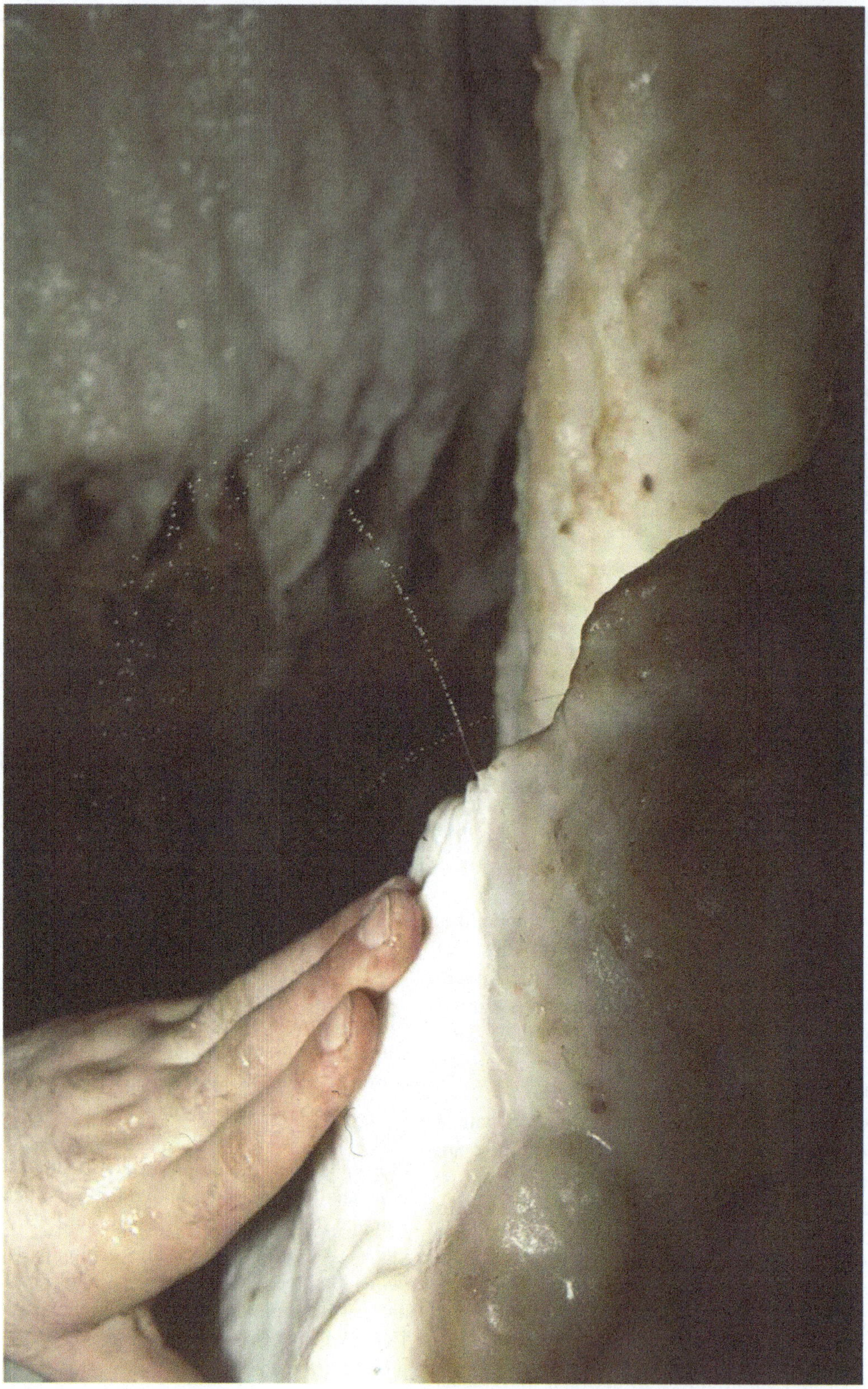

Fig. 10.21 Micro-waterfall in Bezdanjača pod Vatinovcem or Horvatova Cave

Fig. 10.22 Micro-waterfall in Bezdanjača pod Vatinovcem or Horvatova Cave

Fig. 10.23 Circular anemolites from Vrelo Cave near Fužine

Cave (Fig. 10.31) and the discovery of 470 m of new cave channels. These are the forms created by the action of water dripping onto the clay soil. Through their mechanical action, water droplets have shaped clay soils and created "mud forms" that in some places resemble stalagmites, but are completely different from them genetically. They are also subject to the slightest change in temperature and humidity.

In the cave near Bajagić in Dalmatian Zagora, the speleologist Tonči Radja found interesting speleothems in early 2017, which are very likely related to repeated tectonic activity and intense secondary deposition of the speleothems. Part of the stalactite is related to stalagmite, but subsequently the bond with the parent roof of the stalactite ceased (Fig. 10.32).

Crystals on speleothems or single crystals in the Dinaric Karst are much more common in the area of influence of the Adriatic Sea. Specifically, water temperatures above 12–15 ° C allow for faster and more diverse crystallization. Very interesting crystals were found in the caverns in Maslenica junction (Fig. 10.33), the caverns near Stankovci, the cavern in the Bristovac tunnel, etc. These are crystals longer than 10 cm (Fig. 10.34). Such examples are also found in caves in the Dalmatinska Zagora. In the inner, central and outer karst zone, there are negligible groups of such examples. This may be due to the special (warmer) climatic conditions that prevailed in some places.

10.10 Ice Caves

According to the data (Buzjak et al. 2018), so far 321 speleological objects with snow or ice that remain throughout the year or considerable part of the year have been discovered in Croatia. They are mostly located in higher altitudes (above 1000 m.a.s.l.) in the Gorski Kotar, Risnjak, Viševica, Bitoraj, Bjelolasica, Velebit, Biokovo (Fig. 10.35) and Dinara mountains. They also exist in particularly cold and high areas of Lička Plješevica, Ozeblin, Lika and Velika and Mala Kapela. The special shape of such caves or pits allows the flow of air to be re-cooled and maintain an almost constant low temperature. Although global temperature change is also present in speleological objects, ice and snow caves can be considered as a certain rudiments from the last ice age (Malez 1961a).

The ice cave Ledenica in Lomska Duliba was surveyed in 1962, 1973, 1977, 1992–1998 to the depth of −536 m. The length of the cave is 629 m. It was researched by Croatian and Slovak speleologists.

Jelinić (2001) wrote a detailed chronology of the research. The ice cap extends from 50 to 90 m deep, where it crosses a 180 m vertical through an oblique rock-creep. There are three openings through the ice cap today. If there was no ice cap, the entry vertical would be 270 m. The ice melting that led to the formation of these passages (first observed in 1993) continues with varying intensity, which greatly affects the appearance of part of the pit below the ice cap (80–220 m deep). There was a lot of change in this part in recent years with regard to the amount of ice that covers the rocks. These changes prompted Karlovac speleologists to set up a whole new line of descent during the 1995 survey, because it was not possible to descend in the former direction due to ice. Where there were spits in the clear rock, a thick layer of ice was found. This part of the pit is certainly the most dangerous because of the constant slippage of ice and rocks caused by melting. The most serious accident in Croatian speleology happened here, when a landslide of ice and stones severely injured the caver B. Bukovčak from Karlovac.

The results of the ice analyses indicate a relatively small age (450–670 years), but due to the small sampling depth in the ice, the sample is contaminated with dripping water and re-frozen water from the surface. Therefore, more detailed sampling from greater depths in ice should be made.

An interesting result is the determination of the age of the speleothem from the hall at about 50 m depth, which is 301,000 + −55,000 years (Horvatinčić 1996).

The first speleological research of Ledenica in Bukovi Vrh on Velebit Mountain was carried out on 22 of August 1980. Survey showed that the cave is situated at 1325 m.a.s.l. and was formed in middle Triassic dolomites and limestones. This 189 m long and 87 m deep cave hosts a lot of ice (Figs. 10.36, 10.37 and 10.38). New explorations (Garašić 2014) showed that the thickness of ice in last chamber is greater than 17 m (45 × 25 m). Temperature of air in this cave chamber is −1 °C.

Recent research on the age of ice from Lukina Jama was published in Hydrological Processes (Paar et al. 2019). The results showed the age of about 3,500 years. The age of ice of approximately 3,500 years in Vukušić Ice Cave was mentioned by Kern et al. (2008, 2011, 2018). Based on the results of some other analyses that estimated the age of the ice in Dinaric Karst to be about 60,000 years old, it can be concluded that such findings are possible in Velebit.

10.11 Underground Bifurcation

The pit under Debela Glava

Two watercourses were found in the Pit under Debela Glava (Garašić 1990a). One of them gravitates to Glina River Basin and the other one to Glina River Basin and the Korana River Basin. This means that underground bifurcation is present here. The cave is complexed, branched and multi-level, with a vertical entrance. It is known for its large entrance hall (40 × 50 m), one of the largest in the Kordun area. The length of the known channels is 781 m, and the depth is 68 m (Fig. 10.39).

Tracings between the Zrmanja River Basin and the Krka River (Ravni kotari) showed a clear connection (Terzić et al. 2014).

Miljacka is a karst spring situated in the Krka River canyon, in Dalmatia, Croatia (Fritz and Pavičić 1982, 1987). Its catchment is mostly built of karstified carbonate rocks, and it is a part of Dinaric karst. During the last few years, a bulk research programme was performed in the Miljacka spring catchment area. Achieved results comprise all previous conclusions, together with some new aspects. The main reason for this research was establishment of sanitary protection for this strategically important karst spring. Sanitary protection zones were proposed, but because of some new understandings of this complex karst system, they will have to be extended in future with parts of the Krka River catchment area. Although the Miljacka spring is situated in the Krka River canyon, majority of its water flows from quite distant Zrmanja River, which was proven by dye tracer tests in the past. During the recent research, hydrochemical investigation of the Miljacka proven that it is only partially true, and some portion of the spring water originates from the Krka River as well. Therefore, significant parts of the terrain will have to be added to the Miljacka spring catchment area and to sanitary zones as well. The most of the investigated area is built of karstified carbonate rocks, of Eocene–Oligocene (Promina rock mass), Cretaceous, Jurassic and Triassic. Complexity of this karst system is especially presented with the main characteristics of the Zrmanja River, which usually dries out in the swallow-hole zones during the summer dry season. These swallow-hole zones are active throughout the year and water that infiltrates the karst underground in that area discharges mostly at the Miljacka spring. This "hanging" part of the Zrmanja River was proven with boreholes in the past years, but that phenomenon should be investigated in detail. The main problem for any hydro-researcher of the Miljacka spring is shortage of the measured discharge data. This can be avoided in future either by construction of the direct measuring place or by measurements on the Krka River profile, but very close to

Fig. 10.24 Circular anemolites

Fig. 10.25 Circular anemolites

Fig. 10.26 Air circulation measurement in Vrelo Cave

the Miljacka (upstream and downstream), because the possibility of existence of swallow-holes in river bed is quite high.

Miljacka 2 Spring Cave is located on the right bank of the Krka River, two hundred metres downstream of Miljacka Waterfall, at an altitude of 115 m. This cave is the longest topographically recorded speleological object in the area of the Krka National Park. It was surveyed in the length of (3,500) 2,800 m. The last exploration of the main channel in 2010 ended in a large "hall," measuring 30 × 20 m, behind the fifth syphon. There is a sixth syphon in the hall, so it is assumed that the main channel continues. Based on the collected speleomorphological data, the cave is categorized by morphological type into a multi-storeyed and branched speleological object. It was formed within Promina deposits along the fault zone, which extends in the NW–SE direction and the transverse fractures that influenced the formation of wider sections in the main channel. It results from the erosion of water along the fault zone, which extends in the NW–SE direction. According to the genetic type, it is a tectonic–erosion (polygenetic) speleological object. It is a valuable part of the Croatian geo-heritage. The entrance area is dominated by conglomerates and marl limestones, and the end channel is completely formed within marl limestones.

10.12 Radon in the Caves of the Dinaric Karst in Croatia

The concentration of radon 222 isotope gas in speleological objects in the Dinaric Karst of Croatia has been measured since 2005. The first measurements were made in the Vrelo Cave near Fužine and in the caverns in the tunnels Sveti Rok, Vrata, Puljani and Sveti Ilija. In the cavern of the Vrata tunnel, measurements were carried out continuously for 15 years, while at other locations, measurements were made only during speleological surveys. Significant discrepancies were observed in radon gas concentration measurements within speleological objects with natural entrance and caverns. The measurement results deviate 100 or more times between the cave with the natural entrance and the cavern. The biggest differences were observed in the caverns closed by the tunnel door. In these caverns, almost identical micro-climatic conditions were established as before their discovery. It is interesting to compare radon gas concentrations in seismically active areas with those that are not, or to compare radon concentrations between caverns formed in the neotectonic uplift area and the neotectonic sinking region of rock blocks.

10.13 Perspectives for Further Exploration of Large Speleological Systems in the Dinaric Karst of Croatia

Great potential for further successful speleological research is expected in areas that are already partially explored, but also in areas of the Dinaric Karst in Croatia, where previous studies have been rare or have not been performed at all for various reasons. In the Dinaric Karst of Croatia, cave systems that are more than 1,500 m deep and over 100 kms long are yet to be discovered. However, it is this author's opinion that they most certainly exist. Only the most important areas where the potentials of deep or/and long

Fig. 10.27 Special type of conulites (Modrič Cave)

Fig. 10.28 Deaf hall in Horvatova Cave

Fig. 10.29 Blue cave–Modra špilja on the island of Biševo

Fig. 10.30 Cave diving in Varićakova Cave and Panjkov ponor Cave

speleological objects are located geologically will be mentioned here.

For further speleological explorations in the already discovered caves, speleological diving techniques, climbing techniques and techniques for widening narrow sections of cave channels, i.e. carving, digging and blasting, will be mostly used in the near future.

Some perspective areas in Croatian Karst are as follow.

The area of Lika and Gacka River swallow-holes, Crnopac area in south Velebit (Figs. 10.40 and 10.41), Čelavac area (southern Velebit), Kamešnica River spring area, Cetina River spring area beneath Dinara, the area of Mazin anticline and the springs of the river Una, areas of Mala and Velika Kapela, Bitoraj and Višnjevica Mt, Drežnica polje, areas of Pazin swallow-hole towards the river Raša, North Velebit area, swallow-hole areas of Mala Kapela–Drežničko polje–Zagorci, Stajnica–Blata, swallow-hole area of Ogulinska Dobra River, Biokovo mountain, Ličko Sredogorje area, etc.

The Ličko Sredogorje is generally situated between the mountains of Lička Plješevica and Ozeblin on the northeast side and the Velebit massif on the southwest side. To the northwest, it slightly crosses into the Kapela area. It is separated from the aforementioned mountains in the northeast by the Krbavsko and Gračacko fields, in the southwest by the Ličko field, and in the northwest by the Gacko field. In some parts of Ličko Sredogorje, there is a dividing line between the Adriatic and Black Sea basins, as well as smaller regional or local river basins and streams. These are Mesozoic carbonates (limestones, dolomitic limestones, dolomitic limestones and sometimes dolomites. Paleo-and neotectonics are intense. All these have affected the genesis of speleological objects in the area.

Fig. 10.31 Lower Cerovac Cave

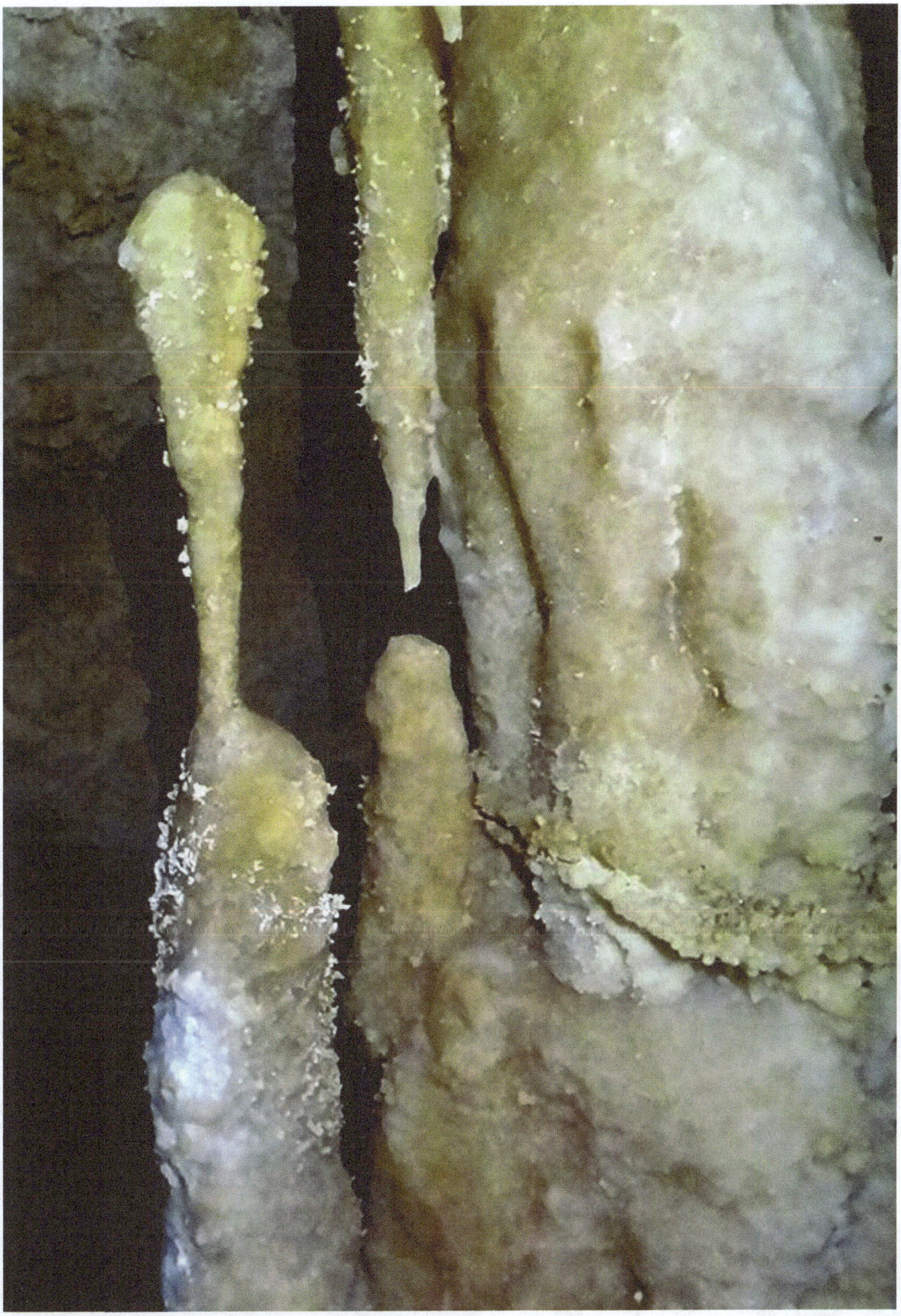

Fig. 10.32 Speleothem in a cave near Bajagić

Fig. 10.33 Cavern near Maslenica junction

Fig. 10.34 Crystals in cavern near Maslenica junction

Fig. 10.35 Mucića ledenica on Biokovo Mountain

Fig. 10.36 Cave Ledenica on Bukov Vrh on Velebit Mountain

Fig. 10.37 Ice in Ledenica on Bukov Vrh

Fig. 10.38 Ice in Ledenica on Bukov Vrh

Fig. 10.39 Pit under Debela Glava—Main Chamber

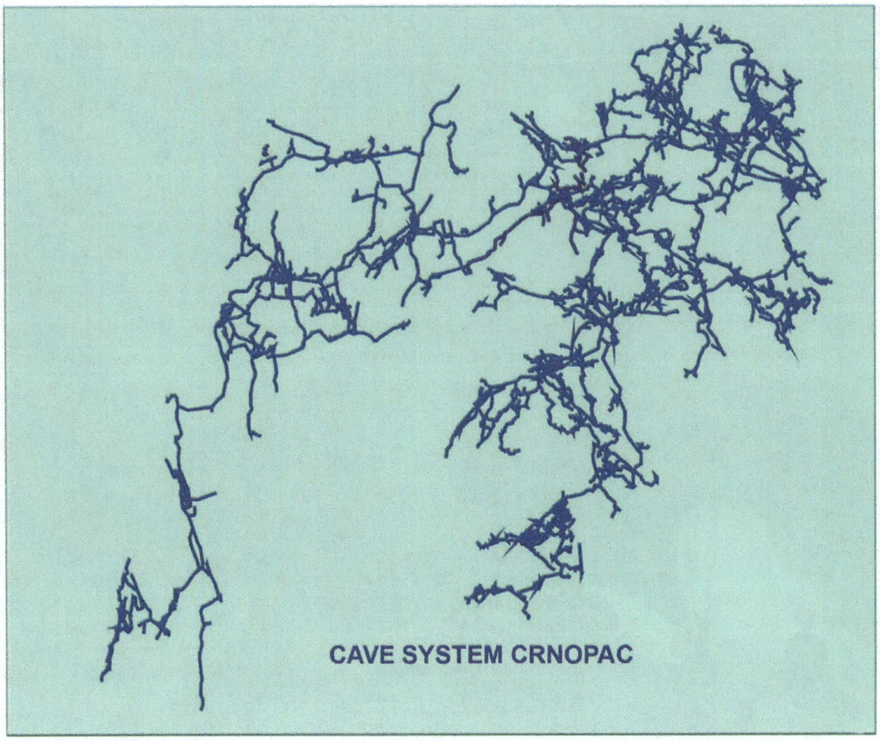

Fig. 10.40 Part of the plan of Crnopac Cave System—the longest cave in Croatia (over 56 km long) by Teo Barišić

Fig. 10.41 Upper Cerovac Cave is located near Crnopac Cave system

10.14 Possibilities of Hypogene Speleologenesis in Karst of Croatia

Hypogene speleogenesis is increasingly observed in the world for the last 20 years (Klimchouk 2001, 2007). Even in places where potential conditions for its existence are not present today, but paleogenetic conditions were favourable for hypogene speleogenesis. There are indications that such objects could exist in the karst of Croatia. However, these assumptions are yet to be proved by research.

Cave Speleometry in the Dinaric Karst of Croatia

11

Abstract

By August 2021, about 12,500 caves and pits had been documented in the Dinaric Karst of Croatia. About 1200 were found during construction works. The number of new caves or the dimensions of previously known caves and pits are constantly updated and show significant activity of speleologists in Croatia. Four pits deeper than 1000 m and caves that are almost 50 km long with a great potential for connection indicate that a significant change in the speleometry of this karst area can be expected soon.

The number of speleological objects on a certain surface, the altitude of their entrances, the dimensions of entrances, halls and channels, the total or partial depth and length of individual morphological or hydrogeological parts within the speleological objects, etc., the volumes and other numerically expressed features can be called speleometric parameters. In a broader sense, speleometry could include various data on measurements of air, water, ice and soil temperatures, relative and absolute humidity, percentages of gases in the composition of speleological objects, air flow rate, natural radiation, etc.

However, the number or frequency of occurrence of speleological objects is the basic information that describes a karst area. Unfortunately, this number has changed a great deal and probably can never or very rarely come close to absolute scientific truth. This is due to the number of information that follows from the exploration of the area, the source of the information and the purpose of the information. For example, in Croatia, up to 12,500 speleological objects are known to speleologists and geologists to date, through published and institutional unpublished literature. Some sources mention an unrealistically small number of only 8500 known speleological objects in Croatia. In the official list of the speleological cadastre of speleological objects of the Republic of Croatia, which was created in 2015 in April 2021, it listed 3401 completely processed caves. Each year, hundreds of new speleological objects are included in the cadastre with complete new documentation. The previously known information is checked and updated. There is also a cadastre called “Clean Underground” which, by 15 August 2021, listed more than 800 speleological objects that were contaminated. In 51 of them, debris was taken to the surface. However, it is most realistic that to date there are about 12,500 speleological objects known in the Croatian part of the Dinaric Karst, following the publication of data from 1910 to 2020.

11.1 The Longest and Deepest Speleological Objects in the Dinaric Karst of Croatia

In the tables is list of the longest caves in Croatia (longer than 2000 m) based on the data collected before 15 August 2021 (Table 11.1) and the list of the deepest caves in Croatia (deeper than 300 m) based on the data collected before 15 August 2021 (Table 11.2).

M. Garasic, *The Dinaric Karst System of Croatia*, Cave and Karst Systems of the World,
https://doi.org/10.1007/978-3-030-80587-6_11

Table 11.1 Caves in Croatia longer than 2000 m

	Cave name	Location	Length (m)
1	Cave system Crnopac	Crnopac, Southern Velebit Mt	54,709
2	Cave system Đulin ponor—Medvedica	Ogulin, Gorski Kotar	16,396
3	Cave system Panjkov ponor—Varićakova špilja	Nova Kršlja, Kordun	13,218
4	Munižaba	Southern VelebitMt	9911
5	Cave system Tounjčica	Tounj, Kordun	9104
6	Cave system Jopićeva špilja-Bent	Brebornica, Kordun	6710
7	Cave system Cavern in tunnel Učka—Zračak nade 2	Učka Mt, Tunnel Učka, Hrvatsko Primorje	6596
8	Slovačka Jama	Mali kuk, Middle Velebit Mt	6416
9	Veternica	Medvednica, Zagreb	6128
10	Jama (ponor) kod Rašpora	Račja Vas, Istria	6184
11	Kotluša	Kijevo, Dalmatia	4843
12	Donja Cerovačka špilja (Lower Cerovac Cave)	Gračac, Lika	4058
13	Gornja Cerovačka špilja (Upper Cerovac Cave)	Gračac, Lika	4035
14	Cave system Lukina Jama—Trojama	Hajdučki kukovi, NP Northern Velebit Mt	3741
15	Miljacka II	Kistanje, Dalmatia	3365
16	Cave system Velebita	Rožanski kukovi, NP Northern Velebit Mt	3346
17	Gospodska špilja	Vrlika, Dalmatia	3060
18	Cave system Vilinska—Ombla	Dubrovnik, Dalmatia	3063
19	Kusa II	Krupa, Southern Velebit Mt	3010
20.U	Vrulja Modrić	Modrič,Dalmatia	2896
21	Nedam	Hajdučki kukovi, NP Northern Velebit Mt	2873
22	Mandelaja	Oštarije, Ogulin	2835
23	Cave system Ponorac—Jovina pećina	Rakovica, Kordun	2834
24	Kusa I	Dobarnica, Southern Velebit Mt	2500
25	Miljacka I and V	Kistanje, Dalmatia	2456
26	Špilja za Gromačkom vlakom	Dubrovnik, Dalmatia	2439
27	Atila	Middle Velebit Mt	2410
28	Markov ponor	Lipovo Polje, Lika	2405
29	Klementina I	Middle Velebit Mt	2403
30	Lubuška Jama	Hajdučki kukovi, Sjeverni Velebit	2395
31	Gojak spring	Gojak, Kordun	2312
32	Golubnjača u Grulovićima	Kistanje, Dalmatia	2189
33	Provala	Bučari, PN Žumberak Samobor	2161
34	Ponor Bregi	Pazin, Istria	2110
35	Cavern in tunnel Plat	Dubrovnik, Dalmatia	2010
50.U	Velika Vrulja	Starigrad Paklenica, Dalmatia	1554
58.U	Vrulja Zečica	Modrič, Dalmatia	1210
U	Majerovo vrilo	Sinac, Lika	942
U	Bakovac sping	Bakovac, Lika	520

U—underwater cave

Table 11.2 Caves in Croatia deeper than 300 m

	Cave name	Location	Depth difference (m)
1	Cave system Lukina Jama—Trojama	Hajdučki kukovi, NP Northern Velebit Mt	1431
2	Slovačka Jama	Malikuk, NP Northern Velebit Mt	1324
3	Nedam	Hajdučki kukovi, NP Northern Velebit Mt	1250
4	Cave system Velebita	Rožanski kukovi, NP Northern Velebit Mt	1026
5	Jama Njemica	Biokovo Mt, Dalmatia	954
6	Mokre noge	Biokovo Mt, Dalmatia	831
7	Cave system Crnopac	Crnopac, Southern Velebit Mt	797
8	Jama Amfora	Biokovo Mt, Dalmatia	788
9	Meduza	Rožanski kukovi, Northern Velebit Mt	706
10	Cave System Vilimova Jama (A-2)—A-1	Biokovo Mt, Dalmatia	589
11	Patkov gušt	Gornji kuk, Northern Velebit Mt	553
12	Olimp	Begovački kuk, Northern Velebit Mt	537
13	Ledena Jama u Lomskoj dulibi	Northern Velebit Mt	536
14	Ponor na Bunovcu	Southern Velebit Mt	534
15	Lubuška Jama	Hajdučki kukovi, Northern Velebit Mt	529
16	Crveno jezero (Red Lake)	Imotski, Dalmatia	528
17	Munižaba	Crnopac, Southern Velebit Mt	510
18	Jama pod Kamenitim vratima	Biokovo Mt, Dalmatia	499
19	Stara škola	Biokovo Mt, Dalmatia	497
20	Fantomska Jama	Visočica, Middle Velebit Mt	477
21	Cavern on km 1 + 637, Tunnel St. Ilija	Tunnel Sveti Ilija, Biokovo Mt	459 (−144/+305)
22	Cave System Cavern in tunnel Učka—Zračak nade 2	Učka Mt, Tunnel Učka, Hrvatsko Primorje	430
23	Stupina Jama	Fužine, Gorski Kotar	413
24	Sirena	Northern Velebit Mt	401
25	Paž	Kita Gavranuša, Northern Velebit Mt	400
26	Biokovka	Biokovo Mt, Dalmatia	389
27	Ovčica	Gromovača, Northern Velebit Mt	379
28	Nova Velika Jama	Biokovo Mt, Dalmatia	361
29	Zečica	Biokovo Mt, Dalmatia	355
30	Jama (ponor) kod Rašpora	Račja Vas, Istria	351
31	Punar u Luci	Pusto Polje, Lika	350
32	Ponor Pepelarica	Klanjeva Ruja, Middle Velebit Mt	348
33	Gnat	Southern Velebit Mt	343
34	Klementina III	Klementa, Middle Velebit Mt	333
35	Podgračišće II (Titina Jama)	Pražnice, Insel of Brač	329
36	Xantipa	Rožanski kukovi, Northern Velebit Mt	323
37	Klanski ponor	Klana, Hrvatsko primorje	320

(continued)

Table 11.2 (continued)

	Cave name	Location	Depth difference (m)
38	Puhaljka	Medak, Southern Velebit Mt	320
39	Klementina IV	Klementa, Middle Velebit Mt	319
40	Duša	Crnopac, Southern Velebit Mt	318
41	Zaboravna Jama	Biokovo Mt, Dalmatia	311
42	Treći svijet	Jarmovac, Gorski kotar	310
U	Una River spring	Neteka, Lika	248
U	Sinjac	Plavča Draga, Lika	203

U—underwater cave

11.2 Constantly Submerged Speleological Objects in the Dinaric Karst of Croatia

There are much more permanently submerged speleological objects or those parts of which are always submerged in the Dinaric Karst of Croatia that is evident from land surveys. Namely, a large number of such caves have entrances located under the sea, rivers or lakes. Sometimes such speleological objects are discovered by diving in the sea or construction work in tunnels. For example, many vruljas and sea caves were discovered by diving in the sea (Modrič, Zečica, Dubci, Katedrala, Banjole, Modra Spilja, etc.). Some were discovered while tunnelling (tunnels Pećine and Ombla). However, there are also cases where some of their entrances are above groundwater level, and only further exploration finds extensions of permanently submerged speleological objects (Podstražišće Cave). All permanent karst river springs (Una, Kupa, Gacka, Krupa, Krnjeza, Dubanac, Ličanka, etc.) can be included in such speleological objects (Kovačević 2015).

The depths of the continually submerged speleological objects that man has explored so far in the Dinaric Karst of Croatia are fascinating. There are several springs more than 200 m deep, and only fifteen are known worldwide. This is due to the intense paleo- and neotectonics that are dominant in the speleogenesis of these areas. Diving techniques today are limiting new knowledge about submerged caves. The author of this book strongly believes that in the future several karst springs deeper than 400 m, without the influence of thermal or mineralized waters from greater depths, will be discovered.

11.3 Volumes of Speleological Objects

Large underground areas are the result of geological processes of mechanical removal or chemical dissolution (Dreybrodt and Kaufmann 2007) of rocks (erosion and corrosion) with the help of tectonic predispositions and hydrogeological conditions. Some of these are extremely large spaces. For example, in the submerged part of Crveno Jezero (528 m deep) near Imotski in Dalmatian Zagora, the volume is estimated at 16 million m^3 (Garašić 2001). The volume of Munižaba Cave (9715 m long, 510 m deep) in southern Velebit is over 2 million m^3, and of the Vrtlina Cave (about 900 m long, 205 m deep) it is over 2 and a half million m^3. Some large underground halls in caverns found in tunnels in Velebit Mountain have a volume of 500,000 to 1 million m^3. The largest volumes of underground spaces were recorded in the so-called Jelar deposits (E, Ol) within limestone breccias (Fig. 11.1), Jurassic Lias limestone deposits (J1) and Upper Cretaceous Senonian (K_2^3) rudist limestone deposits. Although the longest cave system in the Croatian part of the Dinaric Karst is over 40 km long, it does not have the largest volume. Crveno Jezero has the largest known underground volume so far. The volume depends on groundwater level, but it is certainly greater than ten million m^3. However, Crveno Jezero is a vertical

Fig. 11.1 Largest volume of cavities is formed in so-called Jelar deposits within limestone breccias (Velebit tunnel)

speleological object that has a depth (528 m) larger than the largest entrance dimensions (420 m) and is considered a unique pit, not a collapsed sinkhole. Its total volume is certainly well over 20 million m^3 and is one of the largest in the world. Because Crveno Jezero is not technically a hall, it is not on the list of largest cave hall volumes.

Research Methods

12

Abstract

Through more than 140 years of active research of caves in the karst of Croatia, many research methods were used, from the oldest ones to the recent most advanced ones. There was always an effort in Croatian scientific community to use proven and generally accepted research methods.

When researching the speleological objects of the Dinaric Karst in Croatia, from the beginning to the present day, all the most up-to-date and proven speleological research methods relevant to individual periods were used. Speleologists in this area were sometimes ahead of researchers in other parts of the world in terms of inventiveness. For example, in the study of deep vertical speleological objects, they used advance techniques such as ropes (1955–1965), while ladders were still in use in other parts of the world. An example is found in the Balinka cave (1954), which was later explored up to −283 m, and up to −110 m was explored only with the help of hemp climbing ropes and Prusik knots. The first cave diving in speleological objects in the interior of Croatia began in 1959. Diving with partial entry to speleological objects by the sea (land and islands) began ten years earlier. They were conducted by professional and military divers. The first documented cave diving in a syphon was conducted 1952 by diver Karlo Baumann from Croatia in Postojna (Slovenia). It was also the first dive with which the Postojnska Jama and Planinska Jama Caves were to be connected. In June 1956, two divers from Rijeka (Croatia) dived a few dozen metres into the syphon and stated that it continues further in depth rather than towards the surface. Today, even 2 kms beyond the first dives in 1956, the cave diving efforts to connect Postojnska Jama and the Planinska Jama continue. Today, dives use the most advanced trimix and rebreather techniques while diving and have achieved kilometres of dive in the Dinaric Karst in Croatia.

Speleological equipment and techniques changed over time and matched the needs, capabilities and technological level of time in which it was created. Great advances have been made in speleological lighting, which has reached acetylene lamps through wax candles, through torches and lanterns. Such lighting was mandatory for speleologists until 2008–2010, when it was completely replaced by various electric LED lamps that are still in use today. Most of them can work flawlessly at great depths underwater for a long time. In addition, Cyalume Auxiliary Chemical Lighting is used.

Helmets, i.e. head protection, have come from specially made hats, leather mining helmets through climbing helmets, to specially made speleological helmets that are very lightweight, durable and adapted for longer stay underground or under water, with the possibility of wearing speleological lighting on them.

Although it seems that speleological protective suits, gloves and shoes have not made much technological progress in the last fifty years, the experience says otherwise. Today, special undersuits are used to preserve body heat, and from the outside protective overalls are made of special waterproof or semipermeable materials resistant to heavy wear. For special warming of the hands and feet, chemical agents are used in special bags, which in the evoked reaction heat up and transfer heat to the body and prevent fever or hypothermia. Shoes have also made technological advances in their shapes, materials and weight. In particular, these are materials that do not slip on muddy or wet rocks (Fig. 12.1).

Ropes and equipment for vertical descents and climbs are being refined and modified continuously. New materials, weight reduction and new shapes are often present. However, it should be noted that the safety of speleologists when exploring underground depends directly on them (ropes, harnesses, ascenders, descenders, rope anchoring equipment, etc.).

M. Garasic, *The Dinaric Karst System of Croatia*, Cave and Karst Systems of the World,
https://doi.org/10.1007/978-3-030-80587-6_12

Fig. 12.1 Bivack in Siničića cave

All speleological equipment before use must be tested according to international standards prescribed by the UIS - World Speleological Union or certain internationally accepted standards.

Explorations of large vertical channels (Fig. 12.2), such as Patkov gušt in the Velebit, where verticals are larger than 500 m and are among the largest in the world, a combined method of rope anchoring, alteration and double rope safety was used.

The method of expanding the channel with explosives was also sometimes used, starting in 1965. So far, such a method and technique has not been used in channels permanently submerged with water.

Scientific data collection methods are constantly present in this field, be it physical, chemical, biological or geological methods and techniques. Their results in correct interpretations give new insights into the origin, condition and presumed future of speleological objects and bio/geo inventory in them.

DNA studies on the bones found in cave sediments in the Vindija Cave in Zagorje have provided completely new and unknown data on humans over 500,000 years old (Prüfer et al. 2017).

The development of the latest surveying methods will provide an even better understanding of the layout, abundance and characteristics of the speleological objects in the area. Laser 3D and LIDAR imaging on the surface, and especially in the underground, will show the peculiarity of this karst system.

In general, research methods could be divided to those conducted in speleological objects and those subsequently implemented in institutions, cabinets, laboratories, museums, etc. The data compilation and processing are used by the most modern computer programs to create the most realistic model for solving speleological problems in karst of Croatia. In some speleological objects in Croatia, measurements are carried out continuously, for example, absolute displacements of the fault wings or changes in temperature

Fig. 12.2 Vertical channels in pits of Velebit (Šlapice Pit)

and humidity, radon gas concentration, changes in groundwater levels, seismographic tests, etc. are monitored.

All these research methods help to create a more realistic view of the origin, shape and condition of speleological objects in the Dinaric Karst of Croatia.

Croatian speleologists are very active in caving and rescue activities and are recognized as one of the bearers of these activities in Europe and beyond, using state-of-the-art equipment and techniques, i.e. ways of using special speleo rescue equipment.

The highest scientific institution in Croatia—the Croatian Academy of Sciences and Arts (HAZU)—has, since its founding in 1861, through its members, taken care of Dinaric Karst research and speleological activity. Here are some of the most important members of the Croatian Academy of Sciences and Arts who left their mark in Croatian speleology or knowledge of karst: Spiridon Brusina, Đuro Pilar, Mijo Kišpatić, Dragutin Gorjanović Kramberger, Josip Poljak, Grga Novak, Branimir Gušić, Marijan Salopek, Milan Herak, Mirko Malez, Josip Roglić, Ante Polšak, Antun Magdalenić, Zlatko Pepeonik, Ivan Gušić, Branko Sokač, Milan Meštrov, Josip Tišljar, Franjo Fritz, Mladen Juračić, Goran Durn, Igor Vlahović, etc.

Within the Croatian Academy of Sciences and Arts, the Committee for the Karst (Milan Herak, Mirko Malez and Antun Magdalenić) was established in the 1960s, which was renovated in 2012. The members of the Committee are scientists in the field of karst research, led by Ivan Gušić and Mladen Garašić.

International Activity of Speleologists from Croatia

13

Abstract

The international activity of Croatian speleologists is noticeable in international speleological organizations and international speleological expeditions. With research results from Croatia and around the world, Croatian speleologists are well recognized among the international caving community.

From the earliest beginnings of speleology in Croatia, there was a need for joint or international action. This was manifested through personal contacts and visits or explorations of the cave by the individuals or groups who came to Croatia. In 1893, Alfred Eduard Martel, a famous French caver visited and explored the Pazin Cave and some other caves in the Istria area. Many Austrian, Czech, Italian, etc. scientists researched in the area of the Croatian Karst during the Austro-Hungarian monarchy and between the two world wars (Penk, Kraus, Grund, Katzer, Absolon, Bertarelli, Boegan, etc.).

Immediately after World War II, for various political reasons, the international cooperation of Croatian speleologists was reduced to publishing speleological works in foreign journals. However, since the beginning of the International Speleological Gatherings since 1953 in Paris, the international cooperation of speleologists from Croatia has been renewed and upgraded. Croatian speleologists have actively participated in all the speleological congresses held so far on all continents.

Thus, in 2 April 1954, the Croatian Speleological Society (SDH) was founded in Zagreb, whose first president was geologist and the first Croatian doctor of speleology, Dr. Josip Poljak (from 1954 to 1956). He was followed as the President by Academician Josip Roglic (1958–1962), prof. Vladimir Blašković (1962–1966), academician Mirko Malez (1966–1990), Prof. Ph.D. Mladen Garašić (1990–2010), Ph.D. Neven Bočić (2010–2013), Matej Mirkac (2013–2015), Ph.D. Nenad Buzjak (2018–2021). In 1991, the Croatian Speleological Society (SDH) changed its name to the Croatian Speleological Society (HSD). With the change of the Statute and the establishment of several new speleological associations in 1998, HSD was re-registered with the Croatian Speleological Association (HSS). The first president of the Croatian Speleological Association (1998–2010) was Mladen Garašić, and the first secretaries were Tihomir Kovačević (1998–2002), Nebojša Anić (2002–2006) Neven Bočić (2006–2010).

With its independence (1991), the Republic of Croatia represents its full membership in the UIS, which it acquired as one of the founders of the UIS since 16 September 1965. The Chairman of the Organizing Committee of the 4th World Speleological Congress at which UIS was founded in 1965 was Academician Grga Novak from Zagreb, and the first Vice-President of UIS, B.Sc. Stjepan Mikulec also from Croatia. Through its national speleological organization (HSS), Croatia became a full member of the FSE (European Speleological Federation) on 24 September 2009.

From 2013 to 2017, one of the adjunct secretaries at the UIS was from Croatia (Mladen Garašić), who was re-elected to the post (2017–2021). From 2015 to 2019, Prof.dr. Mladen Garašić from Croatia was elected Vice-President of the FSE (European Speleological Federation), and thus during the same period, a Croat was in the presidency of the UIS and the FSE, which never happened before. This is in a way a recognition of the international activity of speleologists from Croatia.

The first speleological international expedition to explore Croatia was the British expedition to the Balinka Cave in 1964 and 1966 (Fig. 13.1). Then there were many international expeditions in Croatia, among which we can enumerate the research of Bunovac pit in 1977 with Swiss cavers, 1980. exploration of the Ledenica Cave with German cavers, the Kamensko 84 expedition with American, British

M. Garasic, *The Dinaric Karst System of Croatia*, Cave and Karst Systems of the World,
https://doi.org/10.1007/978-3-030-80587-6_13

Fig. 13.1 Winch at the entrance of Balinka pit during the international speleological expedition in 1966

and French cavers, cave diving expeditions to the Crveno Jezero, springs of Gacka, Una, Kupa, Krnjeza, etc. with participants from Italy, France, Great Britain, the Czech Republic, Poland, Slovenia, BiH, etc. To date, over fifty international expeditions have been organized in Croatia.

The first expedition organized by Croatian speleologists abroad was in 1976 in Holloch in Switzerland, followed by France (1980), USA (1981), Morocco (1983), etc. To date, speleologists from Croatia have organized over one hundred international expeditions on all continents (Algeria, Madagascar, Cameroon, Namibia, Tanzania, South Africa, Russia, South Korea, Malaysia, Philippines, New Guinea, Thailand, Laos, Vietnam, India, Iran, Lebanon, Turkey, Ukraine, Cyprus, Canada, USA, Mexico, Puerto Rico, Cuba, Brazil, Argentina, Bolivia, Chile, Peru, Australia, New Zealand, Israel, Cambodia, Europe, etc.) except Antarctica. In international speleological organizations, Croatia takes a significant active position, and in the Union International Speleology (UIS)—International Union of Speleology) and the FSE (European Speleological Federation) even the presidency of these organizations (Mladen Garašić).

The leaders and founders of European Cave Rescue Association (ECRA) are also from Croatia (Darko Bakšić, Dinko Novosel). Representatives from Croatia participated in founding the World Speleological Union—UIS in 1965 in Postojna.

Conclusions

14

There is a long tradition of exploring karst and caves in Croatia. The fourth speleological association in the world (after Vienna, Trieste and Postojna) was founded and established here: in Perušić in Lika in 1886 and in 1892 in Kršlja in Kordun. It is rare that an area of the world can boast of such a speleological history. Even today, Croatian cavers are very active and recognized in the cave world. They were among founders of the World Union of Speleologists—Union Internationale de Speleologie (UIS) in 1965. The Dinaric Karst, with all its peculiarities caused by the geological predisposition of lithostratigraphy, tectonics and hydrogeology, has been occupying the interest of both domestic and foreign scientists and experts for centuries. Beginning with the basic theories of water movement in the karst two centuries ago, up to the latest explorations of deep caves on Velebit or large vrulja in the Adriatic Sea.

Out of more than 12,500 known speleological objects in the karst of Croatia, only about 7000 have been mentioned or described in more detail in domestic or foreign literature. The data about the other caves and pits are in the archives of the individuals and institutions that explored them. This book for the first time presents data about some caverns that have never been published before. Unfortunately, some of the caverns mentioned here no longer exist, because the necessary construction work has disintegrated them. But fortunately, their existence, the form of the estate, has been documented. The well-known and significant speleological objects have only been mentioned, and the literature where more information about them can be found is listed. At the time of writing this book, four caves deeper than one kilometre were known (the deepest Lukina Jama is 1431 m, and Crnopac cave systems longer than 56 kms). According to the activity, skills and knowledge of speleologists in Croatia so far, the discovery of new cave channels that will merge already known or completely unknown caves into even larger cave systems with depths of over 1.5 kms and/or lengths over 70 kms can very soon be expected. The potential for learning about new cave channels is even greater, especially in fully or partially submerged caves (springs, sinkholes, estavela and vrulja) (Fig. 14.1).

Many rare occurrences observed in the caves of the Dinaric Karst in Croatia were mentioned for the first time in this book, and it is expected that future generations will continue with their research. The book is intended for a wide range of readers, from those who are discovering speleological secrets for the first time to those who are experienced speleologists, but some of these facts have not been sufficiently known to them so far.

The Dinaric Karst in Croatia has a great potential for new discoveries that will surely leave a mark in the field of karst hydrogeology and especially speleology. The verticality of the speleological phenomena, their genesis, the amount and the type of groundwater in them have characteristics linked specially to this or some less known karst areas of the world.

Results of cavern (research conducted in Croatian Karst in the last thirty years) show that total length of more than a thousand (newly discovered) caverns is over 120 km and the total depth is about 25 km (Fig. 14.2). The longest cavern is over 9 km long, and the deepest one is around 450 m deep.

The deepest discoveries of caverns in the Dinaric Karst in Croatia were found at a depth of 3150 m. There are speleological objects on Dinara at heights above 1800 m. This means that the rock mass in which caves are developed in this area is at least 5 kms thick, which is a rare world phenomenon.

M. Garasic, *The Dinaric Karst System of Croatia*, Cave and Karst Systems of the World,
https://doi.org/10.1007/978-3-030-80587-6_14

Fig. 14.1 More than a thousand (newly discovered) caverns are explored in tunnels, quarries (tunnel Sveti Rok—Velebit Mt)

Fig. 14.2 One of specificum of Dinaric Karst in Croatia are caves at the Adriatic coast (Kamenjak Cave, Istria)

Bibliography[1]

Alfirević S (1969) Adriatic submarine springs in the water system of the Dinaric karst littoral and their problems. Krš Jugoslavije 6:183–200

Alfirević S (1969) Uloga i značaj opskrbe vodom u ekonomskom preobražaju jadranskog otočnog pojasa. Carsus Iugoslaviae 6:381–302

Aljinović B (1981) Depths to the basement of sediments along the line Brač- Palagruža. Geološki Vjesnik 34:121–125

Aljinović B (1986) Karta dubina Mohorovičićeva diskontinuiteta na području Jugoslavije (Depth map of the Mohorovičić discontinuity in the Yugoslavian region). Nafta (zagreb) 37(3):127–130

Babić Lj, Kučenjak MH, Ćorić S, Zupanič J (2007) The Middle Eocene age of the supposed late Oligocene sediments in the flysch of the Pazin Basin (Istria, outer Dinarides). Natura Croat 16(2):83–103

Bahun S (1968) Geološka osnova hidrogeoloških odnosa krškog područja između Slunja i Vrbovskog. Geološki vjesnik, vol 21, pp 19–82. Zagreb

Bahun S (1970) Geological basis of the Ogulin-Plaški karst depression. Carsus Iugoslaviae 7:13–20

Bahun S (1974) Tektogeneza Velebita i postanak Jelar-naslaga (The tectogenesis of Mt.Velebit and the formation of Jelar-deposits). Geološki Vjesnik 27:35–51

Bahun S (1985) Trijaske naslage i Jelar-formacija u dolini Une između Srba i Brotnje (Hrvatska) (Triassic deposits and Jelar Formation in the Una Valley between Srb and Brotnja (Croatia). Geološki Vjesnik 38:21–30

Bahun S (1991) O postanku Crvenog i Modrog jezera. Geološki vjesnik, vol 44, Zagreb.

Bajo P, Stroj A, Paar D, Ujević M, Bakšić D, Lacković D, Čop A, Radolić V (2009) Results of exploring the Munižaba cave (Velebit, Croatia) with special reference to measurements of physico-chemical and microclimatic parameters. In: Abstracts of presentations, 17th international Karstological School "Classical Karst", Karst Research Institute ZRC SAZU, Postojna, pp 51–51

Bajo P (2006) Speleogeneza špilje u kamenolomu Debeljača (Lovinac, Lika). Hrvatski geografski glasnik 68(1):99–111. Zagreb.

Bakalowicz M (2005) Karst groundwater: a challenge for new resource. Hydrogeol J 13(1):148–160

Bakran-Petricioli T, Petricioli D (1999) Life in the sea—submerged springs of the Velebit Channel. Ekološki Glasnik 4:17–23

Bakran-Petricioli T, Petricioli D (2008) Habitats in Submerged Karst of Eastern Adriatic Coast—croatian natural heritage. Croat Med J 49(4): 455–458

Bakšić D (1994) Istraživanja u Lukinoj jami tijekom 1994, Velebiten, br 16, str 25–27. Zagreb

Bakšić D (1998) Patkov gušt (-553m), La deuxieme verticale du Monde. Spelunca No 70. Paris

Bakšić D (1998) Istraživanja Slovačke jame u 1999. godini. Speleolog 46/47:13–16. Zagreb

Bakšić D (2001) Speleološka istraživanja Markovog ponora 1999. i 2000. godine. Velebiten 35:15–22. Zagreb

Bakšić D (2004a) Velebita – nova duboka jama na sjevernom Velebitu, Subterranea Croatica br 2, str 2–5, SK Ursus spelaeus, Karlovac

Bakšić D (2004b) Jamski sustav Velebita—najveća podzemna vertikala na svijetu. Subterranea Croatica br 3, str 2–7. SK Ursus spelaeus, Karlovac

Bakšić D, Lacković D (2002) Jama Amfora, -614 m – najdublja jama Biokova. Velebiten 36:16–22. Zagreb

Bakšić D, Paar D (2006) Croatia and the Deep Caves of Northern Velebit. In: Oliphant TJ (ed) Alpine Karst. Cave books, Dayton, USA, pp 105–124

Barišić T (2019) Jamski sustav Crnopac među 100 najduljih u svijetu. Helop 16:84–85. Šibenik

Baučić I (1984) Jame i spilje na otoku Braču. Brački Zbornik 14:23–34

Benac Č, Juračić M, Bakran-Petricioli T (2004) Submerged tidal notches in the Rijeka Bay NE Adriatic Sea: indicators of relative sea-level change and of recent tectonic movements. Mar Geol 212(1–4):21–33

Benac Č, Rubinić J, Ožanić N (2003) The origin and evolution of coastal and submarine springs in Bakar Bay. Acta Carsologica 32(1):157–171

Bertarelli L, Boegan E (1926) Duemilla Grotte. Touring Club Italiano, Milano, p 332

Biondić B (1988) Tapping and protection of underground water in the Adriatic karst region related to the new conception of structure of Dinarides. In: Proceedings of 21st IAH congress, Guilin, China, pp 187–193

Biondić B, Biondić R, Dukarić F (1998) Protection of karst aquifers in Dinarides in Croatia. Environ Geol 34:309–319

Biondić B, Biondić R, Meaški H (2010) The conceptual hydrogeological model of Plitvice Lakes. Geol Croat 63(2):195–206. Zagreb.

Biodić B, Biondić R (2014) Hidrogeologija Dinarskog krša u Hrvatskoj. Geotehnički fakultet, pp 1–341, Varaždin

Bockovac Ž (1999) Jama Amfora. Speleo'zin 12(7):14–15. Karlovac.

[1]Only significant papers and books which were used in the creation of this book, are mentioned in the literature list. In addition, hundreds of works (mainly not publicly printed professional reports) were used, but because of their abundance and other reasons, they could not all be included in the literature review. I believe that this will be done during the compilation of the entire speleological and karst bibliography in Croatia.

M. Garasic, *The Dinaric Karst System of Croatia*, Cave and Karst Systems of the World,
https://doi.org/10.1007/978-3-030-80587-6

Bočić N, Baćurin Ž (2002) Speleološka istraživanja poluotoka Pelješca u 2002. Speleo'zin 16(2003):7–13 Karlovac
Boegan E (1930) Catasto Italiane delle Grotte. Fasc. l. Grotte delle Venezia Giulia. Trieste
Bögli A (1980) Karst hydrology and physical speleology. Springer, Berlin, p 284
Bognar A (2006) The geomorphological evolution of Dinarides. In: Field trip guide Adria 2006 international geological congress on the Adriatic area. Istituito di Science della Terra, UniversitB di Urbino, Urbino, Italy, pp 23–26
Bognar A, Faivre S, Buzjak N, Pahernik M, Bočić N (2012) Recent landform evolution in the Dinaric and Pannonian regions of Croatia in book. In: Miloš Stankoviansky M, Kotarba A (eds) Recent landform evolution—the Carpatho-Balkan-Dinaric Region. Springer, pp 313–344
Bojanić L, Ivičić D, Božičević S (1980) Inženjerskogeološka problematika dovodnog tunela HE Zakučac II (Engineering geology explorations for the HP Zakučac II tunnel). In: Proceedings of the 6th Yug symposium hydrogeology and engineering geology, Portorož, Book 2, pp 191–200
Bonacci O (1987) Karst hydrology with special reference to the Dinaric karst. Springer, Berlin, p 184
Božičević S (1964) E Senj, Speleološki objekti HE Sklope-II. dio, elaborat, Arhiva geol. Institute, 3721/2, Zagreb
Božičević S (1965a) Poljakova pećina (The Poljak cave). Geol. Vjesnik 18/1(1964):141–157., 4 Tab., 1 crt., Zagreb
Božičević S (1965b) Poljakova pećina in Lika. Bull Sci Cons Acad Yougoslav (A) 10(10):320. Zagreb
Božičević S (1968) Hidrogeologija glavnih ponora rijeke Like. Geološki vjesnik 21:317–328. Zagreb
Božičević S (1984) Hydrogeology of the lost river Gacka in the Dinaric karst after completion of hydroelectric power plant Senj. Krš Jugoslavije 11(2):13–70. Zagreb
Božičević S (1985) Morfogeneza speleoloških pojava Istre i njihova zavisnost o geološkim i hidrogeološkim uvjetima [Morphogenesis of speleological phenomena of Istria and their connection with geological and hydrogeological conditions]. Ph.D. Thesis. Zagreb, pp 1–166
Božičević S (1966) Pećine, jame i ponori s vodom u području Dinarskog krša. Krš Jugoslavije, vol 6, str 137–154, JAZU, Zagreb
Božičević S (1982) HE Krčić. Speleološki radovi. Fond stručne dokumentacije IGI. Zagreb
Božičević S (1993) Kaptaža u kaverni ispod tunela Učka. Pomorski zbornik 31(1993)1:565–586
Božičević S (2012) Dr. Josip Poljak. Prvi Hrvatski speleološki kongres. Zbornik, Speleološki radovi. 1:27–42, Zagreb
Božić V (2014) Ilustrirana povijest speleologije u Hrvatskoj, Zagreb, pp 1–80
Brlek M, Korbar T, Cvetko Tešović B, Glumac B, Fuček L (2012) Stratigraphic framework, discontinuity surfaces and regional significance of Campanian slope to ramp carbonates from central Dalmatia, Croatia. Facies. https://doi.org/10.1007/s10347-012-0342-0
Buzjak N (2002) Speleološke pojave u Parku prirode "Žumberak–Samoborsko gorje" (Speleological features of "Žumberak–Samoborsko gorje" Nature park). Geoadria 7(1):31–49
Buzjak N (2006) Speleomorfološke i hidrološke značajke Đurovića špilje (Čilipi, južna Dalmacija). Hrvatski geografski glasnik 68(2):57–72. Zagreb
Buzjak N (2016) Prilog povijesti speleoloških istraživanja šireg prostora gorskog masiva Risnjaka i NP Risnjak od 19. stoljeća do 2005. godine. Subterranea Croat 14(2):2–16. Karlovac
Buzjak N, Kovač Konrad P, Jalžić V (2013) Recent research of submarine springs Zečica and Modrič (North Dalmatia). In: Proceedings of 3rd scientific meeting quaternary geology in Croatia, pp 14–15
Buzjak N, Bočić N, Paar D, Bakšić D, Dubovečak V (2018) Ice caves in Croatia. In Persoiu A, Lauritzen S (eds) Ice caves. Elsevier, Amsterdam, pp 335–369
Casati L, Garašić M (2016) Speleologia subacquea in Croazia—immersioni nelle sorgenti di Sinjac e Vrelo Une (Inquadramento geografico e geostrutturale del fiume Una). Speleologia 37(75):39–42
Cerkvenik R, Kranjc A, Mihevc A (2018) Karst Wetlands in the Dinaric Karst. In: Finlayson C, Milton G, Prentice R, Davidson N (eds) The wetland book. Springer, Dordrecht (on line)
Cucchi F, Forti F, Ulcigrai F (1994) Lowering of karst surface by corrosion. Acta Carsologica 3:55–62
Cucchi F, Forti F, Stefanini S, Ulcigrai F (1985) Mesures de degradation de roches carbonatees et d'accroissement de stalagmites. In: Act Coll 16 Congress National Speleology, Nancy-Metz, (1985), Spelunca Mem 14:87–90
Cukrov N, Lojen S, Mihelčić G, Dolenec T, Branica M, Papesch W (2003) Isotopic characteristics of tufa in Krka river, Croatia: relation to climate conditions. In: 22nd IAS meeting of sedimentology—Opatija 2003 Abstracts Book 37
Cukrov M, Jalžić B, Omanović D, Cukrov N (2006) Traces of metal in the water column of Urinjska cave. Subterranea Croat 7:25–30
Cukrov M, Ozimec R, Cukrov N, Bedek J, Jalžić B (2012) Sumporače—preliminary results from research into unique anchiline sulphur caves in Croatia. Nat Croat 21(1):36–42. Zagreb
Culver D, White B (eds) (2005) Encyclopedia of caves. Elsevier Academic Press, Amsterdam, New York, 1–654 pp
Cvetko B, Tešović B, Gušić I, Jelaska V, Bucković D (2001) Stratigraphy and microfacies of the Upper Cretaceous Pučišća Formation, Island of Brač, Croatia. Cretac Res 22:591–613
Cvijić J (1893) Das Karstphänomen. Pencks Geogr Abh 5(3):217–330
Cvijić J (18950 Karst. Geografska monografija, Beograd, Štamaparija Kraljevine Srbije
Čalić J (2011) Karstic uvala revisited: toward a redefinition of the term. Geomorphology 134(1–2):32–42
Čepelak M (1977) Novija i buduća speleološka istraživanja u Veternici. Speleolog 24–25:1–8. Zagreb
Čepelak M (1981) Neka objašnjenja uz nacrt Jopičeve špilje. Speleolog 28–29:20–21. Zagreb
Čepelak M (1986) Špiljski sustav Đula-Medvjedica. Speleolog 1984–1985 (XXXII–XXXIII):2–24. Zagreb
Ćosović V, Marjanac T, Drobne K, Moro A (2008) Eastern adriatic coast—external dinarids. In: McCann T (ed) Geology of Central Europe, vol 2. Geological Society London., pp 76–80
Dadić Ž (1984) Osvrt Nikole Gučetića u 16. stoljeću na dvije špilje u okolici Dubrovnika. 9. jug. speleološki kongres (Zbornik predavanja), pp 741–746. Zagreb
Denić-Jukić V, Jukić D (2003) Composite transfer functions for karst aquifers. J Hydrol 274(1–4):80–94
Denić-Jukić V, Jukić D, Vego G (2007) Karst springs discharge modeling by using the composite transfer functions: example of the Žrnovnica and Jadro Springs. In: Proceedings of the XXXV congress of the international association of hydrogeologists "Groundwater and Ecosystems"
Dreybrodt W, Kaufmann G (2007) Physics and chemistry of dissolution on subaerialy exposed soluble rocks by flowing water films. Acta Carsol 36(3):357–367

Durn G, Ottner F, Tišljar J, Mindszenty A, Barudžija U (2003) Regional subaerial unconformities in shallow-marine carbonate sequences of Istria: sedimentology, mineralogy, geochemistry and micromorphology of associated bauxites, palaeosols and pedo-sedimentary complexes. In: Vlahović I, Tišljar J (eds) Evolution of depositional environments from the Palaeozoic to the Quaternary in the karst Dinarides and the Pannonian basin, Field trip guidebook of 22nd IAS meeting of sedimentology—Opatija 2003, 17–19 Sept 2003,. Opatija, Institute of Geology, Zagreb and Croatian Geological Society, pp 209–254, Zagreb

Ereš M, Garašić M, Vukelić S, Bajzek Z, Hranjec M (1999) Iskop tunela "Sveti Rok". Ceste i mostovi, br. 3–4, str. 243–248. Zagreb

Faivre S (2000) Landforms and tectonics of the Velebit mountain range (External Dinarides, Croatia). Ph.D. thesis, Zagreb University, Faculty of Science, Department of Geography, Croatia/Université Blaise Pascal, Clermont-Ferrand II, France, p 360

Faivre S, Bakran-Petricioli T, Horvatinčić N (2010) Relative sea-level change during the late Holocene on the Island of Vis (Croatia) Issa harbour archaeological site. Geodin Acta 23(5–6):209–223

Faivre S, Reiffsteck P (2002) From doline distribution to tectonic movements—example of the Velebit Mountain Range Croatia. Acta Carstol 31 (3):139–154

Fleury P, Bakalowicz M, de Marsily G (2007) Submarine springs and coastal karst aquifers: a review. J Hydrol 339(1–2):79–92

Ford D, Williams P (2007) Karst hydrogeology and geomorphology. Wiley & Sons, p 562

Fortis A (1776) Viaggio in Dalmazia, Venezia

Fritz F, Bahun S (1997) The morphogenesis of submarine spring in the Bay of Kaštela Croatia. Geol Croat 50(1):105–110

Fritz F, Pavičić A (1982) Hydrogeologically "hanging" parts of the Zrmanja and Krka rivers. In: Proceedings of "VII jugoslavenskog simpozija o hidrogeologiji i inženjerskoj geologiji". Novi Sad, in Croatian, pp 115–121

Fritz F, Pavičić A (1987) The Miljacka karst spring catchment area in Krka river valley. In: Proceedings of "IX jugoslavenskog simpozija o hidrogeologiji i inženjerskoj geologiji", Priština, in Croatian, pp 97–101

Gajić-Čapka M, Perčec Tadić M, Patarčić M (2003) Digitalna godišnja oborinska karta Hrvatske (Digital map of the precipitation distribution in Croatia). Hrvatski Meteorološki Časopis 38:21–33

Gams I (1974) Kras. Zgodovinski, naravoslovni in geografski oris. Slovenska Matica, Ljubljana

Gams I (2003) Kras v Sloveniji v prostoru in času. Založba ZRC, ZRC SAZU, Ljubljana, p 516

Garašić M (1975) Istraživanja donje etaže u Jopičevoj spilji. *Speleolog,* 22/23:2–3. Zagreb

Garašić M (1976) Mijatova jama. Sedmi Jugoslovenski Speleološki kongres, Herceg Novi, Zbornik radova, SSJ, pp 135–142, Titograd

Garašić M (1978) Fotografija kao pomoćno sredstvo za rješavanje nekih geoloških problema pri detaljnim istaživanjima speleoloških objekata u kršu. Simpozoj o fotodokumentaciji krša. Zbornik, SSJ, Postojna, pp 19–22

Garašić M (1980) Neotektonika u speleološkim objektima. In: VIII Jugoslovenski speleološki kongres, Zbornik radova, Borsko jezero, str 51–58, fot 2, sl 4. Beograd.

Garašić M (1984) Neotectonical activities—the basis of genesis and morphology of one of the Largest Cave System in Croatia (Yugoslavia). In: The Ninth Yugoslav Congress of Speleology, proceedings, Zagreb, pp 457–466

Garašić M (1985) Proučavanje tektogeneze u spiljama. Priroda 73:2021. Zagreb.

Garašić M (1986) Hydrogeology and morphogenesis of the Caves in Croatian Karst. Ph.D. Thesis, University of Zagreb, Faculty of Geology, Zagreb, pp 1–161

Garašić M (1989) The latest research of speleological objects in Dinaric karst area (Yugoslavia) from the geological point of view. In: 10th International congress of speleology: proceedings, vol 1, UIS, Budapest, pp 223–233

Garašić M (1990) Jama pod Debelom glavom—speleološki objekt na razvodnici slivova rijeke Korane i Gline. Spelaeologia Croat 1:9–14. Zagreb.

Garašić M, Kovačević T (1990) Speleohydrogeology of Velika Pećina near Blata in Lika region. Spelaeologia Croat 1:15–19. Zagreb

Garašić M (1991) Morphological and hydrogeological classification of speleological structures (caves and pits) in the Croatian karst area. Geološki Vjesnik 44:289–300

Garašić M, Kovačević T (1991) Speleogeologija Špilje kod Šušnjara—jedinstvenog speleološkog objekta formiranog u laporovitim stijenama (Banija). Spelaeologia Croat 2:31–36. Zagreb

Garašić M (1992) 1892-the year of estamblishing the first organization connected with speleological activities in Croatia. Spelaeologia Croat 3:53–55. Zagreb

Garašić M, Zanoškar D (1992) Ponor kod Vrata i hidrogeologija bliže okolice Fužina. Spelaeologia Croat 3:23–26

Garašić M (1995) Speleogeneza u okviru hidrogeologije krša i procesa karstifikacije. 1. Hrvatski Geološki Kongers, Opatija, Zbornik radova, Proceedings, pp 177–182, Zagreb

Garašić M (1997a) Stupina jama—najdublja jama Gorskog kotara. Speleo'zin, vol 7, str 7–11, Karlovac

Garašić M (1997b) Some types of speleogenesis and speleohydrogeology in Dinaric karst area (Croatia, Europe). In: Proceedings of the 12th international congress of speleology, vol 1. UIS, La Chaux-de-Fonds, pp 147–150

Garašić M (1997c) Contribution to the Knowledge about Speleology in Croatia. Acta Carsol 26(2):49–54. Postojna

Garašić M (1997d) Karst water tracing in some of the speleological features (Caves and Pits) in Dinaric Karst Area. Tracer hydrology. Balkema CRC Press, Rotterdam, pp 229–236

Garašić M (2001) New speleohydrogeological research of Crveno Jezero (Red Lake) near Imotski in Dinaric karst area (Croatia, Europe): International speleodiving expedition 'Crveno jezero 98'. In: Procceedings of 13th international congress of speleology, Speleo Brazil, pp 555–559

Garašić M (2004) Istraživanje kaverni u tunelu Sveti Rok (cave exploring in Sveti Rok tunnel), Karlovac. Speleozin 12:21–27

Garašić M (2005a) Some new speleological research of Caverns in route of the Highways in Croatian Karst. In: 14th International congress of speleology: Procceedings: Union Internationale de Speleologie, pp 221–224, Athens

Garašić M (2005b) Zanimljiv nalaz siga pod morem. 6. skup speleologa Hrvatske sažeci SDK, Kamanje, str 25–25

Garašić M (2006a) Izvješće o speleološkom istraživanju, fotografskom i topografskom snimanju, te hidrogeološkim opservacijama u novootvorenoj jami (kaverni) u tunelu Vrata (desna tunelska cijev), na stacionaži km 31+292, na trasi autoceste Zagreb—Rijeka, dionica Oštrovica—Vrata kod Fužina u Gorskom Kotaru. Građevinski Fakultet Zagreb 100/06

Garašić M (2006b) Pronađeni i snimljeni najdublji speleothemi u moru. Spelaeologia Croat 7:58

Garašić M (2006c) Novije spoznaje o speleogenezi i kršu, Zbornik radova. Akademik Josip Roglić i njegovo djelo, Hrvatsko geografsko društvo; Split, Zadar, Zagreb, str. 105–111

Garašić M (2007) Sveti Rok—Velika kaverna u cestovnom tunelu na autocesto Zagreb-Split. Hrvatska vodoprivreda XVI, br 180, str 56–60. Zagreb

Garašić M (2009a) Izvješće o speleološkom istraživanju, fotografskom i topografskom snimanju, te hidrogeološkim opservacijama u novootvorenoj jami (kaverni) u tunelu "Sveti Ilija" (servisna tunelska cijev), na stacionaži km 1+193, na trasi spojne ceste Zagvozd—Baška voda u Dalmaciji. University of Zagreb, pp 1–39, Zagreb

Garašić M (2009b) Izvješće o speleološkom istraživanju, fotografskom i topografskom snimanju, te hidrogeološkim opservacijama u novootvorenoj jami (kaverni) u tunelu "Sveti Ilija" (servisna tunelska cijev), na stacionaži km 1+415, na trasi spojne ceste Zagvozd—Baška voda u Dalmaciji. University of Zagreb, pp 1–43, Zagreb

Garašić M (2009c) Izvješće o speleološkom istraživanju, fotografskom i topografskom snimanju, te hidrogeološkim opservacijama u novootvorenoj jami (kaverni) u tunelu "Sveti Ilija" (servisna tunelska cijev), na stacionaži km 1+637, na trasi spojne ceste Zagvozd—Baška voda u Dalmaciji. University of Zagreb, pp 1–55, Zagreb

Garašić M (2010) Prinos poznavanju i novi tipovi speleogeneze proučavane prilikom istraživanja spilja, jama i kaverni u Hrvatskom kršu. In: 4th Croatian congress od geology: Abstracts Book. Hrvatski geološki institut, Zagreb, pp 328–329

Garašić M, Kovačević T (1992) Neki speleološki objekti na trasi autoceste Karlovac—Rijeka, dionica Oštrovica-Delnice i kaverne u tunelima "Tuhobić" i "Vrata. Spelaeologia Croat 3, str 5–14. Zagreb

Garašić M, Vivoda B (2003a) Speleološke pojave (kaverne) u trasi autoceste Zagreb-Rijeka. Ceste i mostovi 49, br 5–6, str 104–111, Zagreb

Garašić M, Kovačević MS, Jurić-Kaćunić D (2010) Investigation and remediation of the cavern in the Vrata tunnel on the Zagreb-Rijeka highway (Croatia) Postojna. Acta Crsologica 39(1):61–77

Garašić M, Cvetković M (2012) Izvješće o rezultatima speleološke prospekcije novootvorenih kaverni I rezultati geofizičkih ispitivanja stijenske mase u neposrednoj blizini kaverni. Faculty of Civil Engineering, University of Zagreb, pp 1–55, Zagreb

Garašić M, Garašić D (2013) Some information about the deepest Caves known in Croatian Karst Area. In: 16th International Congress of Speleology: proceedings, vol 3. Czech Speleological Society, Praha, pp 72–77

Garašić M (2014) New research in cave Ledenica in Bukovi Vrh on Velebit Mt in Croatian Dinaric Karst. In: NCKRI symposium4, international workshop on Ice Caves VI, pp 31–32

Garašić M, Garašić D (2015) Problems with caves during tunneling in Dinaric Karst (Croatia). In: Lollino G, Manconi A, Guzzetti F, Culshaw M, Bobrowsky P, Luino F (eds) Engineering geology for society and territory, vol 5. Springer, Cham

Garašić M, Garašić D (2016) Some new and interesting caverns in tunnels and along stretches of the Croatian Highways (Dinaric Karst, Croatia). Environ Earth Sci 75:5

Garašić M, Garašić D (2017a) The new insights about deep karst springs in the Dinaric karst of Croatia. In: Proceedings of the 17th international congress of speleology, vol 2, July 22–28, Sydney, NSW Australia, pp 30–33, Sydney

Garašić M, Garašić D (2017b) Some new caves under the Dubrovnik city airport in the Dinaric karst of Croatia. In: Proceedings of the 17th international congress of speleology, vol 2, pp 26–29, Sydney

Garašić & Getvaj (2020) Circular anemolites in Cave Vrelo in Croatia. (in print)

Gjurašin K (1942) Prilog hidrografiji primorskog krša. (Contributionto the litoral karsthydrography). Tehnički Vjesnik 59:107–112. Zagreb

Glavaš I (2011) Jama kod Rašpora (Žankana jama, Abisso Bertarelli). Zbornik općine Lanišće 2010, Lanišće

Gorjanović-Kramberger D (1899) Krš Zagrebačke gore. Hrvatski Planinar 11:164

Gorjanović- Kramberger D (1911) Vijesti geološkoga povjerenstva za kraljevine Hrvatsku-Slavoniju za godinu 1910, Hartman, Zagreb

Gorjanović-Kramberger D (1914) Einige Beobachtungen im untern Laufe des Mrežnica-Flusses (in Croatian and German). Berichte der Geologischen Kommision der Königreiche Kroatien und Slavonien für Jahre 1912 und 1913, Kraljevska zemaljska tiskara, Zagreb, pp 107–111

Grandić S, Veseli V, Kolbah S (2002) Hydrocarbon potential of Dugi otok basin in offshore Croatia. Nafta (zagreb) 53(6–7):215–224

Grund A (1910) Beiträge zur Morphologie des dinarischen Gebirges. Geograph Abhandl (Penck) 9(3): 236

Gučetić VN (1584) Sopra le metheore d' Aristotile, Venezzia

Gunn J (ed) (2004) Encyclopedia of Caves and Karst Science, Fitzroy Dearnborn, Taylor & Francis Group. New York, London, 1–902 pp

Gušić I, Jelaska V (1990) Upper Cretaceous stratigraphy of the Island of Brač within the geodynamic evolution of the Adriatic carbonate platform. Djela Jugosl Akad Znan Umjetn 69:1–160

Herak M (1984) Tektonski okvir speleogeneze. Deveti jug. speleo. kong., Karlovac, str. 111–130, Zagreb

Herak M (1986) A new concept of geotectonics of the Dinarides. Acta Geol 16/1, Prirod. istr. 53, JAZU, pp 1–42, Zagreb.

Herak M (1987) Relationship between adriatic and dinaric carbonate platforms. Mem Soc Geol It 40:289–293

Herak M (1991) Dinarides—mobilistic view of the genesis and structure. Croat Acad Sci Art Acta Geol 21:35–83

Herak M, Bahun S, Magdalenić A (1969) Pozitivni i negativni utjecaji na razvoj krša u Hrvatskoj (Positive and negative influences on the development of the Karst in Croatia). Krš Jugosl., akad. znan. i umjetn Zagreb 6:45–78

Herak M (2002) Povijesni temelji hrvatske geologije, HAZU, Zagreb, pp 1–181

Herak M, Stringfield VT (1972) Karst—important karst regions of the northern hemisphere. Elsevier, Amsterdam

Herak I, Magaš B, Sarić A, Habe F (1976) Prilog bibliografiji krša Jugoslavije 1666–1974 (Contribution of Bibliography of Karst of Yugoslavia from 1666–1974). Carsus Iugoslaviae 9(3):81–149, Zagreb

Herak M, Bahun S (1979) The role of the calcareous breccias (Jelar Formation) in the tectonic interpretation of the High Karst Zone of the Dinarides. Geol Vjesnik 31:49–59

Hlaj A, Poropat B (2008) Vodič kroz speleološke objekte Bujštine. Acta Bullearum II, Pučko otvoreno učilište Buje, pp 108–116

Horvatinčić N (1996) Izotopna mjerenja u Ledu, jama Ledenica, Velebit. Zbornik radova trećeg simpozija Hrvatskog društva za zaštitu od zračenja, pp 297–301, Zagreb

Hudec M, Božičević S, Bruna V, Blajvajs R (1980) Stabilizacija svoda kaverne u tunelu Učka. Saopćenje 5 Jugoslovenskog simpozija iz mehanika stijena i podzemnih radova, knjiga 2, Split, pp 64–70

Jalžić B (2007) Jamski sustav Lukina jama—Trojama, pp1–156, HPS, Zagreb

Jelinić I (2001) Ledena jama Lomskoj dulibi. Speleolog 46/47, str. 17–22, Zagreb

Juračić M, Bakran-Petricioli T, Petricioli D (2002) Cessation of karstification due to the sea-level rise? Case Study of the Y-Cave, Dugi otok, Croatia. In: Grabovšek F (ed) Evolution of karst: from prekarst to cessation. Založba ZRC, Postojna, Ljubljana, pp 319–326

Jurkovšek B, Cvetko Tešović B, Kolar-Jurkovšek T (2013) Geologija Krasa = Geology of Kras—Geološki zavod Slovenije, p 205, Ljubljana

Katzer F (1909) Karst und Karsthydrographie. Zur Kunde der Balkanhalbinsel, Sarajevo, pp 1–94

Kern Z, Bočić N, Horvatinčić N, Fórizs I, Nagy B, László P (2008) Palaeoenvironmental records from ice caves of Velebit Mountains—Ledena Pit and Vukušić Ice Cave, Croatia. In: Kadebskaya O, Mavlyudov BR, Pyatunin M (eds) 3rd International workshop on Ice Caves proceedings, Kungur, pp 108–113

Kern Z, Széles E, Horvatinčić N, Fórizs I, Bočić N, Nagy B (2011) Glaciochemical investigations of the ice deposit of Vukušić Ice Cave, Velebit Mountain, Croatia. Cryosphere 5:485–494

Kern Z, Bočić N, Sipos G (2018) Radiocarbon-dated vegetal remains from the cave ice deposits of Velebit mountain, Croatia. Radiocarbon 60 (5):1391–1402

Klimchouk A (ed) (2001) Speleogenesis. Evolution of Karst Aquifers, p 1–528

Klimchouk A (2007) Hypogene speleogenesis: hydrogeological and morphogenetic perspective. National Cave and Karst Research Institute, Special Paper no.1, Carlsbad, pp 1–106

Korač N, Barudžija U (2017) Geološke značajke Jame kod Rašpora na Ćićariji. Subterranea Croat 15(1):14–18, Karlovac

Korbar T (2009) Orogenic evolution of the external Dinarides in the NE Adriatic region: a model constrained by tectonostratigraphy of Upper Cretaceous to Palaegone carbonates. Earth Sci Rev 96(4):296–312

Kovačević T (2006) Đurovića jama-špilja novi speleološki biser u turističkoj ponudi "Lijepe naše". Subterranea Croat 4(6):16–22. Karlovac

Kovačević T (2010) Međunarodna speleološka ekspedicija "Panjkov ponor 2010". Speleolog 58:42–53. Zagreb

Kovačević T (2013) Sedam veličanstvenih vrila našeg Dinarskog krša. Hrvatska vodoprivreda XX(198):114–119. Zagreb

Kovačević T (2015) MSE "Kamena galaksija—Zrmanja 2016." Scubalife 6(20):72–81

Kranjc A (2004) Dinaric karst. In Gunn J (ed) Encyclopedia of caves and karst science. Fitzroy Dearborn, New York, London

Kranjc A (2010) Short hystory of research. In: Prelovšek Z-H (ed) Mihevc. Introduction to the Dinaric karst, Postojna, pp 9–13

Kuhta M, Božičević S, Kapelj S, Miko S (1999) Protection and utilization study of the natural environment of the Modrič Cave and its surroundings. Technical report. Croat Geol Surv, Zagreb

Kuhta M, Novosel A (2000) Hydrogeological exploration of the vruljas in the Jurjevska Žrnovnica Bay. In: Proceedings of 2nd Croatian geological congress, pp 653–658

Kukuljan L, Glavaš I (2012) Rezultati najnovijih istraţivanja u Jami kod Rašpora. Subterranea Croat 14:42–46. Karlovac

Kukuljan L, Reš D, Rubinić A, Grozić D, Trinajstić N, Pavlek M, Bregović P, Delić T, Bedek J, Vitas B, Randić M (2018) Šverda—Speleološka istraživanja zaleđa izvora Rječine. Subterranea Croat 16(1):4–292. Karlovac.

Kuželički R, Ružić, D (2008) Tunel Pećine. Građevinar 60(6):529–542. Zagreb.

Kwokal Ž, Cukrov N, Cuculić V (2014) Natural causes of changes in marine environment: mercury speciation and distribution in anchialine caves. Estuarine Coastal Shelf Sci 151:10–20. Elsevier.

Labaš V (1987) Neke specifičnosti građe podzemlja dijela Centralne zone dinaridskog gravimetrijskog minimum. Nafta 38(10):555–559. Zagreb

Lacković D (1989) Špilja u kamenolomu Tounj. Hrvatski Planinar 81:233–234. Zagreb

Lacković D, Glumac B, Asmerom Y, Stroj A (2011) Evolution of the Veternica cave (Medvednica Mountain, Croatia) drainage system: insights from the distribution and dating of cave deposits. Geol Croat 3:213–221

Lacković D, Šmida B, Horvatinčić N, Tibljaš D (1999) Some geological observations in Slovačka jama cave (-1268 m) in Velebit mountain Croatia. Acta Carsologica 28(2):113–120

Lončar N, Bar M, Bar-Mathews M, Ayalon A, Faivre S, Surić M (2015) Izotopni sastav siga iz speleoloških objekata kao pokazatelj promjena u paleookolišu. Knjiga sažetaka: Hrvatsko geografsko društvo, pp 35–35. Zagreb

Lugomer D, Bubanović B, Redžepagić M (1999) Iskop sjeverne strane tunela Sveti Rok (stručni rad). Građevinar 8(1999):535–545

Malez M (1955) Đulin ponor u Ogulinu. Geolv vjesnik 8–9:153–172. Zagreb

Malez M (1956a) Pećina Drenovac (Die Hoble Drenovac) Zagreb. Speleolog 3–4:52–56

Malez M (1956b) Istraživanje pećina Gorskog kotara. Ljetopis JAZU LXIII:347–354. Zagreb

Malez M (1960) Pećine Ćićarije i Učke u Istri. Acta Geol 2, Priorod.istr. JAZU 29:163–263, Zagreb

Malez M (1961a) Speleološki objekti jugozapadne Like. Acta Geol 3, vol 31, str.107–241, JAZU, Zagreb.

Malez M (1961b) Pećina Veternica kao paleolitsko nalazište s tragovima kulta medvjeda. Drugi jugoslavenski speleološki kongres, 123138, Zagreb

Malez M (1963) Kvartarna fauna pećine Veternice u Medvednici. Paleontolo Iug JAZU 5:1–193. Zagreb

Malez M (1965) Cerovačke Pećine/die Höhlen von Cerovac/The Cervac Caves/Les Grottes de Cerovac, Published as Izdanja Speleoloskog drustva Hrvatske, No 1. SB. 41, p 21 full page B&W photos, 2 large folding maps in the rear

Malez M (1984) Povijest speleoloških istraživanja u Hrvatskoj. Zbornik predavanja 9. jugosl. spel. kongresa, Karlovac, pp 73–102

Malez M (1987) Quaternary vertibrata fauna from Modrić cave near Rovanjska. Rad Yug Acad Sci Art 432:141–154. Zagreb

Malez M, Božičević S (1964) Medvjedja pećina (Bärenhöhle) auf der Insel Lošinj als ein Beweis für Postwürmsche Transgression in nördlichen Teil des Adriatischen Meeres. Bull Sci JAZU v9 110. 4–5:105–106. Zagreb

Malez M, Božičević S (1965) The Medvjedja Pećina (Bear Cave) on Lošinj Island, a rare case of submerged cave. Prob Speleol Res, pp 211–216, Prague

Magdalenić A (1971) Hidrogeologija sliva Cetine. Krš Jugoslavije 7/4

Malinar H (2018) Geneza špilje Krušcice u Lici i moguća daljnja istraživanja. Speleolog 66:40–47, Zagreb

Marjanac LJ, Marjanac T (2004) Glacial history of the Croatian Adriatic and Coastal Dinarides. In: Ehlers J, Gibbard PL (eds) Quaternary glaciations—extent and chronology. Elsevier, vol 2, pp 19–26

Marijanović A (2018) Krški fenomen—Rijeka Vrljika. Hrvatska vodoprivreda XXVI, 225:33–36. Zagreb

Markovinović D, Garašić M (2010) Implementacija gravimetrije u speleologiji i istraživanju kaverni. In: The first croatian speleological congress with international participation. Abstracts, Hrvatski speleološki savez, Zagreb, pp 86–87

Martel EA (1896) Sur de la Foiba Pisino (Istria). Compt. Rend. de Acad. des Sciences, 123, 28th December 1896th Paris.

Martel EA (1897) La Foiba de Pisino. La Nature, Paris, p 25

Matas M (2009) Krš Hrvatske—geografski pregled i značenje. Geografsko društvo Split, pp 1–264

Matičec D (2009) Krški Dinaridi. Tektonski pregled. In: Velić I, Vlahović I (eds) Tumač Geološke karte Republike Hrvatske 1:300000, Croatian Geological Survey, pp 105–106, Zagreb

Maucci W (1952) L' ipotesi dell "Erosione inversa. Bolletino dellea Societa Adriatico di Scienze Naturali, vol XLVI, Trieste.

Mihevc A, Prelovšek M (2010) Geographical position and general overview. In: Mihevc A et al (eds) Introduction to the Dinaric Karst. ZRC Publishing, pp 6–8, Postojna

Mihevc A (2010) Geomorphology. In: Mihevc A, Prelovšek M, Zupan Hajna N (eds) Introduction to the Dinaric Karst. Karst Research Institute at ZRC SAZU, Postojna, pp 30–43

Mihevc A, Zupan N (1996) Clastic sediments from dolines and caves on the route of highway at Divača. Acta Carsologica 25:169–191

Miko S, Kuhta M, Kapelj S (2002) Environmental baseline geochemistry of sediments and percolating waters in the Modrič Cave Croatia. Acta Carsologica 31(1):135–149

Milanović P (1981) Karst hydrogeology. Water Resources Publication, Littleton, p 434

Milanović P (1996) Ombla spring, Croatia. Environ Geol 27:105–107

Milanović PT (2006) Karst istočne Hercegovine i Dubrovačkog priobalja (Karst of Eastern Herzegovina and Dubrovnik Littoral). Asocijacija speleoloških organizacija Srbije, Beograd, p 362

Mojsisovic E, Tietze E, Bittner E (1880) Geologie von Bosnien. Jahrb Geol Reichsanstalt, pp 2–272

Nicod J (2003a) Understanding environmental problems in Dinaric Karst. Dela 20:27–41

Nicod J (2003b) Les karsts Dinariques—paysages et problèmes: Slovénie, Croatie, Bosnie-Herzégovine, Monténégro Karstologia. Mémoires 10:1–183

Otte M (1968) The ponor of the Pazinčica and its role in the underground hydrological system in central and southern Istria. Geogr Bull 30:61–77

Paar D, Mance D Stroj A, Pavić M (2019) Northern Velebit (Croatia) karst hydrological system: results of a preliminary 2H and 18O stable isotope study. Geol Croat J Croat Geol Surv Croat Geol Soc 72(3):205–213

Palmer A (2007) Cave geology. Cave Books, Dayton, pp 1–454

Pamić J, Gušić I, Jelaska V (1998) Geodynamic evolution of the Central Dinarides. Tectonophysics 297:251

Perica D (1998) Geomorfologija krša Velebita (Karst geomorphology of the Velebit Mountains). Ph.D. dissertation, PMF, Geografski odsjek, Zagreb, p 220

Petricioli D, Bakran-Petricioli T, Kodba Z, Jalžić B (1995) Basic biological properties of temporary submarine springs (vrulje) in Modrič and Zečica coves (Adriatic Sea, Croatia). Paklenički Zbornik 1:195–198

Poljak J (1922) Periodička jezero Blata. Glasnik Hrv. priorodoslovnog društve 34:71–82. Zagreb

Poljak J (1934) Pećina Veternica u Zagrebačkoj gori. Priroda 5:133–139

Poljak J (1935) Pećine okolice Ogulina, Velike Paklenice i Zameta. Rasprave Geološkog instituta Kraljevine Jugoslavije, br 5. Beograd

Prelogović E, Aljinović B, Bahun S (1995) New data on structural relationships in the Northern Dalmatian Dinaride Area. Geol Croat 48(2): 167–176

Prelogović E, Pribičević B, Ivković Ž, Dragičević I, Buljan R, Tomljenović B (2004) Recent structural fabric of the Dinarides and tectonically active zones important for petroleum-geological exploration. Nafta 4:155–161

Prüfer K, de Filippo C, Grote S, Mafessoni F, Korlević P, Hajdinjak, M, Vernot B, Skov L, Hsieh P, Peyrégne S, Reher D, Hopfe C, Nagel S, Maricic T, Fu Q, Theunert Q, Rogers R, Skoglund P, Chintalapati M, Dannemann M, Nelson B.J, Key F.M, Rudan P, Kućan Ž, Gušić I, Golovanova LJ, Doronichev V, Patterson, N, Reich D, Eichler E, Slatkin M, Schierup M, Andrés A, Kelso J, Meyer M, Pääbo S (2017) A high-coverage Neandertal genome from Vindija Cave in Croatia. Science 358(6363):655–658

Roglić J, Baučić I (1958) Krš u dolomitima između Konavoskog polja i morske obale. Hrvatski geografski glasnik 20(1):129–137. Zagreb

Roglić J (1974a) Contribution to Croatian karst terminology. Beograd, Savez Geografskih Institucija Jugoslavije

Roglić J (1974b) Odnos izmedju površja i podzemlja Dinarskog krša. Acta Carsologica. In: 6th Yugoslav speleological congress, Sežana-Lipica, 10–15 oktober 1972, vol 6, pp 11–19, Ljubljana

Rubinić A, Kukuljan L, Glavaš I, Rubinić J, Rutić I (2013) Cave explorations and application of hydrological model in Rašpor cave (Istria, Croatia) . In: 16th international congress of speleology, Czech Republic, Brno, Proceedings, vol 2, pp 369–375. Brno, Czech Republic

Slišković I, Ivičić D (2000) Hidrogeologija Imotskog polja [Imotski Polje hydrogeology]. Rudarsko-Geološki Glasnik 4:59–67

Sakač K (1984) Dr. Josip Poljak, prvi hrvatski speleolog. Zbornik Devetog Jugosl Spel Kongr Karlovac, str 817–825. Zagreb

Sokač B (1973) Geologija Velebita (Geology of Velebit Mt). Dissertation, PMF, University of Zagreb, pp 1–151

Stroj A, Paar D (2019) Water and air dynamics within a deep vadose zone of a karst massif: Observations from the Lukina jama–Trojama cavesystem (−1, 431 m) in Dinaric karst (Croatia). Hydrol Process 33:551–561

Surić M (2002) Submarine karst of Croatia—evidence of former lower sea levels. Acta Carsologica 31(3):89–98

Surić M (2006) Late Pleistocene-Holocene paleoenvironmental changes-records from submerged speleothems from the Eastern Adriatic Sea (Croatia). Doctoral thesis. University of Zagreb, Faculty of Science, p 213, Zagreb

Surić M, Juračić M, Horvatinčić N, Krajcar BI (2005) Late Pleistocene-Holocene sea-level rise and the pattern of coastal karst inundation: records from submerged speleothems along the Eastern Adriatic Coast (Croatia). Mar Geol 214(1–3):163–175

Surić M, Jalžić B, Petricioli (2007) Submerged speleothemes—expect the unexpected. Examples from the Eastern Adriatic Coast (Croatia). Acta Carsologica 36(3):389–396. Postojna

Surić M, Lončarić R, Lončar N (2010) Submerged caves of Croatia—distribution, classification and origin. Environ Earth Sci 61(7):1473–1480

Surić M, Korbar T, Juračić M (2014) Tectonic constraints on the late Pleistocene-Holocene relative sea-level change along the north-eastern Adriatic coast (Croatia). Geomorphology 220:93–103

Surić M, Lončarić R, Buzjak N, Schultz S, Šangulin J, Maldini K, Tomas D (2015) Influence of submarine groundwater discharge on seawater properties in Rovanjska-Modrič karst region (Croatia). Environ Earth Sci 74(7):5625–5638
Šegota T (1968) Sea level changes in holocen and younger Wurm. Geogr Bull 30:15–39
Šegota T, Filipčić A (2003) Köppenova podjela klima i hrvatsko nazivlje. Geoadria 8(1):17–37. Zadar
Štefanac D (2011) Gradusa Cave. Determination of cave jewelry. Mountaineering Association of the University "Velebit", Velebiten
Terzaghi K (1913) Beiträge zur Hydrographie und Morphologie des kroatischen Karstes. Mitt, Jahrb Kgl Ungar Reichsanstalt 8(6):256–369
Terzić J, Marković T, Lukač RJ (2014) Hydrogeological properties of a complex Dinaric karst catchment: Miljacka Spring case study. Environ Earth Sci 72(4):1129–1142
Turner S (1960) Bojanje rijeke Like na Markovom ponoru u Lipovom polju, Arh. "Elektroprojekta", Zagreb
Vasseur F (2019) Ekspedicije speleoronilačke udruge Plongee Sout u Hrvatskoj 2013–2015. Subterranea Croat 17(1):2–11. Karlovac
Valvasor JW (1689) Die Ehre deß Herzogthums Crain. I. Th., W. M. Endter, p 696. Laibach, Nürnberg
Velić I, Tišljar J, Vlahović I, Matičec D, Bergant S (2003) Evolution of the Istrian Part of the Adriatic Carbonate Platfgorm from the Middle Jurassic to the Santonian and Formation of the Flysch Basin during the Eocene: main events and regional compression. In: Vlahović I (ed) Evolution of depositional environments from the Palaeozoic to the Quaternary in the karst Dinarides and the Pannonian basin, Field trip guidebook of 22nd IAS Meeting of Sedimentology, Opatija 2003, 17–19 Sept 2003, Opatija, Institute of Geology Zagreb and Croatian Geological Society, pp 3–17, Zagreb
Velić I, Šparica M, Vlahović I (2009) Klastične i karbonatne naslage karbona i perma. In: Velić I, Vlahović I (eds) Tumač geološke karte Republike Hrvatske 1:300000. Croatian geological Survey, Zagreb, pp 20–23
Velić I, Vlahović I (2009) Karbonatna platforma. In: Velić I, Vlahović I (eds) Tumač geološke karte Republike Hrvatske 1:300000. Croatian geological Survey, Zagreb, pp 44–52
Vlahovic I, Tišljar J, Velic I, Maticec D (2002) The Karst Dinarides are composed of relics of a single Mesozoic platform: facts and consequences. Geol Croat 55(2):171 183
Vlahović I, Tišljar J, Velić I, Matičec D (2005) Evolution of the Adriatic carbonate platform: palaeogeography, main events and depositional dynamics. Palaeogeogr Palaeocl 220(3–4):333–360
White WB (1988) Geomorphology and hydrogeology of karst terrains. Oxford University Press, Oxford, p 464
Zaninović K (ed) (2008) Klimatski atlas Hrvatske (Climate atlas of Croatia 1961–1990/1971–2000). Državni hidrometeorološki zavod, Zagreb, p 200
Zlokolica-Mandić M, Jekić M (2017) Pećina Piskovica – najdulja hrvatska pećina u flišu. *Subterranea Croatica*, vol.23,p 29–33, Karlovac.
Zupan Hajna N (2010) Geology. In: Introduction to the Dinaric Karst, Mihevc, Prelovšek, Zupan Hajna. Inštitut za raziskovanje krasa ZRC SAZU, pp 14–19. ISBN 978-961-254-198-9, Postojna
Žic V, Truesdale V, Cuculić V, Cukrov N (2011) Nutrient speciation and hydrography in two anchialine caves in Croatia: tools to understand iodine speciation. Hydrobiologia 677(1):129–148

The manufacturer's authorised representative in the EU is Springer Nature Customer Service Centre GmbH, Europaplatz 3, 69115 Heidelberg, Germany. If you have any concerns regarding our products, please contact ProductSafety@springernature.com

Printed and bound by CPI Group (UK) Ltd, Croydon, CR0 4YY
30/09/2026
02225516-0012